HIGHWAYS

Volume 1
Highways and Traffic

(*Courtesy of Highways Design and Construction*)

The interchange between the M6 Motorway and the Birmingham City Expressway at Gravelly Hill, Birmingham

This textbook is dedicated to

Nuala

who helped much

with 'heart an' hand'

HIGHWAYS

Volume 1
Highways and Traffic

C. A. O'Flaherty

B.E.(N.U.I.), M.S., Ph.D.(Iowa State), C.Eng., F.I.Mun.E., F.C.I.T., M.Inst.H.E., M.I.E.I.
Professor of Transport Engineering and Director of the Institute for Transport Studies, University of Leeds

Edward Arnold

First published in 1967 by
Edward Arnold (Publishers) Ltd,
25 Hill Street, London W1X 8LL

Second Edition 1974

Boards edition ISBN 0 7131 3322 8
Paper edition ISBN 0 7131 3323 6

Printed in Great Britain by Butler & Tanner Ltd, Frome and London

Preface to First Edition

In recent years highway engineering has undergone a profound change. Not too long ago it was considered by many—educationalists not excluded—that the recipe for the successful building of a highway simply involved a little art, a dash of science and a large helping of intuition. The practical result was only too often the construction of a facility which was inadequate and unable to meet the traffic demands placed upon it.

The dramatic rise in living standards since World War II with its consequent demand for higher travel standards has caused this approach to road-making to be questioned and discarded. Now it is recognized that highways can play an indispensable part in making the daily chores of life more enjoyable, as well as more convenient. Their importance is indicated by the great numbers of motor vehicles which ply from the home to the school, between working places and from town to town. Thus the need for road facilities to carry these vehicles safely and efficiently has brought to the fore the demand for highway engineers with outlooks and training which are, in many instances, different from the traditional ones. The result has been that today's highway engineer is now expected not only to be conversant with the traditional engineering skills but, in addition, he must have a reservoir of knowledge with regard to planning, economics, statistics, physics and chemistry, as well as computer technology.

Because of the magnitude of the knowledge which he is now expected to absorb, the tendency in recent years has been for the young highway engineer to specialize in some particular aspect of his subject, e.g. traffic planning, geometric design, pavement materials and design, traffic management. As a result, many erudite textbooks have been, and are being, written to cater for these specialisms. What is often forgotten, however, is that the great majority of highway engineers are not just specialists in only one branch of their subject; in the course of their careers they may have to deal with problems ranging from the planning, design and construction of a motorway to the comparatively simple regulation of traffic movement through an intersection. Perhaps more important is the fact that the young engineer who is about to centre his career in highways must have a fundamental grasp of all the aspects involved before he can decide in which branch of the subject he wishes to specialize—or, indeed, whether he wishes to specialize at all.

It is to meet the needs of this latter group that this textbook has primarily been written. Its purpose is to present an integrated picture of current thought with regard to the theory and practice of highway engineering.

Thus, the basic philosophy has been to formulate into a unified matrix the basic concepts relative to highway planning, design, construction and traffic management. The aim in so doing has been to present the material in a simplified manner, with emphasis upon the 'why' of particular highway procedures.

Much of the material presented in this book is the outgrowth of lectures to both undergraduate and postgraduate students in highway engineering. Its outline therefore follows the form in which it is felt desirable that a student should develop his interest in highways.

The text first exposes the young engineer to the historical development of the road and to the manner in which the relevant British legislative, financial and administrative procedures have come about. The planning and economic factors which underlie the rationalization relative to the types of road that are needed and where they should be put are next considered, and these are then followed by discussions regarding the physical surveys which influence the exact locations of highway facilities.

Following upon the decisions as to where the highways are to be built, the student is next introduced to the features which affect the geometric and structural design of highways. With regard to geometric design, this requires an understanding of how the road configuration can influence vehicle speed, flow and safety, so that the detrimental aspects of these influences can be minimized throughout the design. In respect of structural design, this requires that emphasis be placed on recent research which reflects current thoughts as to how pavements behave under traffic; this enables the adequacies/inadequacies of the various pavement design procedures to be considered in some detail.

Because of the controlling influence which they exert over pavement performance, special attention is paid to the materials which form pavements and to the soils upon which they rest. In the discussions on soils and soil stabilization, the emphasis is placed upon the physical and chemical features and mechanisms which influence stability, rather than assuming that soils are structural materials which simply require some mechanical tests to be carried out on them in order to determine their properties.

Once a highway has been built, the highway engineer is still faced with the problem of ensuring that the traffic is able to move on it in an orderly, efficient and safe manner. Thus it is that in the latter parts of the text the reader is introduced to the basic concepts of traffic management and control.

It is not intended that this text should just be about roads and roadmaking in Britain. While, of course, special attention is paid to the highway research and developments which have taken place in this country in recent years, considerable reference is also made to work in other countries. Since the greater part of the world's highway research is carried out in the United States, this has necessitated drawing heavily upon American highway technology. Irrespective of the source of data, however, the intention throughout the writing of the text has been to provide the young engineer, no matter where he may reside, with basic knowledge on how to plan, locate, design, build and manage the traffic on highways.

It has been stated that one picture says as much as a thousand words and hence diagrams illustrating fundamental concepts and practices are used extensively. Where appropriate, also, example problems have been integrated throughout the text. References are provided at the end of each chapter; these have been selected not only to corroborate items of recent or controversial origin but also to provide fruitful sources of definitive information for further study. Thus, while the text has been designed primarily to meet the needs of undergraduate students in Universities and Colleges of Technology, it is hoped that it will also serve as an introductory text for highway research workers and as a reference book for practising engineers and planners engaged in the very fascinating task of building highways.

Since the work was begun a number of revised Standards, Codes, Specifications, Road Notes, etc. have been issued and are in current use. Inevitably some of the referenced extracts used in the book differ slightly from the latest editions of these publications and the reader is strongly advised always to consult the most up-to-date versions. This is especially important where the tables in the book have been abbreviated and are shown mainly for illustrative purposes.

Preface to Second Edition

The primary aim of this work is as described for the first edition, to provide a basic text book for the young engineer about to centre his career in highways, whether it be in their planning, design, construction, or operation. The approach therefore is similar—except that, for economic and handling reasons, it has been necessary to divide the material between two volumes. Thus, Volume 1: Highways and Traffic covers the general administrative aspects of highways together with those other chapters of particular interest to the traffic engineer. Volume 2 deals with the 'hard' side of highways, i.e. their physical location and structural design, as well as the materials used in their construction. (The contents of Volume 2 are set out on p. xiv.)

There are a number of other changes in the new edition. First, and most important, the text has been updated. Thus every chapter has been modified in some way or another, particularly the chapters on Highway Planning, Economics, Geometric Design, and Traffic Management.

A complete new chapter on Parking has been added. As it has special relevance to traffic planning in urban areas, this chapter has been included in the text just after Highway Planning—even though it also contains material of relevance to design and operations.

The opportunity has also been taken to 'metricate' the text. This has meant in certain circumstances that formulae empirically derived in imperial units have had to be changed into metric units. While it is hoped that the likelihood of error has been minimized, the writer is nevertheless conscious that mistakes may have crept in, and will welcome any being brought to his attention. Meanwhile, the reader is reminded that the factual material in the text is primarily given for demonstration purposes, and is urged to seek out the original material (references are given in most instances), and to consult the most up-to-date 'official' versions of codes, recommended standards, specifications, etc, when engaged in actual planning, design and field-work.

COLEMAN O'FLAHERTY

Institute for Transport Studies
and Department of Civil Engineering,
University of Leeds,
Leeds 2.

Feb. 12, 1973.

Acknowledgments

I am indebted to the following for their courtesy in allowing me to reproduce diagrams and tables from their publications in this and its complementary volume. The number in the title of individual tables and figures indicates the reference at the end of the chapter where the source is to be found.

American Association of State Highway Officials, Washington, D.C.
Asphalt Institute, College Park, Maryland
Association of Asphalt Paving Technologists, New Orleans
Australian Road Research Board
British Road Tar Association, London
British Standards Institution, 2 Park Street, London W.1. (from whom copies of the latest edition of the complete Standard or Code can be obtained)
California Division of Highways, Sacramento, California
Cement and Concrete Association, London
Department of the Environment, Transport and Road Research Laboratory, Crowthorne, Berks.
Greater London Council
Her Majesty's Stationery Office, London (All figures from H.M.S.O. publications are Crown copyright and reproduced by permission of the Controller)
Highway Research Board, Washington, D.C.
Institution of Civil Engineers, London
Institution of Highway Engineers, London
Institution of Municipal Engineers, London
McGraw-Hill Book Co., Inc., New York
Permanent International Association of Road Congresses, Paris
Proceedings of the International Conferences on the Structural Design of Asphalt Pavements
Public Roads (Washington, D.C.)
Public Works and Municipal Services Congress Council, London
Quarry Manager's Journal, London
Roads and Road Construction, London
Royal Society of Arts, London
Town Planning Review (Liverpool University)

Traffic Engineering (Institute of Traffic Engineers, Washington, D.C.)
Traffic Engineering and Control, London
U.S. Department of Defence, Department of the Army, Washington, D.C.
D. Van Nostrand Company, Inc., New York
John Wiley & Sons, Inc., New York
World Touring and Automobile Organization, London

I should also like in particular to again thank Professor R. H. Evans for his constant encouragement throughout the preparation of the first edition of this text. I am very grateful to my colleagues D. Andrews, E. Bennett, J. Cabrerta, G. Collins, M. Dyer, J. Fox, J. Garner, W. Houghton-Evans, G. Leake, P. Mackie, J. Rawcliffe, G. Rainnie and P. Sutcliffe from within the University, and to B. Hutton, M. Mateos, H. Roberts and M. Shaw from without, all of whom at some time or other read some part of the draft manuscripts of the first or second edition and commented critically upon them. Not only I, but the publishers also are indebted to Mrs. R. Holmes and Mrs. G. Hancock for their excellent interpretation of my scribbles when typing the manuscripts. Last, but far from being least, I should like to thank my colleagues, B. Firth and, most particularly, J. Higgins, who prepared all the diagrams in both editions, often working only from scrappy notes and sketches.

COLEMAN O'FLAHERTY

Contents

		Page
	Preface to first edition	vii
	Preface to second edition	x
	Acknowledgments	xi
1	The road in perspective	1
2	Highway legislation, classification, finance and administration	22
3	Highway planning	53
4	Parking	129
5	The economics of road improvements	188
6	Geometric design	216
7	Traffic management	287
8	Highway lighting	366
	Index	383

Complementary to this Volume

HIGHWAYS, Volume 2: HIGHWAY ENGINEERING

1 Highway location, physical surveys, plans and specifications
2 Surface and sub-surface moisture control
3 Pavement materials
4 Soil engineering for highways
5 Soil stabilization
6 Flexible pavements
7 Bituminous surfacings
8 Rigid pavements
Index

1 The road in perspective

Roads can be considered as much a cause as a consequence of civilization; they both precede and follow it. History testifies that the provision of roadways is necessary to draw a country out of a state of barbarism, but that civilization is not attained until communications between neighbours is made so easy that the local differences which breed narrowness and bigotry are minimized. As civilization is advanced and prosperity increases there is an inevitable demand for better and speedier communication facilities, especially the roads. Indeed it can truly be said that the prosperity of a nation is bound up with the state of its roads, and that the roads act as a palimpsest of a nation's history.

It is fashionable nowadays to heap abuse upon the roads of Britain because of their relative inability to absorb the avalanche of motor vehicles descending upon them. That they are not capable of handling at all times the demands placed upon them cannot be denied; what is not always appreciated however is that the roads of Britain were never purpose-built to meet modern communication needs but that, instead, the motor vehicle was thrust upon a road system already in existence and possessed of a deep association with the history of the country. For this reason therefore, it is desirable to present a picture of how the existing roads evolved; by so doing it is possible to explain the existence of certain obstacles in the way of adapting the inherited roads to meet modern requirements. Included in this historical perspective is some brief account of particular financial and administrative problems affecting the roads since they, as much as the technical limitations of the times, have greatly influenced the development of the roads in Britain.

ANCIENT ROADS

The birth of the road as a formal entity is lost in the mists of antiquity. Most certainly however the trails deliberately chosen and travelled by ancient man and his pack animals were the forerunners of the road as we know it today. As man developed and his desire for communication increased, so inevitably trails became pathways and pathways developed to become recognized travel-ways; no doubt some levelling was done and perhaps soft and marshy patches filled and made firm to ease the movement of man and his beasts of burden.

The invention of the wheel over 5000 years ago made necessary the construction of special hard surfacings capable of carrying concentrated and greater loads than hitherfore. That this was realized by the ancients is evidenced by the multiplicity of sometimes sophisticated but more often crude roadways that have been discovered by archaeologists. It is not feasible, nor indeed is it necessary, to attempt to discuss all of the fragmentary evidence relating to the development of these roads and their methods of construction, and so instead only a few of the more important of these early travelling ways will be described.

Non-British roadways

Perhaps the most monumental of the earliest recorded 'roadways' was the stone-paved sloping causeway constructed at the direction of the Egyptian King Cheops to facilitate the conveyance of the huge limestone blocks used in the building of the Great Pyramid. This causeway consisted of an embankment having a bottom layer of sandy soil beneath a layer of broken limestone, on top of which were placed layers of tiles or bricks; the various constructional layers were contained within two walls of quarried stone which were strengthened by placing sloping embankments of sand against either side. This causeway took 10 years to complete and the finished facility was '5 furlongs long, 10 fathoms wide and of height, at the highest part, 8 fathoms'.[(1)] Indeed it can be said that this construction was in no way inferior to the pyramid itself.

What has been described as the oldest road in Europe, and which might also be called the first dual-carriageway, was the roadway built in Crete about 2000 B.C. to connect Cortina on the south-east of the island to Knossos on the north. Archaeologists have described the method of construction as follows. First of all the earth was excavated to a depth of about 200 mm over a width of about 4·5 m, and the subsoil levelled and compacted. Into this excavation was placed a watertight foundation course of sandstone embedded in a clay-gypsum mortar about 200 mm deep. On top of this a 50 mm layer of clay was laid to form a bed for the pavement. A 300 mm wide pavement composed of 50 mm thick basalt slabs formed the crown of the road and this was flanked by pavements of rough limestone blocks grouted with gypsum mortar. It is believed that the central pavement was for the use of pedestrians while light-wheeled traffic ran on the sides.

A third road of constructional interest is the royal processional route in Babylon which was probably built about 620 B.C. This road was only about 1220 m long but it is notable for the materials utilized in its making. The road was paved to a width of 10–20 m with stoneslabs placed on a foundation of three or more layers of brick jointed with bitumen. The middle part of the surface was paved with limestone slabs while the sides were deemed to be adequately surfaced with red and white breccia slabs set in bitumen.

These few examples illustrate that road construction had achieved a certain sophistication many thousands of years ago. That the *purpose* of a road was also appreciated by at least some of the ancients is illustrated by the character of the route known as the Persian Royal Road. This organized trade-cum-military road is believed to have run from Smyrna eastwards across Turkey and then southward through Persia to the Persian Gulf, for a total length of 2400 km. The territory through which it passed was part of the Persian Empire which was then (520–485 B.C.) ruled by King Darius I. This monarch was determined to keep his empire under control and decided to have the route laid out so that he could move his troops quickly. To facilitate movement he provided resting places at intervals of one day's walking along the way. He further arranged for it to follow the shortest practical route, even to the extent of by-passing some of the largest towns in the empire; he did, however, provide service roads into these towns. His purpose in by-passing the towns was to ensure swift movement unhindered by congestion and other obstacles which were even then prevalent in the built-up areas.

Another who appreciated the importance of the road was the Indian ruler Chandragupta, who in the period 322–298 B.C. constructed a 2400 km road across the sub-continent. This monarch set up a special ministry to organize and carry out maintenance on the route, provide milestones and resthouses, and operate the ferry systems at river crossings. Even then obtaining the finance for road maintenance was a problem and so he obtained a monopoly of the salt trade in order to secure funds to support the roads.

British trackways

As far as is known there were no formal roads constructed in Britain prior to A.D. 43. Certainly the driftways used by the early inhabitants of this country cannot be dignified as roadways, being nothing more really than the customary paths marked out by the frequent passing of man.

The oldest of the British pathways are located on the downs of south and south-east England. These were not haphazard tracks wandering aimlessly about the country, but had instead a very definite sense and purpose. The longest and most definite of these are known as Ridgeways because of their location just below the crests of hills. These tracks, which came into existence about 2800 B.C. in order to join centres of occupation, only descended into the valleys when it became necessary to cross rivers at fordable locations; other than this, they stuck to the hill crests where the underbrush was less dense and there was more freedom of movement.

As time progressed the Ridgeways developed almost imperceptibly into what are now known as Trackways. These included pathways which ran on the lower slopes; they were sufficiently above the bottoms of valleys to ensure good drainage but yet were low enough to obviate unnecessary climbing. The Trackways were primarily trade routes and along them settlements and shrines developed; perhaps the most famous of these is the great sun temple at Stonehenge which was constructed about the beginning of the Bronze Age.

ROMAN ROADS

The Romans are generally given credit for being the first real highway designers. Roads played a vital role in the maintenance of the Roman empire; in fact it can be said that the empire was held together by a well-designed system of roadways extending from Rome into Italy, Gaul, Britain, Illyria, Thrace, Asia Minor, Pontus, the East, Egypt and Africa. These regions were divided into 113 provinces traversed by 372 great roads totalling 52 964 Roman miles.[2] This mileage is equivalent to about 78 000 km.

There is much discussion as to whether or not the Roman roads should be classified as military facilities. They are usually regarded in this way as their construction was motivated by military expediency, it being necessary for occupying military force to be able to travel quickly to any point in the province threatened by attack. They were non-military roads, however, in that their main purpose was not for conquest but to aid administration and quell rebellion *after* a country had been occupied.

The great Roman road system was based on 29 major roads radiating from Rome to the outermost fringes of the empire. The first and perhaps the greatest of these roads was the Via Appia which travelled from Rome in a south-east direction through Capua on to Brundisium in the heel of Italy. This 660 km road, which was started in 312 B.C. and reached Brundisium in 244 B.C., provided improved communication between Rome and North Africa and the East. This roadway—sections of which can still be seen today—was usually constructed at least 4·25 m wide; this enabled two chariots to pass each other with ease and legions to march six abreast. It was common practice to reduce the gradient by cutting tunnels and one such tunnel on the Via Appia was about 0·75 km long.

Two geometrical features of the Roman roads are of particular interest. The first of these refers to the habit of constructing their roads well above ground level. It seems certain that nearly all of the great Roman roads were built on embankments, the major ones averaging some 2 m in height.[3] While the raising of the way had the engineering by-product of serving to keep the surface dry, its main purpose was to give a commanding view of the surrounding countryside so that troops passing along were less liable to surprise attack. From this safety measure has arisen the modern term for a road, that of 'Highway'.

The second feature of the Roman road was its directness. The popular conception of the straightness of the Roman roads is, however, somewhat exaggerated. Certainly the Romans attempted to construct direct routes where possible but, instead of attempting to connect one terminus to another by one major straight road, they planned it so that a number of individual stations along the route could be connected by a series of straight lines. Thus, for instance, Britain's Watling Street appears on the map as a gradual curve, whereas in reality it was composed of 9 separate straight lines connecting points along the way. The adoption of the straight line by the Romans was primarily motivated by the fact that a straight road obviously

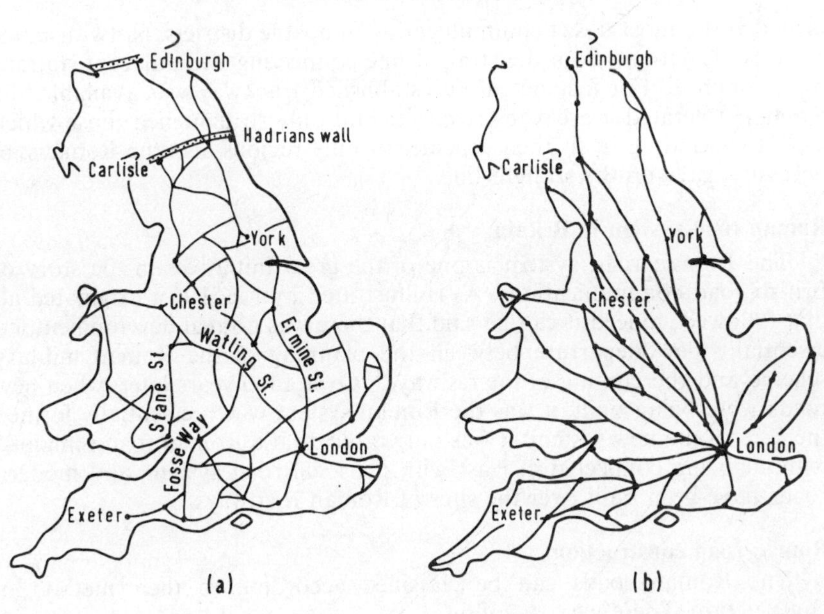

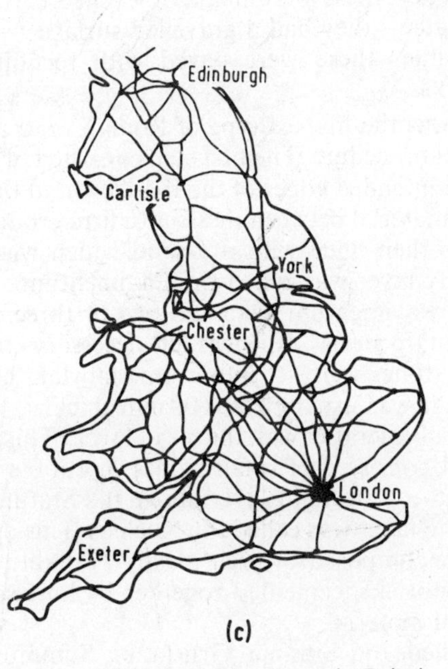

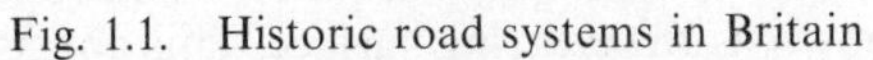

Fig. 1.1. Historic road systems in Britain

(*a*) Some of the more important Roman roads
(*b*) Chief directional roads, 1771
(*c*) Pre-motorway trunk road system

provided the most direct communication in hostile districts. Notwithstanding this, deviations from the straight line connecting two adjacent stations were common. For instance, if an established trackway was available, the Romans tolerated a great many curves and only straightened those which were too winding to fit their scheme. In hilly regions also the Romans, of necessity, gave up the straight line.

Roman road system in Britain

The Roman road system is one of the great initiatives in the story of British road communications. As is illustrated by Fig. 1.1, it originated all that followed; indeed it can be said that there was no real development, no essentially new departure between the planning of the Roman military scheme and the coming of the railway.[4] Over 1400 years later, when new roads were being built, it was the Roman system which essentially formed the core of the new system; it was only rarely that a Roman road remained without being connected at least with the local road system, and modern roads have been built over the sites of Roman roadways.

Roman road construction

The Roman roads can be classified according to their method of construction. These were as follows:

1. Viae terrenae—these were made of levelled earth.
2. Viae glareatae—they had a gravelled surface.
3. Viae munitae—these were paved with rectangular or polygonal stone blocks.

The *viae munitae* was the highest type of Roman road and its construction followed a classical procedure. The first step consisted of cutting two parallel trenches along the intended edges of the roadway and then removing all the top soil and loose material between these until firm ground was reached. The excavated cut was then filled with dry sand which was well rammed into position; this sandy layer was called the Pavimentum.

On top of the Pavimentum were laid two or three courses of large flat stones or small square stones. If it was felt necessary to make this a waterproof layer, these stones were set into a lime mortar. This layer was called the Statumen and it was usually 250–600 mm thick.

The Rudus or Ruderatio was the next layer. This was usually about 225 mm thick and consisted of small stones or rubble set in mortar. This presumably formed a watertight layer above the Statumen.

The penultimate layer was called the Nucleus. This usually consisted of a fine concrete layer composed of such locally available materials as gravel, broken pottery or bricks, cemented together by lime so as to form a hard permanent mass of material.

What was called the Summa Grusta or Summum Dorsum by the Romans is now known as the wearing surface. This consisted of fitted stone blocks embedded in the top of the newly laid nucleus. The thickness of these blocks varied of course, but usually they were between 150 and 300 mm thick.

The type of construction just described was of course varied to suit the needs of a particular location. For instance, when swampy soils were met, it was quite common to drive wooden piles into the soft ground and then to construct the road on these piles—method of construction similar to that employed today.

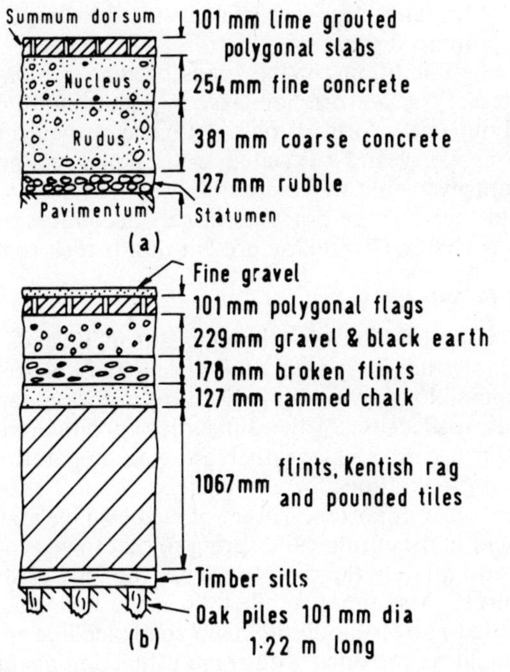

Fig. 1.2. Sections through Roman roads in Britain
(*a*) Fosse Way, near Radstock
(*b*) Through Medway Valley, near Rochester

POST-ROMAN PERIOD TO THE EIGHTEENTH CENTURY

The withdrawal of the Roman legions from Britain in A.D. 407 not only anticipated the death of the Roman empire but it also preceded the complete breakdown of the only organized road system in Europe. Only once or twice were improvements attempted, notably by Charlemagne, who made strenuous efforts to build and maintain good local roads. After his death in A.D. 812 the Roman road system deteriorated still further.

Roads prior to the Middle Ages

It does not appear that following the departure of the legions a single new road was deliberately planned and constructed in Britain until the advent of the 17th century. Although the Roman roads were the main highways of internal communication for a very long period, with time however they began to decay and disintegrate as they were exposed to the wear and tear of natural and man-made agencies. Their decay was not due simply to carelessness and ignorance of maintenance practices, but in reality it had its roots in deep-seated political causes.[5] During the Dark Ages the country was split into a number of small Kingdoms whose rulers preferred the political ideal of isolation rather than co-ordination. Their needs and desires were parochial rather than national and thus they exerted little or no effort to preserve the through roads—except as it suited their internal needs—and so these soon fell into disrepair.

Inevitably the conditions of the roads became deplorable. New roads simply consisted of tracks worn haphazardly according to need. New tracks were made about every form of obstruction, care being usually taken to avoid private property and cultivated land. When sections of an existing road became untraversable, they were simply abandoned and new trackways created around them. These practices largely account for the winding and indeed tortuous character of many present-day British roads and lanes.

Roads during the Middle Ages

Throughout the Middle Ages the main through roads all over Britain were nothing more than miry tracks full of ruts and holes. As a result it was practically impossible to travel long distances in the winter, no matter how urgent the reason. Because of the difficulties encountered in transporting commodities during inclement weather, a town's preparations for the winter often resembled preparations for a long siege.

Fairs were a most important feature of medieval life and the biggest fairs were always held in the autumn and spring so that towns could re-provision. The potholes and mud on the roads were not the only obstruction to travel to the fairs during these times. While they lasted it was customary to post specially mounted patrols along the main roads leading to the fairs so that the traveller could be provided with some protection against the terrors of the robber and the highwayman. Historic evidence of the need for these guards is given in the Statute of Winchester which, in 1285, was the first statute to attempt to deal with the roads. This statute ordained 'That highways leading from one market town to another shall be enlarged where as bushes, woods, or dykes be, so that there be neither dyke nor bush whereby a man may lurk to do hurt within 200 feet [61 m] of the one side and 200 feet [61 m] of the other side of the way'. Unfortunately there is ample evidence to testify that this statute was more noted in its breach than in its observance.

In general, the care and maintenance applied to roads during medieval times were merely a matter of chance. The only systematic approach to road maintenance came about at the instigation of the religious authorities. In keeping with the times, road repairs were considered as pious and meritorious work of the same sort as visiting the sick or caring for the poor. Men saw road repairing as a true charity for the unfortunate travellers.[6] To encourage the faithful to take part in constructing or repairing roads and bridges, the religious authorities often granted indulgences to those who performed such public services.

British roads were in a state of continual decline as the medieval times came to a close. The suppression of the monasteries by Henry VIII completed the road decay by removing the most important class of road-makers and maintainers; unfortunately the new owners of the ecclesiastical estates were not inclined to accept the road obligations imposed by custom on the original landowner .

Coincident with the decline of the religious orders an agricultural revolution took place which substituted pasture for arable land all over Britain. This had the additional effect of lessening the necessity for moving produce by cart and resulted in the generally held consideration that there was less need for taking care of the roads.

These reasons, taken in conjunction with the long and expensive wars to which the country was subjected, resulted in the roads of Britain rapidly reaching a very low state indeed after the turn of the 16th century.

Roads between the 16th and 18th centuries

By the middle of the 16th century the road conditions were so bad that in 1555 Parliament was forced to pass an Act 'For amending the High-ways being now very noisom and tedious to travel and dangerous to all Passengers and Carriages'. This was the first Act which provided by statute for the maintenance of highways. It required the individual parishes to be responsible for the maintenance of their own roads, and laid down rules by which they had to supply all the materials and labour needed for this purpose. In addition each parish had to provide from amongst its own inhabitants an official to superintend, gratuitously, this road maintenance; these officials were usually known as Surveyors of Highways. The lack of esteem in which the office of Surveyor was held is reflected in the fact that the official was rarely either a person of high social position or a man with technical skills. Invariably the post was held by a local farmer or shopkeeper who was both incompetent and unpopular.

At the same time as road conditions were deteriorating, the opportunities for travel were increasing. About the middle of the 16th century the first wheeled coaches appeared in Europe. Until that time most travellers, even the Courts on their royal progresses, still rode on horseback or travelled on foot. The first coach to be seen in Britain appeared in London in 1555, having been specially constructed for the Earl of Rutland. Long distance passenger coach services were commenced from London in 1605, and it is known that hackney coaches were plying the streets of the capital city in 1634.

It is perhaps ironic to note that even during these times the traffic problem in London was considered to be chaotic. In 1635 the situation was considered to be so undesirable that a Royal Proclamation was published which stated 'That the great number of hackney coaches of late time seen and kept in London, Westminster and their suburbs, and the general and promiscuous use of coaches there, were not only a great disturbance to his Majesty, his dearest Consort the Queen, the Nobility, and others of Place and Degree, in their passage through the streets; but the streets themselves were so pestered, and the pavements so broken up, that the common passage is thereby hindered and more dangerous; and the prices of hay and provender, and other provisions of stable, thereby made exceedingly dear. Wherefore We expressly command and forbid, that no hackney or hired coach be used or suffered in London or Westminster or the suburbs thereof. And also, that no person shall go in a coach in the said streets, except the owner of the coach shall constantly keep up four able horses for our service, when required'.

Two features of this proclamation make it notable. The first of these is the fact that little attention was paid to it by the populace, illustrating that even then the people accepted as their right that they should have usage of the roads. More significant perhaps is the fact that it marked the onset of the concept, held to this day by many people in authority, that if the roads were not capable of handling the traffic, then the traffic should be reformed to suit the roads.

As time wore on, the official attitude became that traffic of all types was a nuisance that had to be suppressed if possible and, if not, to be allowed to exist only in accordance with the most rigorous regulations. Thus a 1621 Act forbade the use of four-wheeled wagons and the carriage of goods of more than one ton weight because vehicles with 'excessive burdens so galled the highways and the very foundations of bridges, that they were public nuisances'.(3) This type of legislation continued during the 17th and 18th centuries, becoming increasingly detailed and increasingly incomprehensible.

ROADS INTO THE TWENTIETH CENTURY

In spite of the legislative attempts to curb its growth, the 17th century saw a steady increase in the amount of wheeled vehicular traffic. With the onset of the 18th century, foreign trade became more important to the steadily developing manufacturing industries in Britain, and soon long trains of carts were laboriously dragging coal from the mines to the iron-works, glassworks and potteries, and manufactured goods to the ports and harbours. The state of at least part of the road system at this time is probably best described by this rather graphic description of the Mile End Road in 1756 '...resembled a stagnant lake of deep mud from Whitechapel to Stratford, with some deep and dangerous sloughs; in many places it was hard work for the horses to go faster than a foot pace, on level ground, with a light four-wheeled post chaise'.(7)

Acceptance of the need for good roads

The first half of the 18th century saw the occurrence of two events which signified the growing acceptance of the need for competent road-makers and a viable road system. These events were the formal construction of 400 km of roadway in Scotland by General Wade, and the formation of the Turnpike Trusts in England.

General Wade's roads. The most famous of the British roads of the early part of the 18th century were those constructed by General Wade in the southern highlands of Scotland. These roads, construction of which began in 1726, were military roads in the same sense as the Roman roads were military roads; they were constructed to ensure that the near success of the Insurrection of 1715 was not repeated. They followed from a visit to Scotland by General Wade in 1724, when he was officially asked to determine what measures were necessary for the development of the country. His recommendation, to build a road system, was accepted by the government and he was given 500 soldiers to carry out the construction.

General Wade's roads were planned to make the homes of the clans easily accessible to the forts and garrison stations located at strategic positions throughout the highlands. In addition, they were intended to have the pacific purpose of improving the general conditions in the highlands, which were at that time deplorably backward. The difference between the old travelled ways and their formal replacements is best signified by the inscription on the monument erected to General Wade on the road from Inverness to Fort William:

'Had you seen these roads before they were made, You would lift up your hands and bless General Wade.'

The Turnpike Trusts. At the turn of the 18th century the methods employed in making and maintaining roads in England were very primitive. The usual procedure was to dig loose earth from the ditches and throw it on the roadway in the hope that the traffic would beat it into a hard compact surface. However, these surfaces were not capable of withstanding the stresses imposed on them by the wheeled traffic, and they were soon churned into bands of mud with the onset of wet weather.

Confronted by the abominable state of the highways a new principle of road maintenance became accepted; this was that the road-users should pay for the upkeep of the highways. In 1706 there was passed in Parliament the first of a succession of statutes which created special bodies known as Turnpike Trustees. The number of these bodies came eventually to number over 1100, administering some 36 800 km of road, constructed at the cost of an accumulated debt of £7 million, and raising and spending an annual revenue of more than £1·5 million.[(7)] These Trusts, established by literally thousands of separate (albeit similar) Acts of Parliament from 1706 onwards, were each empowered to construct and maintain a specified length of an important roadway and to levy tolls upon certain kinds of traffic. The powers were given only for a limited term of years, usually twenty-one, since it was the intention to remove the toll-gates as soon as the roads were in sufficiently good shape to do this. However, since this happy state never came to pass, every Trust eventually applied for a new Act extending its existence and so the Trustees became virtually permanent Local Authorities on roads.

Whatever the faults of the Turnpike Trusts—and they were indeed very many—there can be no doubt that they marked a most important stage in the development of the road system in Britain. First of all, the system made it possible for people to choose to travel, whereas before they only travelled when absolutely necessary. More important, they established the principle that the road user should pay his share of the road costs. Most important of all, however, is the fact that the advent of the turnpikes brought about and made definite what is in effect the present-day major road network.

If during the 18th century anybody had bothered to make a turnpike map of England it would have shown, not a system of radiating arteries of communication, but scattered instances of turnpike administration that were unconnected with each other. At first these sections would have appeared as mere dots on the map, but with time it would be seen that the dots were gradually increasing in number and size so that eventually they formed continuous lines. In 1794 it was stated, that they 'so multiplied and extended as to form almost a universal plan of communication throughout the kingdom'. It is this network that finally determined today's network of roads. That this is so can be seen from a comparison of the road system described by Patterson in 1771 and the present system; these are illustrated in Fig. 1.1. The influence of the Roman system is also apparent in this figure.

The Turnpike Trusts were important for another reason. They emphasized the lack of skilled road-makers in the country.

Developments in road construction and administration

Although towards the end of the 18th century considerable emphasis was placed upon the need for improved communications in the country, little engineering skill was applied to the construction of roads. The idea that road-making needed any special knowledge or education was incompatible with the professional thinking of that time, with the unhappy result that the application of engineering principles to road-making was entirely neglected. The 'science' of road construction as understood and practised in Britain in the 18th century was simply to heap more soil on top of the existing mud and hope that traffic would compact it into a hard surface. Indeed so little was known about the repairing of roads that sometimes when ruts were being filled the sides of the travelled way were made higher than the centre. The concept of providing drainage seems never to have been considered in practice and as a result the ruts made by wheeled vehicles were continually filled with water. When it is considered that one report stated that the ruts on certain turnpikes were over 1 m deep, the need for stating the importance of good drainage seems rather superfluous.

Fortunately, however, there came into the scene about this time a number of men who each in their own way contributed significantly to the development of the road and the science of road-making.

Pierre Trésaguet. The father of modern highway engineering can truly be said to be Pierre Trésaguet, the Inspector General of Roads in France between 1775 and 1785. He was the first engineer to fully appreciate the importance of the moisture content of the subsoil and its effect upon the stability of the road foundation. Hence he made provision for drainage and, in comparison with the Romans, designed a relatively light cross-section on the principle that the subsoil should be able to support the load laid upon it.

Trésaguet not only emphasized the importance of drainage but he also advocated a diminished road thickness to meet the varied requirements of traffic. He also enunciated the necessity for continuous organized maintenance instead of intermittent repair if the roads were to be kept usable at all times, and so he divided the roads between villages into sections of such length that an entire road could be covered by maintenance men living nearby.

Robert Phillips. In Britain the man who perhaps might be termed the pioneer of road design is a certain Robert Phillips. As early as 1736 he presented a paper to the Royal Society entitled 'A Dissertation concerning the Present State of the High Roads of England, Especially of those near London, wherein is proposed a New Method of Repairing and Maintaining Them'. In this paper Phillips gave a very comprehensive account of the slovenly practices which were employed in road-making and maintenance. He was concerned with the clay and gravel roads then in use and emphasized that a layer of gravel, if resting upon a well-drained 'sole', would be beaten by the traffic into a solid road surface.

John Metcalf. Unfortunately Phillips' lead was not accepted at the time and road-making remained a lost art in Britain until about 1765 when John Metcalf came to prominence. Known as 'Blind Jack of Knaresborough' this master road-builder was a most extraordinary and fascinating character. Born of humble parentage in 1717, he contracted smallpox and became blind at the age of six. He refused, however, to let his affliction become a disability and in the course of a full and varied life he was in turn a musician, soldier, waggoner, horse-trader and eventually, when he was well over forty years old, a road-maker. In 1765 this Yorkshireman was given the contract to construct three miles of turnpike road between Harrogate and Boroughbridge; the efficiency and stability with which he constructed this section was so impressive that eventually he was given contracts to build some 290 km of roadway in his native county, including all the necessary bridges, culverts and retaining walls.

Two aspects of Blind Jack's work make him particularly notable. The first of these was his insistence on good drainage. In constructing the road he emphasized that it should have a good foundation and used large stones to obtain this quality. On top of this he placed a layer of excavated road material in order to raise the roadbed, and on top of this he placed a surface layer of gravel. The carriageway was arched to throw off the rainwater, which was then drained away by capacious ditches on either side of the road.

Perhaps the most individualistic of his road achievements, however, was his approach to building roads over soft ground. Instead of avoiding boggy terrain as had been the custom until that time, he devised a special type of road to go across it. This consisted of constructing his normal roadway on top of a raft sub-base composed of bundles of heather carefully prepared and placed in position on the soft ground.

Thomas Telford. At the end of the 18th century the condition of the populace in Scotland was so bad that the government had fears that the countryside might become depopulated through emigration. In 1802 Thomas Telford was sent to the highlands to determine methods for their development and improvement. This was a particularly well-informed choice since Telford, in retrospect, can beyond doubt be ranked as one of the very great civil engineers of all time. Born in 1757 this Scotsman was the son of a farmherd and as such he had the very minimum of formal education. Early in life he was apprenticed to a stonemason and afterwards worked at his trade in Edinburgh, London and Portsmouth. In 1786 he was appointed County Surveyor of Shropshire and soon distinguished himself by the stature of his building projects, particularly those in roads and bridges. In 1793 he became engineer to the Ellesmere Canal Company and constructed some remarkable aqueducts at Chirk and Pont Cysylltan. In 1801 his proposal for a new London Bridge of a single span of iron won him considerable acclaim as a designer of vision.

This then was the man who when he went on tour in Scotland found neither roads nor carts north of Inverness, and was able to state in his report that all goods were carried by ponies or women. In the southern highlands he found the stone-covered roads of General Wade to be in disuse, primarily because they had been built to lead to military strategic positions and not to the ports or market centres. He wrote a report containing five proposals, the most important being that there was a great need for improving communications by means of roads and bridges; he stated that 'they will not only furnish present Employment, but promise to accomplish all the leading Objects which can reasonably be looked forward to for the Improvement and future Welfare of the country, whether as regards its Agriculture, Fisheries or Manufactures'.

Thomas Telford's proposals were accepted and in 1803 he was given governmental authority to put them into effect. His industry in so doing is testified by the 1470 km of new roads and 1117 bridges constructed under his personal supervision during the following 18 years. This road system is remarkable for many reasons, not the least of which is that it illustrated how good road communications can revolutionize conditions in a depressed economic area. He perfected the broken stone method of construction for which he is famous, while recognizing that the road should be designed for the conditions prevailing. Thus, he recognized that the wear of the Highland roads by traffic was insignificant and foresaw that the main enemies were the natural ones of storm, flood and frost. Hence he provided for maintenance by appointing road superintendents to ensure that the roads were kept well drained and damage caused by winter frost was promptly repaired.

Thomas Telford's most renowned achievement as a road-maker was the reconstruction of the London–Holyhead road. Following the Union with Ireland in 1805, the need for improving the means of travel between London and Holyhead was raised by both the Irish Members of Parliament, who had to use the route perpetually, and the Postmaster General who considered the way to be in such poor condition as to be unsafe for the mail coaches. In 1810 Telford was asked to survey the road and in 1815 he began its reconstruction.

The construction used on this roadway is illustrative of the thoroughness with which Telford approached road-building. He first laid a foundation layer of hand-packed stones varying in depth from 227 to 177 mm at the centre and from 127 to 76 mm at the haunches. Each stone was placed in position with its broadest end downward. The specification required that the top face of each block be not more than 101·5 mm wide and that the interstices between adjacent stones be filled with fine chippings. The centre 5·5 m of road was then covered with stones in two layers about 101 mm and 51 mm thick, the size of the individual stones being such that each could be passed through a ring 63·5 mm in diameter. This formed the 'working' portion of

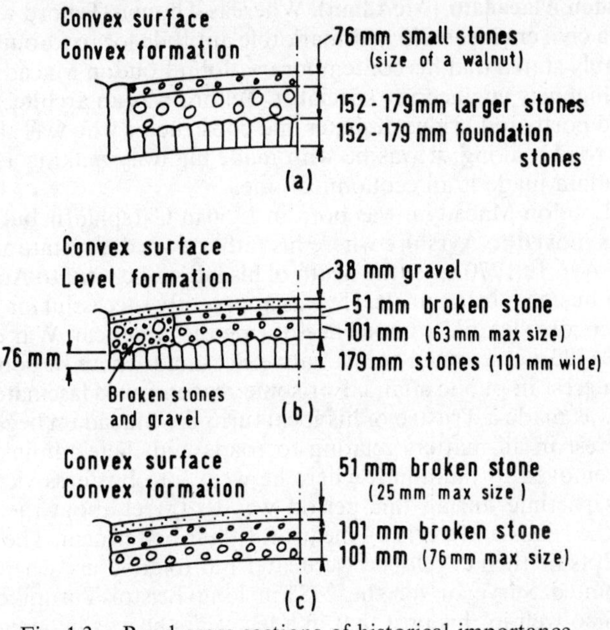

Fig. 1.3. Road cross-sections of historical importance
(*a*) Trésaguet's construction
(*b*) Telford's construction
(*c*) Macadam's construction

the highway and the layers were left to be consolidated by the traffic. The side portions, each 1·8 m wide, were then made up of broken stone or clean gravel, and levelled off to give a cross-fall of not more than 152 mm on a 9·14 m wide section. Over all was then placed a binding layer of gravel about 37·5 mm thick, the watering-in and consolidation being again left mainly to the weather and the traffic. Since the interstices in the foundation layer were sufficiently large to admit water from the surface, cross-drains were often provided beneath this layer at intervals of about 90 m, so that such water could be discharged into side ditches.

Not only did Telford develop a structurally sound method of road construction, but he also paid particular attention to the alignment of roads.

It is recorded that the old road in Anglesey rose and fell a total vertical distance of 1079 m between its extremities 38·5 km apart; its replacement, built by Telford, was 3·22 km shorter and rose and fell 391 m less than the old road. He is said to have taken 32·2 km off the total journey from London to Holyhead, while at the same time easing the gradients so that there was no slope greater than 1 in 20. When completed in 1830 this road was considered to provide by far the best facility for land travelling of its time. Its quality is attested to by the fact that much of this road is incorporated in today's A5 trunk road.

John Loudon Macadam (McAdam). Whereas Thomas Telford was first and foremost a civil engineer with a remarkable aptitude for road-building, it can be justifiably stated that his contemporary, John Loudon Macadam, was the first true highway engineering specialist. Telford was an architect, builder of canals and ports, and bridge designer and constructor who was also a perfectionist in road-making; it was he who made highway-making a respectable art. Macadam made it an economical one.

John Loudon Macadam was born in 1756 at Carsphairn, but early in his life he was moved to Ayrshire where his father owned an estate and founded a bank at Ayr. In 1770, after the death of his father, he went to America to be trained in business by an uncle. He was apparently successful for he was able to amass a considerable competence during the American War of Independence. In 1783 he returned to his native Scotland where he soon began to take an interest in public affairs. For some reason roads fascinated him, and when he was made a Trustee of his local turnpike Macadam began to take a great interest in all matters relating to roads and their administration. In 1798 he removed to Falmouth where he went into business victualling the navy. His pastime during this period was to travel about the south-west studying the roads and devising means for their betterment. The knowledge and concepts he then developed were later put to good use when in 1816 he was appointed Surveyor for the 238 km long Bristol Turnpike Trust. He succeeded so well in this post that in a few years he was in demand all over England as an adviser to other Turnpike Trusts and eventually, in 1826, to the government itself.

Macadam is best known for the method of road construction he advocated and which, in modified form, bears his name today. The two fundamental principles which he emphasized were 'that it is the native soil which really supports the weight of the traffic; that while it is preserved in a dry state it will carry any weight without sinking' and 'to put broken stone upon a road, which shall unite by its own angles so as to form a solid hard surface'. These concepts are as valid today as they were over a century and a half ago.

Under Macadam's system of construction the foundation was shaped to the intended surface camber, thereby giving good side drainage to the foundation as well as a uniform constructional thickness throughout the entire width of the road. The amount of cross-fall which he called for was about half of that required by Telford, being at the very most 76 mm on a 9·14 m wide road. Macadam held little respect for costly stone-paved foundations and instead considered two 101 mm layers of 76 mm broken stone to be quite satisfactory. On top of these he placed a finishing layer of angular stone fragments not greater than 25·4 mm in size, consolidated first by ramming them into the interstices and then by the traffic. Macadam relied upon the road metal to produce by attrition under traffic sufficient fine particles to fill the remaining interstices and bind the whole lot together into a smooth hard waterproof surface.

While Macadam is renowned for his method of construction, his roads were actually in some ways inferior to Telford's. It is no coincidence that many modern roads still exist on their original Telford foundations, whereas it is doubtful indeed whether any of the original Macadam roads have survived. The greatness of his construction lies in the fact that it was a technique which for efficiency and cheapness was an enormous improvement over any method used by his contemporaries.

Macadam's real claim to fame should perhaps be based more legitimately on his eventual destruction of the prevalent hostility to increased traffic, and the advocation of the importance of effective highway administration. He treated with scorn the controversy then very much to the fore about whether the size and loads of vehicles should be regulated. He pointed out that it was possible to construct roads to carry any type of traffic, and that in fact it was the methods of road construction which were actually at fault and not the traffic on the roads. He continually emphasized that the roads must be made to accommodate the traffic and not the traffic regulated to preserve the roads.

Most important were the principles he advocated for producing effective roads and their administration. He persistently urged that the essential feature of any efficient service must necessarily be the trained professional official, paid a salary which enabled him to be above corruption—Macadam said that he must be of good social status—and who would give his entire time to his duties and be held responsible for the success of any undertaking. Macadam also advocated that roads be constructed by skilled labour using the best instruments available. He also urged the creation of a central highway authority which would act in an advisory and monitory capacity on all matters relating to the roads, although at the same time he depreciated any attempt to nationalize the roads on the prophetic grounds that the government would utilize them as a source of revenue instead of looking upon their upkeep as a public service of the first magnitude.[5]

Macadam's proposals were considerably ahead of their time, and it cannot be said that they were accepted in bulk during his lifetime. In general, however, they were regarded with favour and there is no doubt that they initiated concepts and shaped events influencing the future of the whole transportation industry. Whereas to Trésaguet can be given the credit of bringing science to road-making, there is no doubt that John Loudon Macadam was the first Highway Engineer.

The roads and the railways

What all the travel of the coaching days and all the urgings of Macadam and his contemporaries failed to do, the railways did quite incidentally and without effort; they forced the government to reform the administration of the roads. The railways were invented to satisfy a need that the roads proved incapable of meeting, that of keeping pace with the progress of industry. From the time that the first railway—it was known as the Surrey Iron Railway, used horse traction, and ran from Wandsworth to Croydon—was opened for traffic in 1805, the railways never looked back. When the Stockton–Darlington railway was opened in 1825, it became obvious that the new means of communication marked a complete supersession of through passenger traffic by road. The transfer of long distance traffic from road to rail was instantaneous as soon as a railway became open to the towns along its route. This is simply illustrated by the fact that the last stage-coach between London and Birmingham ran in 1839, whereas the railway line was opened only a year previously in 1838.

What Macadam came to describe as the 'calamity of railways' fell upon the Turnpike Trusts in particular between the years 1830 and 1850. Because of the ease of train travel long distance traffic by road was so heavily curtailed that very many of the trusts were brought to a state of chronic insolvency and soon they began to disappear. All projected highway improvements were either abandoned or shelved and during the second half of the 19th century the roads went back to practically the same state as they had been in before the establishment of the toll system. In 1864, conditions were so desperate that a Commission of the House of Commons was appointed to inquire into the Turnpike Trusts; its final report strongly recommended the abolition of the entire toll system. This advice was accepted and acted upon by refusing to grant a renewal when the authorized term of a Turnpike Trust's existence was at an end. While this means of terminating the toll system was a slow one, it was also a thorough as well as economic method. By 1890 there were only two Turnpike Trusts as against 854 in the year 1870. The final Trust, that on the Anglesey portion of the London–Holyhead road, collected its last toll on November 1st, 1895.

From stage-coach to motor vehicle

The abolition of the Turnpike Trusts meant that all roads reverted to the old system of parish maintenance. A measure of the chaos into which these turnpike roads were thrust is reflected in the statistic that at the turn of the latter half of the 19th century there were some 15000 separate and mutually independent highway boards in England and Wales.

The transfer of some 37000 km of turnpike roads naturally caused resentment in those parishes that had to support them out of local funds, especially as these were the 'main' roads and had to receive considerable upkeep. In 1878 the Highways and Locomotive (Amendment) Act abolished the parochial charges, defined main roads, and legislated that half the cost of the main roads be spread over the county. In 1882 the State accepted financial responsibility for aiding the highways and made a grant of one-fourth of the cost of the main roads to the highway authorities; in 1887 this grant was increased to one-half. The Local Government Acts of 1888 and 1894 resulted in changes which essentially stabilized responsibility for the roads by assigning the rural main roads to county councils, county boroughs became responsible for all their own roads, and the rural, urban and borough councils had to look after the remaining roads.

It must be mentioned here that all of these reforms in road administration were undertaken at a time when the roads were practically deserted, unrepaired and almost universally considered useless. The reforms were primarily the outcome of despair over what to do with the miles upon miles of derelict roads, rather than in any constructive and pioneering spirit. Fortunately, however, these reforms came at a time, still unsuspected, when they could be of great use, for by the turn of the 20th century the mixed blessing of the motor car had arrived.

The beginning of the motor age in Britain can perhaps be precisely defined as the year 1896, for this year marked the passing of the Locomotive on Highways Act. Up to that date all road locomotives were limited to speeds of 6·44 km/h in open country and 3·22 km/h in a populous neighbourhood, and each vehicle had to be preceded by a man carrying a red flag. The Act of 1896 gave to the light locomotive, in which was included the motor car, the right to travel upon the highways at a speed not exceeding 22·53 km/h.

The enfranchisement of the motor vehicle in 1896 was celebrated in November of that year by an organized motor run from the Thames Embankment in London to Brighton. Little did these pioneer motorists realize that they were ushering in a transportation revolution in Britain; their trip marked a return of traffic to the roads and the need for a new look at the roads. Engineering for highways was about to be initiated.

THE FUTURE

The foregoing discussion has attempted to outline the major factors influencing the development of the road in Britain. It has stopped short at the beginning of the 20th century for a number of reasons. One of these is that further developments in the British highway system are discussed elsewhere in this text, where they are more fitted. A more important reason however is the fact that, once the era of the motor vehicle is reached, the history of the road is no longer simply the history of the road in Great Britain. Prior to the 20th century it is possible to say that road development in this country was reflective of what was occurring in other nations; after the turn of the century this was not so. Without doubt the emphasis shifted at this stage to the United States where the most striking feature of the American economy and way of life was its developing dependence upon the motor vehicle and highway transportation.

At the present time some 60 per cent of the world's motor vehicle population is located within the United States—this in spite of the fact that under 7 per cent of the world's population live there. According to the latest statistics at hand, almost 90 million cars were operated in America in 1970; this is approximately 1 car for every 2·3 people in the country. In Britain[(8)] the corresponding figures for 1970 were 11·5 million and a ratio of 1 to 4·7.

Coincident with the rise of the motor vehicle in the United States has been the growth of an elaborate road network within the country. At the moment there are about 5·47 million kilometres of road in the United States, ranging from complex urban and rural motorways to 'dirt' roads reaching into every corner of the land. To meet the tremendous demands for highway construction and reconstruction, great research programmes have had to be initiated, not only in the existing field of highway engineering, but in the vast new field of transport planning and engineering. The results of these studies have revolutionized highway engineering practice throughout the world, and either directly or indirectly stimulated many of the investigative programmes being carried out in countries now entering the motor age.

Perhaps the greatest single spur to road development and organized highway research came about with the passing of the American Federal Aid Highway Act of 1944 which authorized the designation of a highway system 'so located as to connect by routes, as direct as practicable, the principal metropolitan areas, cities, and industrial centres, to serve the national defense, and to connect at suitable border points with routes of continental importance in the Dominion of Canada and the Republic of Mexico'. This network, known as the Interstate and Defense Highway System, connects 90 per cent of the American cities with populations of over 50 000, by means of about 66 000 km of motorway constructed at a cost of some 40 billion dollars. About 9650 km of this system are composed of extensions into urban areas and city circumferential urban routes.

The Interstate System, although constituting only a little more than 1 per cent of America's total road and street system, is expected to carry 20 per cent of all the traffic. Most of the routes are 4-lane dual-carriageways, growing to 6 and 8 lanes in and near metropolitan areas. Access to the highways is controlled throughout the entire system, entry being only at carefully selected locations. Traffic interchanges, overpasses, and underpasses eliminate all grade crossings, whether they be highway or railroad. When the system is completed, it will be possible to drive from coast to coast across America without encountering a single traffic light, roundabout or stop sign.

It has been said that one has only to look at the United States today in order to see the Europe of tomorrow. Whether this is so or not is, of course, a matter of fascinating speculation which is incidental to the content of this textbook. It is very probable, however, that the vehicle ownership statistics for the United States are representative of what can be expected elsewhere. If this is so then there can be little doubt that motorway systems equivalent in scope and magnitude to the Interstate System will soon be criss-crossing Europe and, eventually, the less developed and emerging nations of the world.

SELECTED BIBLIOGRAPHY

1. COLLINS, H. J. and C. A. HART. *Principles of Road Engineering.* London, Arnold, 1936.
2. PANNELL, J. P. M. *An Illustrated History of Civil Engineering.* London, Thames & Hudson, 1964.
3. ANDERSON, R. M. C. *The Roads of England.* London, Benn, 1932.
4. BELLOC, H. *The Road.* Manchester, Hobson, 1923.
5. HARTMAN, E. H. *The Story of the Roads.* London, Routledge, 1927.
6. JACKMAN, W. T. *The Development of Transportation in Modern England.* London, Cass, 1962.
7. WEBB, S. and B. WEBB. *English Local Government: The Story of the King's Highway.* London, Longmans, Green, 1913.
8. BRITISH ROAD FEDERATION. *Basic Road Statistics.* London, The Federation, 1972.

2 Highway legislation, classification, finance and administration

The public roads of Britain are a government responsibility. They are supported by taxes raised from the people and maintained and operated by the central government and the local authorities through their public servants. This responsibility is a necessary one, due not only to the magnitude of the highway network, but also because of the unique role that it plays in the national economic effort. Indeed, the influence of the road network is now such that major, and sometimes minor, improvements inevitably affect the lives, and even the fortunes, of the people living and working nearby.

Because of this governmental relationship a particular type of administration has grown up in order to manage the highways and cope with the problems which they create. Since the highways necessitate the expenditure of vast sums of money, while at the same time they often impinge on the basic civil rights of the individual, much legislative effort has been exerted to protect the interests of the public. As a result the young highway engineer entering upon his professional career and putting his special technical qualities to work is often confounded by the multiplicity of administrative and legislative factors which he must take in his stride before he can attain a position of executive responsibility.

To be efficient the road engineer must be fully aware of the nature of his product. While he will usually have qualified non-technical advice and facilities at his disposal, he must nevertheless understand the basis of these facilities so as to make the best use of them. The function of this introductory chapter therefore is to give a brief perspective of highway administration and legislation, and the manner in which they function to provide the highway transportation service. Intimately bound up with administration and legislation are the methods by which the highways are classified and financed, and so these are discussed also.

LEGISLATION

The basis of all highway administration and finance lies in the legislative process which brought them into being. In formulating the laws which enable finance to be raised and allocated, and administration to function, Parliament indirectly as well as directly set out and defined the objectives of performance, thereby outlining the processes now in existence. The following is therefore a brief summary of the legislative history which underlies the operation of the present system. No attempt is made here to refer to all the laws that are of interest; these are literally too numerous and too detailed to cover within the limitations of this text. Instead emphasis is laid on those laws which, it is felt, have had particular significance in setting in motion events or restrictions that are of particular importance to highways and their traffic today.

Finally, it should be mentioned that, although the Highways Act, 1959 and the Road Traffic Act, 1960 together consolidate most of the pre-1960 laws relating to the creation, management and use of highways and bridges, a certain volume of legislation remains scattered in a variety of statutes. It is not possible, due to space limitations, to refer to these Acts here. For quick reference to this legislation, the reader is referred to an excellent aide-memoire that is available in the technical literature on highways.[(1)]

Pre-World War I laws

Prior to 1888 the roads of Great Britain were administered by highway boards, the area of their responsibilities being determined by the grouping of a series of parishes, each of which had the right of electing a member to such boards. The roads under the control of these bodies were all rural in character, while those passing through built-up areas were directly maintained by county borough councils, municipal corporations or boroughs, and local boards which, in addition to having the duties and powers of maintaining the highways, were the local authorities for such purposes as public health and public utility services. At that period relatively negligible grants were made to the highway boards from the central government towards the cost of many local services, including the cost of through roads.

In 1888 and 1889 two most important Acts, the Local Government (England and Wales) Act, 1888 and the Local Government (Scotland) Act, 1889, brought into existence local authorities known as county councils which were given far reaching duties dealing, amongst other matters, with public health, police, lunacy, small holdings, contagious diseases of animals, and 'main' roads. In addition to being the highway authorities for the erstwhile Turnpike Trusts, the county councils had the power to declare as main roads other roads which in their opinion should be so declared by virtue of their through traffic and the total volume of traffic. The obvious intention of these laws was to ensure that the maintenance of main roads should become a county, as against a district or local, charge. Under these Acts, the county boroughs were to maintain all roads within their districts.

The reform in highway administration initiated by the Local Government Acts of 1888 and 1889 was extended by the Local Government Act of 1894. This legislation abolished the highway boards and highway parishes and transferred the duty of maintaining roads other than main roads, in rural areas, to rural district councils.

What may well be described as the legislation which signalled the emancipation of the motor car was passed in 1896 in the form of the Locomotive on Highways Act. This Act abolished the requirement that a road locomotive had to be preceded by a man carrying a red flag, and gave to it the right to travel upon the highway at a speed not exceeding 22·5 km/h.

The year 1909 saw the beginning of official government recognition that special measures were needed on a national level to ensure adequate road facilities. This was signified by the passing of the Development and Road Improvement Funds Act, 1909, which constituted a Road Board with powers to use money raised from vehicle and petrol taxation to build new roads and to carry out schemes to widen and straighten existing roads, to by-pass villages, and to allay the dust nuisance. This Act, for the first time, brought a central government department into being that really had power to define which roads should be given special assistance, and what form the improvements should take. Money for these purposes was raised by a petrol tax and a graduated scale of motor vehicle licence duties based on a horse-power rating defined by the Royal Automobile Club and paid into a fund called the Road Improvement Fund; in time this became known as the Road Fund.

The year 1909 also saw the passing of the Housing, Town Planning, Etc. Act, the purpose of which was to ensure that land in the vicinity of towns should be developed in such a way as to secure proper sanitary conditions, amenity and convenience. This legislation is of significance in that it discloses for the first time the gulf which, unfortunately, exists in many instances to this day between the highway engineer and the town planner. The emphasis in the Act was placed upon securing proper public health considerations, within which the social implications raised by the construction of a new road or road system were not at all taken into account. As a result, and more by default than any other reason, the planning and design of roads were regarded as primarily the responsibility of the engineer and only secondarily as part of the town planning process.

Legislation between the World Wars

In 1919, for the first time in its history, the government created a Ministry of Transport to which the powers and duties of the Road Board, together with its officers and staff, were transferred. The Ministry of Transport Act empowered the Minister of Transport to devise and implement a road classification system after consultation with the Roads Advisory Committee and the local authorities.

Following this legislation the country was divided into administrative divisions, each under the jurisdiction of a Divisional Road Engineer who was expected to study the road and bridge problems in his area, and to pronounce upon proposed future developments as they were submitted by the local highway authorities. Another outgrowth of the Ministry of Transport Act was the selection in the year 1922 of 36 619 km of roadway to be called Class I roads and 23 564 km of Class II roads; these are the origins of the now familiar A and B identification of roadways.

The year 1920 saw the passing of two important pieces of legislation, the

Finance Act, 1920 and the Roads Act, 1920. The Finance Act created a simple system of taxing cars which, in essence, remained in operation until after the second World War. For the first time also, goods vehicles were taxed more severely than passenger cars. The Roads Act established the form of registration and licensing which is in use to this day. The county and county borough councils were made responsible for registering the vehicles and collecting the duties imposed. An important by-product of the Roads Act was that for the first time it was possible to obtain accurate data regarding the numbers of the different types of vehicles in use.

The Roads Act, 1920 also provided that contributions for the maintenance of existing roads could be granted from the newly constituted Road Fund, which took the place of the Road Improvement Fund. As a result of this the Minister of Transport, in 1922, made available grants amounting to 50 per cent of the cost of maintaining and improving Class I roads, and 25 per cent of the same costs for Class II roads, as well as special grants to the improvement of other roads, and to meet special and unexpected contingencies.

The Road Improvements Act, 1925 gave to highway authorities the power to prescribe building lines along roads and, at bends and corners, to control the height and character of walls, fences, hedges, etc., and to restrict the erection of new buildings where adequate sight distances were required. The Act also authorized the Ministry to conduct research and experimental work; this led eventually to the establishment of what is now known as the Transport and Road Research Laboratory.

The Finance Act of 1927, for the first time since mechanically propelled vehicles were taxed, decided to regard a portion of the receipts from ordinary motor car taxation as a luxury tax, and to use that portion as a contribution to the general exchequer expenses of the country.

The Local Government Act, 1929, recognized that the maintenance and repair of roads was in the hands of too many minor local authorities and set about reducing this number. Prior to this Act, by far the greatest amounts of roads, known as 'district roads', were under the control of some 600 rural district councils. With the passing of the Act these bodies were eliminated as highway authorities and their roads, equipment, plant and officers were transferred to the county councils; the maintenance and improvement charges for these roads then became county-at-large charges. In like manner, municipal boroughs and urban districts with populations less than 20 000 also ceased to be highway authorities and the roads in their areas became county roads and the responsibility of the county councils. The larger urban districts in England and Wales were left in undisturbed possession of their roads with one major proviso; where a road not previously designated as a county main road was classified by the Ministry of Transport, then that road became a county main road for the expenditure on which the county council and the Ministry of Transport became responsible. The position of county boroughs remained unchanged. In Scotland, the small burghs were also dispossessed of their road maintenance powers, and the residents of these localities became for the first time county ratepayers for highway purposes.

In 1928 a Royal Commission on Transport, which was established to consider problems arising out of the growth of the motor vehicle population and usage, concluded that some regulation of public transport was necessary in the public interest. The result was the passing of the Road Traffic Act of 1930 under which control of public service vehicles was vested in Area Traffic Commissioners appointed by the Minister of Transport. Their function was to fix and maintain fares, to sanction routes and time-tables, and to eliminate any unnecessary services. Every operator of a public service vehicle was required to obtain an annual licence for his vehicle (this has since been changed to 3 years in the case of stage-carriage licences) and upon renewal he had to justify his services against any rival provider of transport who might object. The net effect of these requirements was to protect established public transport operators; this was at the price of tying them down to fixed lines and thereby reducing their flexibility. This Road Traffic Act also revised the speed limits for certain types of commercial vehicles while it swept away all limits for motor cars and motor cycles, thereby making it possible for the first time for highway engineers to design officially for high vehicle speeds. It made third party insurance obligatory and set standards and penalties for careless and dangerous driving.

The year 1933 saw the passing of the Road and Rail Traffic Act which attempted to regulate with respect to goods vehicle licensing and to ensure that commercial traffic paid a greater share of the cost of the road system. A second Road Traffic Act was passed in 1934 which imposed an overall speed limit of 48 km/h in all built-up areas, i.e. roads in areas where systems of street lighting were in existence. At the time it was considered that this was the universal answer to the road accident problem; this, unfortunately, as is well known today, was not the case. The second major feature of this piece of legislation was that it introduced a driving test for all drivers who did not hold a licence before 1st April, 1934. A special driving licence was also required for drivers of heavy commercial vehicles.

The Road Traffic Act, 1933, also permitted advances from the Road Fund to local authorities for the purpose of erecting traffic signs. In addition the Minister of Transport was empowered to make regulations regarding the use of pedestrian crossings, and the county councils were enabled, under certain conditions, to provide street lighting.

In 1935 the Government awoke to the dangers of ribbon development and its effects upon the free movement of vehicles on the road. As a result it passed the Restriction of Ribbon Development Act which made it unlawful to construct any building within 67 m of the centre of a classified road, or to construct, form or lay out any means of access to or from such a road, without the consent of the highway authority. This Act also incorporated a new system of defining the land needed for future roads and road improvements.

The following year, another most important legislative Act was passed. This was the Trunk Roads Act, 1936, which for the first time made the Ministry of Transport a highway authority in its own right. This was done by giving the Ministry complete financial responsibility for 7249 km of major through routes outside the boundaries of London, the county boroughs in England and Wales, and the large burghs in Scotland. The Act also made provision for delegating the Minister's highway authority powers to the local authorities so that they could act as his agents for these roads.

The final important pre-war piece of legislation was the Finance Act of 1936. This Act required the Ministry of Transport to draw its revenue for highway purposes from sums voted by the House of Commons. Whereas, before, part of the receipts from vehicle taxation automatically went to the Road Fund where, *as a right*, they were devoted to road improvement, now the Ministry of Transport had to budget for its monies in a manner similar to, and in competition with, other ministries.

Post-World War II legislation

The first major road legislation after the war was the Trunk Roads Act of 1947. This Act transferred another 5929 km of Class I roads to the responsibility of the Ministry of Transport, together with all private bridges on the trunk roads. On this occasion the Minister became responsible also for main roads designated as trunk roads which passed through county boroughs and the county of London. In addition, and for the first time, provision was made for the two carriageways of a dual carriageway road to diverge for reasons of engineering or property so that the land in between need not be considered part of the roadway. Furthermore, the Minister was required to keep the national system of through routes under constant review, and to reorganize it as changing circumstances demanded.

In 1947, the Town and Country Planning Act was passed and this had the effect of requiring planning authorities to prepare plans showing the intended use of all land within their areas. This Act had the effect of co-ordinating highway and land use planning, since road proposals now had to be shown on the statutory development plans, and their relationship to land use and development was apparent to all. Furthermore, it meant that road proposals would be kept up-to-date since the development plans were to be revised every five years.

The year 1947 also saw the passing of the Transport Act which set up a British Transport Commission for the purpose of securing an efficient and properly integrated system of certain public transport and port facilities within Great Britain. This Act, which gave considerable powers to the Minister of Transport by enabling him to give directions to the Commission as to the exercise and performance of their functions in relation to matters which appeared to him to be of national interest, marked the first major move toward having a co-ordinated public transport system in Great Britain.

The motorway era in Great Britain can be said to have begun with the passing of the Special Roads Act of 1949. This Act is of major importance in that it marks the departure from the previously held principle that everyone normally had the right of free access along the highway; now it became possible to construct roads reserved for the exclusive use of certain classes of traffic. By a government decision made in 1946 motorways were considered to be reserved for the purpose of facilitating the movement of long distance motor traffic and of connecting some of the main centres of population.

The Transport Act, 1953, provided for the winding-up of the Road Haulage Executive of the Transport Commission, and the transfer of the Railway Executive's powers to area authorities. More important, however, is the fact that this Act signified the enfranchisement of road transport. Since that time the number of goods vehicles on the roads has increased by over 65 per cent.

The first of what were to be two major traffic management legislative measures was passed in 1956 in the form of the Road Traffic Act. Under this Act, compulsory annual tests were introduced for all vehicles aged ten years and older (these have since been reduced to three years). This was supported by stricter regulations on the sale of unroadworthy vehicles and the renewal of driving licences. In a further attempt to improve road safety, the penalties for dangerous driving were increased and extended to cover pedal cyclists, and pedestrians became legally bound to obey the directions of police on traffic duties. In addition the Minister of Transport was empowered to approve local authority schemes fixing parking places and charges, and to sanction experimental traffic management schemes by the police. These latter provisions were brought about by the need to ease traffic congestion within the London metropolitan area.

The second bill of particular importance with respect to traffic management was the Road Traffic and Road Improvement Act, 1960. This Act provided the Minister of Transport with power to take more effective action in dealing with the traffic, parking and roadworks' problems in Greater London. Local authorities were also given new powers to make orders for the purpose of a general scheme of traffic management in a specified area, as well as in connection with the provision of off-street parking facilities. In addition provision was made for the police to appoint traffic wardens, while members of the public were given the option of paying fixed penalties instead of being prosecuted for certain traffic offences.

About this time also, a number of most important measures were taken in order to simplify and bring up-to-date the law with respect to roads and their traffic. These measures involved the passing of a number of consolidating Acts, namely, the Highways Act, 1959, the Road Traffic Act, 1960, and the Vehicle Excise Act, 1962.

The Highways Act codified the existing statute law, the only amendments made being for the sake of clarification, the repeal of obsolete or unnecessary provisions, and the making general of a few clauses in local and adoptive Acts which were accepted as uncontroversial. No radical alteration was included in this Act as it was intended to be the forerunner of general amending legislation which would bring the highway law of the country up to modern requirements.

The Road Traffic Act was a consolidation Act for the purpose of gathering together previous traffic legislation. It contained the following seven parts: 1. General provisions relating to road traffic. 2. Minimum age for driving motor vehicles and licensing drivers. 3. Public service vehicles. 4. Regulation of carriage of goods by road. 5. Licensing of drivers of heavy goods vehicles. 6. Third party liability. 7. Miscellaneous and general.

The Vehicle Excise Act consolidated previous legislation relating to the licensing and registering of mechanically propelled vehicles.

In 1961, the Highways (Miscellaneous Provisions) Act was passed which made it possible for the first time to bring an action against a highway authority in respect of damage resulting from its failure to maintain a highway maintainable at the public expense. In 1962, another Road Traffic Act, which was primarily concerned with promoting road safety, granted more flexible powers to impose speed limits; a penalty system for traffic offences was also presented. Another provision dealt with the control of hovercraft when used on roads.

The Transport Act of 1962 provided for the dissolution of the Transport Commission, and the division of its functions between four separate boards, appointed by the Minister of Transport, for British Railways, London Transport, British Transport Docks and British Waterways. A Transport Holding Company was appointed to take over all the commercial activities, i.e. bus companies, road haulage and travel agencies, run in company form. In addition a Nationalized Transport Advisory Council was set up to advise the Minister of Transport regarding the co-ordination and other aspects of the nationalized transport undertakings.

The Road Safety Act, 1967, was of particular importance in that it clarified the situation regarding drinking and driving. This Act made it an offence to drive or attempt to drive if 100 ml of blood contained more than 80 mg of alcohol, and gave the police power to impose selective roadside breath tests and to require blood or urine samples if the tests were positive. The existing offences of driving or being in charge while under the influence of drinks or drugs were retained. The Act also increased governmental powers with regard to roadside checks on commercial vehicles.

The Transport Act, 1968, marked a major breakthrough with regard to governmental support for public passenger transport. It gave the Minister of Transport and the Secretaries of State for Scotland and Wales the power to designate any area outside London as a Passenger Transport Area, with both an Authority and an Executive to integrate and develop public transport services. The governmental National Bus Company acquired the Transport Holding Company's bus operating and manufacturing companies, except for those vested in the Scottish Bus Group (responsible to the Secretary for State for Scotland). The Act also increased the fuel grant for stage carriage services, provided 50/50 Exchequer/local authority grants for uneconomic bus and ferry services and 25 per cent grants to buy new buses, as well as empowering central government and local authorities to make grants toward approved capital expenditure on public transport facilities. The purposes for which local authorities could make orders regulating traffic were widened to help buses, pedestrian movement and amenity; the authorities were also enabled to restrict parking meters to specific classes of users, and to use surplus meter revenues for purposes connected with public transport or road provision. The Transport Act also gave to the National Freight Corporation and to British Railways Board the right to object to certain goods services (e.g. involving vehicles weighing over 16 tonnes when loaded and doing trips of more than 160 km) if they could be provided wholly or partly by rail.

The Motor Vehicles (Variations of Speed Limit) Regulations of 1962, 1966 and 1971 provided for speed limits for different classes of vehicles on the various types of road. The Zebra Pedestrian Crossings Regulations, 1971 specified a control area within which vehicles are banned from overtaking, parking or waiting on the approach to a crossing, as well as indicating the areas within which pedestrians should not cross the road. The Road Vehicles Lighting (Standing Vehicles) (Exemption) (General) Regulations, 1972, provide that cars, motorcycles and light goods vehicles may park without lights on any road provided that (i) the road is subject to a speed limit of 48 km/h or less, (ii) no part of the vehicle is within 13·75 m of a road junction, and (iii) the vehicle is parked close to the kerb and parallel to it and, except in one-way streets, with its nearside to the kerb.

CLASSIFICATION

Britain has one of the most comprehensive and penetrating roadway systems of any country in the world. The statistics show that there are 334 894 km of maintained roads at this time on this island. Table 2.1 gives the breakdown of the different types of road forming this total.

Trunk roads

Following the Trunk Roads Acts of 1936 and 1947, the central government assumed responsibility for 13 178 km of main road which were to form the nucleus of a national system of routes for through traffic. As can be seen from the figures in Table 2.1, these early trunk roads still form the core of Britain's main road system.

TABLE 2.1. *Composition of the highway system in Great Britain in 1971*[2]

Classification	*No. of km*	*Distribution of costs, %*	
		Central government	*Local government*
Trunk	14 668 (1269)*	100	—
Principal	32 727	75	25
Other	289 380	—	100

*Figures in brackets indicate the amount of motorways within the total.

The term 'trunk road' does not guarantee that a particular level of service will be provided on any highway bearing the title. It is an all-embracing term which includes roadways ranging from very busy motorways to ones less than 5·5 m in width and carrying very light volumes of traffic. In general, however, the level of service provided by any given trunk road is a relative function of its national and regional importance.

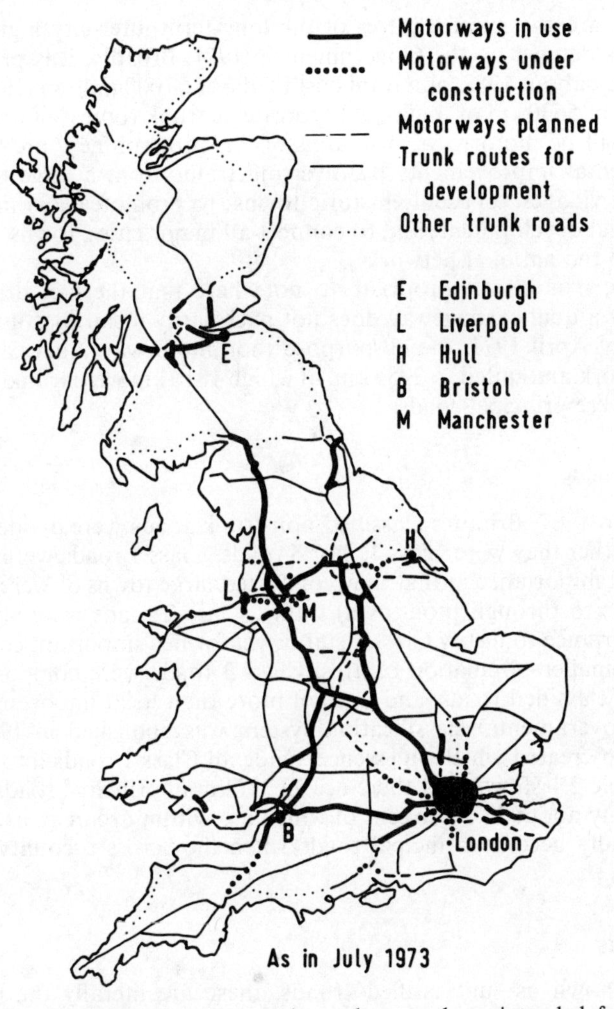

Fig. 2.1. The inter-city strategic trunk road network as intended for the early 1980's.

By means of enabling legislation which became law in May, 1949, special roads can be reserved exclusively for particular classes of traffic, e.g. motor vehicles, or pedestrians, or pedal cycles or other prescribed traffic. The most important of these is the motorway, usage of which is restricted to motor vehicles and certain types of motor cycles. Apart from its high-quality geometric design standards, the most important features of a motorway are that there is complete control of access and complete separation of all conflicting traffic movements. Essentially all motorways built to-date form part of the trunk road network.

Figure 2.1 shows the motorway system as of July, 1972. Also shown in this figure are the main features of the long-term inter-city highway programme as defined by the Government in 1971. Broadly, it is proposed to complete by about 1982, at a total cost of about £3 000 million (1971 prices), a network of 5630 km of 'high quality strategic trunk routes' of which about 3200 km will be motorways. The aims of this programme[3] are to achieve environmental improvements by diverting traffic from a large number of towns and villages, especially historic towns; to promote economic growth and regional development; and to connect all major cities, towns, ports and airports to the national network.

Finally, it is also appropriate to note here that the fact that a given highway is a dual carriageway does not necessarily mean that it is a trunk road. As of April 1971, the all-purpose (non-motorway) dual carriageway road network amounted to 2955 km, of which 1810 km were trunk roads and 1145 km were principal roads.

Principal roads

Prior to 1967, Britain's classified non-trunk roads were divided according to whether they were Class 1, 2 or 3 roads. Class 1 roads were primarily of regional importance in that they connected large towns or were of special importance to through (non-town) traffic. Class 2 roads were not of sufficient importance to justify Class 1 status, yet formed important connections with the smaller population centres. Class 3 roads were composed of the remaining classified roads, and were of more than local importance.

This governmental classification system was abolished in 1967, and a new system created which, in essence, made all Class 1 roads into Principal roads. Table 2.1 shows that there are 32 727 km of principal roads in Great Britain, only about 11·5 per cent of which are within urban areas, i.e. trunk roads usually become principal roads when they cross a county borough boundary.

Other roads

Also known as 'unclassified' roads, these are literally the remaining highways within the national (urban and rural) network. They include what used to be known as Class 2 and 3 roads, as well as previously unclassified roads.

Highway identification

As will have been gathered the *classification* of a highway is related to its general importance within the national system; it also has implications with regard to the manner in which roads are funded. It is important not to confuse classification with *identification* which simply refers to the numbering system by which given roads are known.

A glance at any road map will show that all roads other than motorways are first divided according to whether they are A-roads or B-roads. The

A-roads—which tend to be the old trunk and classified roads—are then generally sub-divided into three, indicated by the number of digits after the letter, whilst all the B-roads and some A-roads have four digits after the letter.

There are nine one-digit primary roads, six of which are centred in London, and the remaining three on Edinburgh. London-based roads are the A1 to Edinburgh, (and proceeding in a clockwise direction) the A2 to Dover, A3 to Portsmouth, A4 to Bristol, A5 to Holyhead, and A6 to Carlisle. The Edinburgh-based roads are the A7 to Carlisle, A8 to Glasgow, and A9 to Wick.

Between each of the primary roads there are other important roads which are given double numbers starting with the initial primary number. Thus, these roads between, say, the A3 and A4 start with the number 3, e.g. the A30 between London and Penzance.

Other lesser A-roads have three digits after the letter, and these start with the appropriate figure for the sector, e.g. the A483 between the A4 and A5, and joining the A48 with the A5. All B-roads and some A-roads of lesser importance have four figures, starting with the primary digit of the district in which they are located, e.g. A3072 between Bude and Bickleigh, or the B2080 between Tenterden and New Romney.

FINANCE

Highway grants

The roads of Britain can be divided into three main groupings from the point of view of financial support, and the amount of money which any particular road receives is dependent on its grouping (see Table 2.1).

The central government (via either the Secretary of State for the Environment in England, the Secretary of State for Scotland, or the Secretary of State for Wales) is the highway authority responsible for all the trunk roads, and as such it provides all the funds for the planning, design, construction and maintenance of these facilities. The decision as to whether or not a road is classified as a trunk road is made by the central government itself.

Prior to 1967, the central government paid for 75, 60, and 50 per cent of the capital costs of major, and certain minor, improvements on Class 1, 2 and 3 roads. In 1967, the decision was taken to change the system so that the central government would pay for 75 per cent of the cost of improvements to (including new sections of) principal roads; all other expenditures including maintenance costs on principal roads as well as maintenance and improvement costs of non-principal and non-trunk roads, must be paid for out of local authority funds. To alleviate hardship which this change might cause to local authorities, the central government simultaneously introduced a non-specific or block grant in support of the local rates, i.e. a calculated sum is given to the local authority, from the national exchequer, on the basis of approved expenditure estimates borne by the rates, but the local authority may then divide this grant between its local services as it sees fit.

A major advantage of the above change is in relation to principal roads, particularly in urban areas. Although most of the traffic on urban principal roads might be local traffic, very many of the local authorities could not find the finance under the old system to improve these roads; under the new system, the great burden is shifted to the central government—which also means that a higher quality improvement can be provided than might otherwise be the case. Furthermore, as a *quid quo pro*, the central government is released from all tedious problems dealing with the allocation of funds for the (now) unclassified roads—problems which are better dealt with by local government anyway.

Sources of revenue

As is indicated in the previous discussion, the two main sources of highway revenue are the central government funds and the local government rating funds. The following discussion is therefore primarily concerned with the financial means by which these monies are raised.

Central government revenues. The road grants provided by the central government are financed from the general revenue of the state and, following allocation by Parliament, they are distributed by the Secretaries of State. In 1971 the amount of money spent in this way was equivalent to 3·0 per cent of the national expenditure.

The general revenue of the state comes, of course, from many sources, not the least of which is the taxation revenue from motor vehicles. In 1971 the motor taxation contribution to the national exchequer was equivalent to 12·3 per cent of the national revenue. Since the income from motor taxation by far exceeds the national expenditure on roads, it will be assumed for the purpose of this discussion that the money raised by this taxation is in fact the money used to finance the road system.

As indicated by Fig. 2.2, there are three main forms of motor vehicle taxation. These are (1) Motor fuel taxes, (2) Vehicle and licence duties, (3) Purchase taxes.

A *Motor fuel tax* was first levied upon motor fuels in Britain in 1909, when a rate of tax of 3d per gallon was imposed on light oils used for motoring purposes, a 50 per cent rebate being given to commercial vehicles. In 1916 this tax was raised to 6d on the gallon and this was levied until 1921 when the motor fuel tax was repealed entirely. In 1928, however, a new petrol tax at the rate of 4d per gallon was again established, and this has since been raised to its current value of 4·95p per litre.

Following the Development and Road Improvement Funds Act, 1909, the same graduated scale of *Vehicle Registration Duties* was imposed on all motor vehicles. In 1921, the duties for cars and goods vehicles were separated and this practice has continued ever since. The current duties

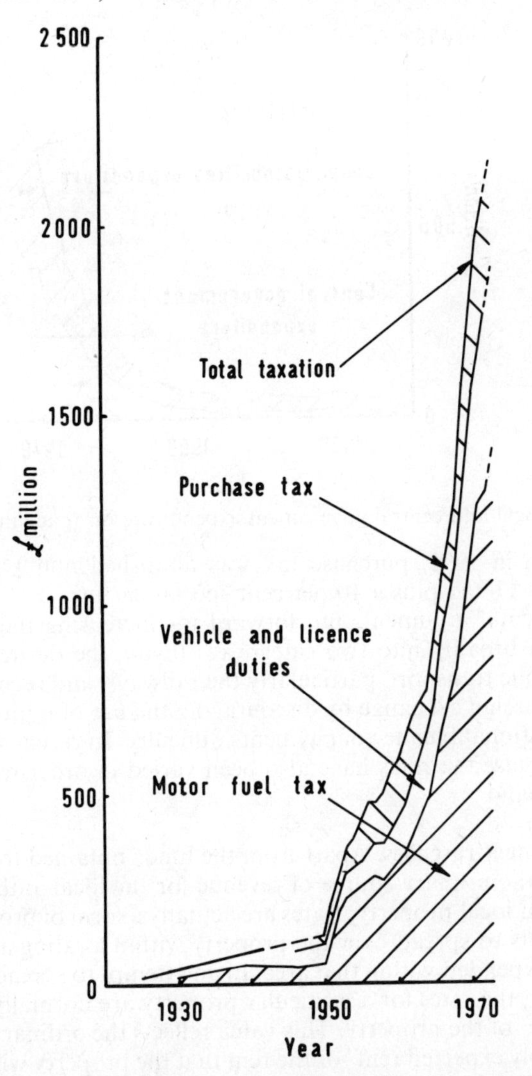

Fig. 2.2. Motor taxation revenue in Great Britain[2]

(imposed in 1968) are a flat rate of £25 per year for all cars, and a graduated rate for goods vehicles varying from £24 to £135 (+£13·50 per $\frac{1}{4}$ ton over 4 tons).

A *Purchase Tax* of 33$\frac{1}{3}$ per cent was first imposed on cars in 1940, and since then the tax rate has varied at irregular intervals, depending upon national economic needs (e.g. 66$\frac{2}{3}$ per cent in 1951, 25 per cent in 1972). Other than during the period 1950–9, purchase tax has never been levied on goods vehicles. With the introduction of the *Value Added Tax* (*VAT*) in

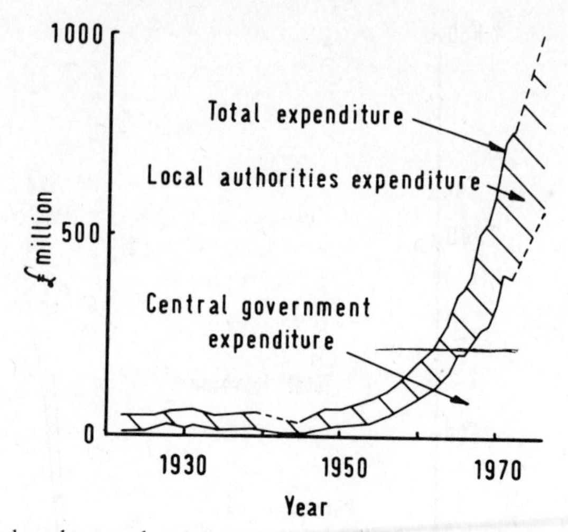

Fig. 2.3. Local and central government expenditure on roads in Great Britain[2]

Great Britain in 1973, purchase tax was abolished and replaced with a 10 per cent VAT tax plus a 10 per cent special car tax.

Governmental arguments put forward for increasing the fuel tax have tended to fall broadly into two categories: firstly, the desire to encourage the use of public transport, particularly the railways, and secondly, the need to conserve foreign exchange by discouraging the use of imported materials in times of national balance of payments difficulty. In recent years, both the fuel and purchase tax rates have also been varied in order to help regulate domestic demand.

Local government revenues. Apart from the funds obtained from the central government, the primary source of revenue for the local authorities are the rates levied on local property. Rates are actually a form of property taxation which attempts to spread over the property within a rating area the cost of the services expended within that area. In an attempt to spread these costs in a fair manner, the rates for a particular property are normally based on the 'annual value' of the property. This value reflects the ordinary letting value, the 'reasonably expected rent' or the rent that the property will command in the open market. This rent is estimated by a local valuation officer who is a member of the valuation department of the Board of Inland Revenue. When the local authority is informed of the annual value of the property, it then levies an assessment against the property and this is called the rate for the property.

The monies raised from the rates are allocated by the elected representatives of the people to the various services, including roads, according to the needs in relation to the total expenditure.

As well as loans, another minor method of financing particular stretches of new roadway is to tax frontagers for the cost of constructing the new facilities. Thus, for instance, if a new housing estate is being constructed the roads will be built by the estate contractors and the cost of so doing is then added on to the cost of the houses serviced by these roads.

Brief discussion relative to highway revenue

Roads cost a lot of money to construct and maintain; of this there is no doubt. What is very much in doubt, however, is the equitableness of the presently used methods of raising the required revenue. If road costs were to be charged in direct accordance with the benefits received, then the expenditure on each road or road system should ideally be assessed against the following three main beneficiaries: (1) The general public, (2) The adjacent properties, (3) The motorists. That this principle is accepted is reflected in the form in which highway finance is now carried out. Unfortunately, however, the manner in which the costs are allocated between the beneficiaries is not based on any scientific evaluation but rather has grown up as a result of practical expediency and need.

Payments by the general public. This concept requires payments to take place in the form of grants from the central government towards the construction and maintenance of highways that are in the national or general interest. Under this concept the funds allocated for this purpose would come from finance raised by non-motor vehicle taxation sources.

Roads falling into this national category are undoubtedly the rural motorways and trunk roads, since their function is to serve national economic needs for long-distance vehicle movement. Central government support for urban motorways should probably also be included since, in a densely populated island country the size of Britain, it can probably be said with justification that in cities where urban motorways are necessary their influence is indirectly, if not directly, national; motorways in such cities as London or Birmingham, to name only two, can certainly be included within this user classification.

It must be pointed out, however, that not since 1932 has the general public as a whole apparently contributed directly towards the cost of the highway system in Britain. A simple comparison of motor vehicle taxation figures and those for central government highway expenditure during recent years shows quite clearly that no non-motor vehicle taxation monies have been used for road purposes but that in fact it is the property owners/occupiers and the motoring public who have contributed the entire financial support to the road system.

Payments by property owners/occupiers. The second group of people to benefit from the road system are the property owners/occupiers who use it to obtain access to their homes or places of business. Roads falling into the category for which the ratepayers should be primarily responsible can be divided into two types; (*a*) roads which are used for the direct purpose of providing access to houses and land—the most obvious example is a cul-de-sac, and (*b*) roads which provide internal communication within the local community, e.g. the streets of a town. The former type of road should ideally be financed by means of direct assessments upon the owners/occupiers of the homes or land being serviced, while the funds to support the latter roads must be provided by the community's properties as a whole, i.e. from local government taxation.

The financing system in Britain generally attempts to apportion the road costs to the ratepayers according to the benefits derived. Whether the present apportionment to the ratepayers is the correct one, is, of course, not known. However, there is no doubt that the current apportionments reflect the general order of importance of these roads to the local communities who pay for their upkeep.

Payments by the motorists. It follows that since streets and highways exist for the primary purpose of allowing vehicles to travel upon them, then the road-users should contribute a substantial amount towards their construction and upkeep. This principle is accepted not only by most highway administrators but also by the motorists themselves. It is at this stage that agreement ends, however, since it is not possible to say, firstly, just how much of the cost of the road system should be borne by the motorists and, secondly, how the portion to be borne by them should be allocated amongst the various users of the highways.

It is doubtful whether a formula based on scientific investigation and evidence will ever be derived that will equitably apportion the road costs between the motorists and the other beneficiaries of the road system; inevitably this apportionment will be—and indeed perhaps it should be—dictated by national policy needs. What is more likely, however, is that eventually sufficient evidence will become available to permit the allocation of costs in such a way that those vehicles which require highly designed facilities, or which cause the greater wear and tear, have to pay a proportionately greater part of the costs than those smaller vehicles which cause less wear and tear and can be adequately served by more lightly constructed facilities.

Various theories have been suggested as to how highway costs should be allocated between different classes of road users. The following are the most widely known:

1. *Differential-cost concept*. This concept, also known as the increment-cost concept, seeks to apportion the costs on an increment-weight basis. Thus the first costing is made on the basis of what the highways would cost if all the vehicles were passenger cars and light commercial vehicles. The additional costs needed to adapt the highways to meet the structural and geometrical requirements of larger commercial vehicles and buses, e.g. thicker pavements, wider carriageways, more gentle hill slopes, additional maintenance, etc., are next calculated and these extra costs are then allocated to the vehicles falling into this category. Finally the extra costs involved in adapting the highways to meet the heaviest vehicles are calculated and allocated.

The total costs are then apportioned in the following manner: all vehicles pay the first or basic increment of cost; all vehicles except the passenger cars and light commercials also pay the second increment; while the final increment is paid only by the heaviest vehicles. In this way the total costs allocated to the motoring public are distributed in an equitable manner.

2. *Weight-distance concept*. This theory assumes that the cost of using the highways should be directly related to the weights of the vehicles and the distances which they travel. The weight relationship is used to reflect the amount of wear and tear and the need for sturdier facilities, while usage is reflected in the distance tally. Thus a passenger car weighing 1 tonne and travelling 5000 km/year would be expected to pay only one-tenth of the taxation paid by a 5 tonne commercial vehicle which does 10 000 km in the year.

This method of taxing commercial vehicles has been used in several states of the United States with, in general, unsuccessful results. To determine the assessment for each vehicle a daily log of the miles travelled had to be maintained by the operator. This met with considerable opposition from the operators who were adversely affected, and it is said that not all of them kept honest records.

3. *Operating-cost concept*. This taxation theory proposes that vehicles should be charged for using the highways on the basis of scientifically determined knowledge of their operating costs. Thus the fees paid by commercial vehicles could be considerably greater than those paid by passenger cars since their operating costs are greater. Proponents of this method of charging argue that the operating cost of a vehicle is proportional to the value of the benefits derived from its use on the highway.

This method actually results in a lower assessment on commercial vehicles than is levied by the weight-distance method. While the cost of operation of a commercial vehicle increases as the weight of the vehicle increases, it actually decreases on a weight-distance basis as the load increases.

4. *Differential-benefits concept*. This theory proposes that motor vehicle taxation can be determined on the basis of estimates of the savings incurred by the different classes of vehicle due to the highway facilities being improved. Thus, prior to every taxation period, an economic analysis would be carried out on the effects of the current improvement programme on the operating costs of the different classes of vehicles, and taxes would then be levied in direct proportion to the savings incurred by each class.

With the present state of knowledge regarding road-user costs, there is no doubt that this method is not a practical proposition at this time.

5. *Area-occupied concept*. Another name for this might be the 'congestion tax theory' since it attempts to assign taxation on the basis of the carriageway space occupied by the different classes of vehicles. Thus commercial vehicles would again have to pay a higher rate of taxation than private cars due to their larger bulk.

This method of taxing road-users has its most fruitful application in urban areas where the principal problem is one of motor vehicle congestion. Details of how a tax of this type might be used to reduce congestion in built-up areas are given in the chapter on Traffic Management and so will not be further described here.

Present system. It is often suggested that motor vehicle taxation in Britain is unjust. Criticisms of the existing system can be divided into those which refer to the total share of the highway costs borne by the motorists, and those which refer to the manner in which the motorists' allocation is shared between the various classes of road-users.

1. *Total motor vehicle taxation.* At the present time the motorist appears to be taxed very heavily in comparison to the highway benefits received in the form of direct road expenditure. Figs. 2.2 and 2.3 show that the direct contributions to the national exchequer in the form of motor vehicle taxation are, and have been for some time, substantially in excess of not only the central government expenditure but also the total national expenditure on the roads. In fact, the motorists' share is even greater than is illustrated by the central government taxation revenue since road-users are, more often than not, property owners/occupiers as well and hence they have to contribute an additional amount in the form of rating towards the cost of supporting the non-trunk roads.

In discussing the contributions to the national exchequer, the major consideration is whether the motoring taxes are levied for the primary purpose of supporting the highways, or in order to help finance general government expenditure. If this latter consideration is the one accepted—and in this country it is so—then the government is entitled to raise revenue by whatever legal means are available in order to provide the country with the necessary defence and social services. From this aspect the motor vehicle is just another fruitful source of taxation, just as are cigarettes and drinks. Hence, it can be argued that there is as little justification for the entire revenue derived from motor taxation to be used for the construction and maintenance of highways as there is for all the beer and spirit taxes to be used to build bigger and better public houses.

Even if this latter approach is accepted there is still a strong case for having in existence some consistent relationship between the benefits received by the motorist in the form of improved road facilities and the finance which he provides in the form of motor vehicle taxation. From the highway administrator's viewpoint this is a desirable concept, since highway improvements are best carried out on the basis of a planned orderly schedule coupled with the assurance of a steady income.

2. *Differential apportionment.* It is indeed very difficult to say what should be an individual motorist's rightful contribution to the motor taxation revenue. For instance, there is no doubt that, as a means of charging for the use of the roads, purchase tax at least is indefensible.[4] It can really only be considered as a form of luxury tax which bears no relationship whatsoever to the use that a vehicle makes of the highways. This is most obviously demonstrated by the fact that in Great Britain commercial vehicles, which cause the greatest amount of wear and tear to the highway system, paid no purchase tax at all. A further disadvantage of the purchase tax mechanism is

that it encourages the road-user to keep his vehicle on the road for a longer period of time in order to get better value for money. In fact the only beneficial function of the purchase tax is the short-term negative one of restricting the entry of new vehicles on to the roads until better roads can be provided.

The motor-fuel tax is a more equitable method of charging for highway usage since it at least partially reflects road maintenance costs due to wear and tear, as well as congestion costs. The more a vehicle uses the highway to move about, the greater is its fuel consumption and hence the greater its contribution in the form of taxation. In addition, when a vehicle is travelling on a congested roadway it is forced to move at a low speed which, when coupled with numerous stopping delays, also increases the rate of fuel consumption. Nevertheless, the fuel tax in its present state can only be considered as a rather haphazard method of relating the charge to the usage. There are some very obvious inequalities, in that a car with a high rate of fuel consumption pays more than one of similar size with a lower consumption rate, or a petrol-driven goods vehicle pays more than a diesel-powered one. Also petrol consumption on poor roads is greater than on high quality ones so that the anomaly occurs where the charge for using the low quality roads is actually greater than that for the better ones.

Apart from its usage as a means of raising further revenue, the registration of vehicles is necessary for police and other social purposes. The administrative cost of providing for the registration of a vehicle is so small that it leaves a considerable margin for the Treasury when the existing rates are applied. What these rates should be is literally anybody's guess. They can only be regarded as service charges for using the highway facilities; hence they are not truly reflective of the wear and tear caused to the roads or the costs of congestion in particular locations. Thus, for instance, the 'Sunday driver' doing perhaps 5000 km per year pays the same tax as the business supporting a commercial traveller's car on the road for say, 30 000 km in the year. While the registration tax on goods vehicles at least attempts to differentiate between the various types, there are still many other anomalies within the system which make it unsatisfactory.

In summary, therefore, it can be said that the present methods of raising revenue by means of motor vehicle taxation are not at all reflective of the extent to which vehicles make use of the roadway. It is also true, however, that practice in this country in this respect is not at variance with that in most other countries. Certainly an equitable method of motor taxation for highway purposes has yet to be devised and proven in practice.

HIGHWAY ADMINISTRATION

It can be said with some definitiveness that the system of highway administration in Great Britain was not specifically created to direct and manage the roads and their traffic. Rather it is the outgrowth of the need to handle the problems created by the roads at various intervals in history. Some of these have already been chronicled in the Road in Perspective chapter, as well as in the brief summary of highway legislation, and so will not be further described here. Instead, the brief discussion will be primarily concerned with highway administration as it exists today, although some explanation of important changes now in the offing will be also given.

Highway authorities

In 1970 the total number of highway authorities in Great Britain was 1195. Their composition was as follows[2]: England—839, Scotland—233, Wales—123.

As might be expected, this multiplicity of highway authorities has meant that there is a certain inefficiency built into the system of highway administration. Too much will not be said about this, however, as the system is in the process of drastic change as a result of the local government reorganization in Great Britain which is scheduled for implimentation in 1974. While the full extent of the changes in local government structure have yet to be defined, there is no doubt whatsoever but that there will be a very significant reduction in the number of highway authorities. One measure of the likely change is that instead of there being, as at the time of writing (1972), 78 County Boroughs and 46 County Councils (including the Greater London Council) in England, there will be substituted six Metropolitan County Councils, 38 non-metropolitan County Councils, and the Greater London Council. As well as cutting down on the number of highway authorities, the reorganization should also ensure that the new larger authorities will be able to attract and sustain staff of high calibre, and have better resources for dealing with roads and their problems.

Department of the Environment. On November 12, 1970 a new central government department was brought into being; this was the Department of the Environment. The head of this department, the Secretary of State for the Environment, is a politician appointed by the Crown on the advice of the Prime Minister. He heads a department staffed by three other Ministers of the Crown and a staff of permanent civil servants at whose head is a Permanent Secretary (see Table 2.2a). Of particular interest to highway engineers are the responsibilities vested in the Ministers for Transport Industries and Local Government and Development.

The Secretary of State for the Environment, who bears final responsibility including all statutory powers, is primarily concerned with the strategic issues of policy and priority (including public expenditure) which determine the operations of the Department as a whole. He also takes personal charge of the Department's co-ordinating work on environmental pollution.

The Minister for Local Government and Development is concerned with matters of local government; regional land use and transport planning; the countryside and conservation; roads; road passenger transport; water, sewerage and refuse disposal. The Minister is assisted generally by two Parliamentary Under-Secretaries of State and by a Secretariat of civil servants (see Table 2.2b).

The Minister for Transport Industries is responsible for all matters relating to ports; general policy on the nationalised transport industries; railways; inland waterways; the Channel Tunnel; freight haulage; international aspects of inland transport; road and vehicle safety, and licensing; sport and recreation. In this he is assisted by a Secretariat (see Table 2.2c) (partly as for the Minister for Local Government), and a parliamentary Under-Secretary of State who also reports directly to the Secretary of State for the Environment on matters of environmental pollution relating to clean air and noise.

(There is one major point which should be made here with regard to the organizational set-up shown in Table 2.2. Government departments are constantly reorganizing in order to meet continually changing demands—thus the central organization shown in the table should not be taken as fixed in any way, but rather as a guide to the overall framework as it exists.)

The Department of the Environment also has regional offices located across England, each under the charge of a Regionnnal Director (a civil servant of Under-Secretary rank) who generally deals with the detailed problems of his region. (While the Road Construction Unit Director reports directly to London, he consults with the Regional Director of the DoE on major road projects.) Each Regional Director has working with him two Regional Controllers, one on Roads and Transportation, the other on Housing and Planning. It is through the office of the Regional Controller (Roads and Transportation) that most of the dealings with local authorities regarding highways take place. There is also very close liaison between the two Regional Controllers where developments in the transport field are seen to obviously interact with planning aspects, e.g. when land use transport demand studies are to be carried out.

Specifically, the Regional Directors have wide responsibilities relative to the planning, design and supervision of construction of trunk road schemes costing less than £1 million, as well as for all day-to-day aspects of road maintenance and management, road lighting and traffic aspects of planning applications for all trunk roads within their areas (of which there are 10 in England). In addition, they work closely with local authorities on all aspects of traffic management on principal and unclassified roads.

Planning and design of major trunk roads. Although the Department of the Environment bears the responsibility for trunk roads, it does not, in practice, carry out any physical work on them but instead delegates the responsibility for improvement and maintenance. In the case of maintenance and the design of improvement schemes costing less than £1 million, this responsibility is delegated to a local authority which acts as the Department's agent authority at an appropriate fee. When the improvement scheme—which can be a new motorway or simply a realignment of an old trunk road—is a large one, i.e. costing more than £1 million, the responsibility for its design and supervision of construction is delegated to a *Road Construction Unit.*

Prior to 1967, County Councils or consulting engineers acted as agent authorities for all trunk road improvement programmes. This did not prove

TABLE 2.2. *Organization of the London Headquarters of the Department of the Environment on November 12, 1970*

(a) Distribution of duties at ministerial and top official level

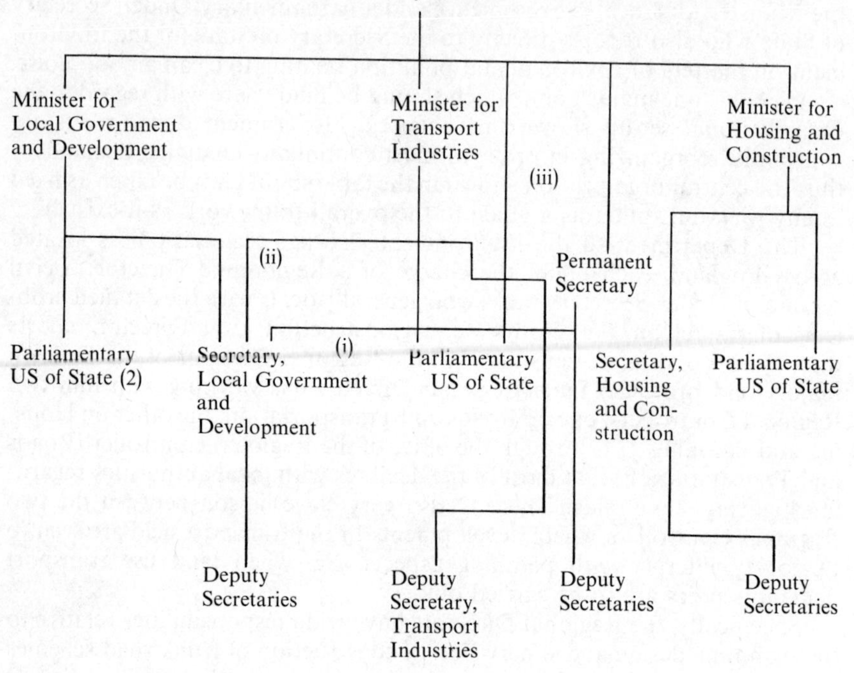

Notes: (i) Report line for sport only. (ii) Report line for road safety and environmental pollution only. (iii) Report line for environmental pollution only.

Table 2.2 — *cont.*

(b) *Secretariat* (*Local Government and Development*)

- Permanent Secretary
  - Secretary (Local Government and Development)
    - Deputy Secretary
      - Chief Scientific Officer Pollution Unit
      - Under-Secretaries Local Government (3)
    - Chief Engineer
      - Deputy Chief Engineers (2)
      - Chief Alkali Inspector
      - Chief Inspector of Audit
      - Principal Finance Officer and Accountant General
    - Chief Planner
      - Deputy Chief Planners (3)
      - Chief Estate Officer
    - Deputy Secretary
      - Under-Secretaries Planning (7)
      - Chief Housing and Planning Inspector
    - Deputy Secretary
      - Under-Secretary Passenger Transport and Urban Planning
      - Under-Secretary London
      - Deputy Chief Engineer, Urban and Regional Professional Group
      - Under-Secretary Road Safety and Vehicle Safety (Reports eventually to the Minister for Transport Industries)
      - Chief Mechanical Engineer
    - Director General Highways
      - Chief Highway Engineer
        - Deputy Chief Engineer
      - Deputy Director General 1
        - Under-Secretary
      - Deputy Director General 2
        - Deputy Chief Engineer
      - Under-Secretary Highway Planning and Economics
      - Under-Secretary Highway Lands and Contracts
    - Chief Information Officer

Table 2.2—*cont.*

(c) *Secretariat* (*Transport Industries*)

Permanent Secretary
- Deputy Secretary (Transport Industries)
  - Chief Information Officer
  - Under-Secretary Ports
  - Under-Secretary Railways
  - Under-Secretary Nationalized Transport
  - Under-Secretary Regional and International Aspects of Inland Transport
  - Under-Secretary Freight
  - Under-Secretary Centralized Licensing
- Under-Secretary Finance
- Deputy Secretary
  - Directors of Economics (2)
  - Director of Statistics
  - Under-Secretary Strategic Planning and Briefing
  - Director of Scientific Studies
  - Director of Road Research
- Deputy Secretary, Director of Organization and Establishments
  - Under-Secretaries (3)
  - Director of Management Services
- Solicitor
- Advisors

too satisfactory for the larger schemes, partly because some County Authorities lacked the expertise to plan, design and supervise the construction of motorways, and partly because central government, although publicly accountable for road construction, lacked sufficient personnel with experience of motorway construction who had the capability of carrying through appraisal of the schemes put forward. To overcome this, the government established, between April 1967 and March 1968, six Road Construction Units in England with the avowed intention of achieving three main objectives: A. To concentrate scarce design and engineering resources into larger units. B. To streamline administrative decision-taking through improved delegation to regional offices. C. To provide opportunities for central government personnel to gain first-hand experience of the design and supervision of road construction schemes.

Within each designated area there are a number of RCU sub-units (or sub-areas). The reason for this is that, in practice, it was found that the personnel with the most relevant expertise were already employed by the County Councils and so it was necessary for central government to reach agreement with the Councils *re* the basis on which these personnel could be made available to the Road Construction Units. The end result of this negotiation was that certain well-staffed counties were designated as 'Sub-Units' with the following characteristics: 1. The County Surveyor of the designated county, e.g. West Riding of Yorkshire, is the Chief Engineer of the Sub-Unit. 2. Personnel from each selected county are assigned to the Road Construction Sub-Unit but they remain, legally, employees of the County Council. 3. Sub-Units are assigned responsibility for the design and supervision of construction of road programmes in counties which are not Sub-Units. 4. All costs associated with the activities of the Sub-Unit are borne by the Department of the Environment. The situation now is that although the Director of each regional unit is a civil servant, over 90 per cent of the staff of the Sub-Units and of the regional RCU headquarters are seconded county employees.

It might also be noted here that the Road Construction Units have no responsibilities for the design and construction of roads in county boroughs. It is not known yet what the effect local government reorganization will have on the Road Construction Units.

Road research. Because of the very strong and active influence which highway research has upon both governmental decision-making and road design, it is appropriate to comment here on the manner in which it is carried out in Great Britain. Without doubt, the greater part of this research is undertaken by the Transport and Road Research Laboratory, which is part of the Department of the Environment. Other important research is carried out by the Greater London Council and at certain major universities.

In 1930, the then Minister of Transport established a small experimental station at Harmondsworth, Middlesex at which research could be carried out into highway engineering, soil mechanics, and bituminous and concrete technology. Over the years this organization has grown tremendously, and today the Laboratory has a staff of nearly 1000 (of whom about half are scientists or engineers) with an annual expenditure of about £4 million. Its main activities are concentrated at Crowthorne, Berkshire, but there is a smaller Scottish branch at Livingston at which highway construction problems peculiar to Scotland are studied.

The range of topics studied at the Transport and Road Research Laboratory (the name was changed from that of Road Research Laboratory in January 1972) are reflected in the organization chart shown in Table 2.3. The results of its researches are made known by the publication of Technical Papers, Road Notes (containing summarized recommendations of immediate practical value) and, particularly, via a series of research reports coded as belonging to the L.R. series. In addition, each year papers are published by members of the Laboratory in the journals of the learned societies or the technical press. The library contains most of the important literature dealing with highway planning, design and construction; the resources of this library are available to road engineers and other interested workers.

Scottish Development Department/Welsh Office. Since 1956 the Secretary of State for Scotland (via the Scottish Development Office), and since 1965 the Secretary of State for Wales (via the Welsh Office) have had powers and responsibilities in Scotland and Wales, respectively, which are comparable with those of the Secretary of State for the Environment in England.

TABLE 2.3. *Organization of the Transport and Road Research Laboratory (September 197*

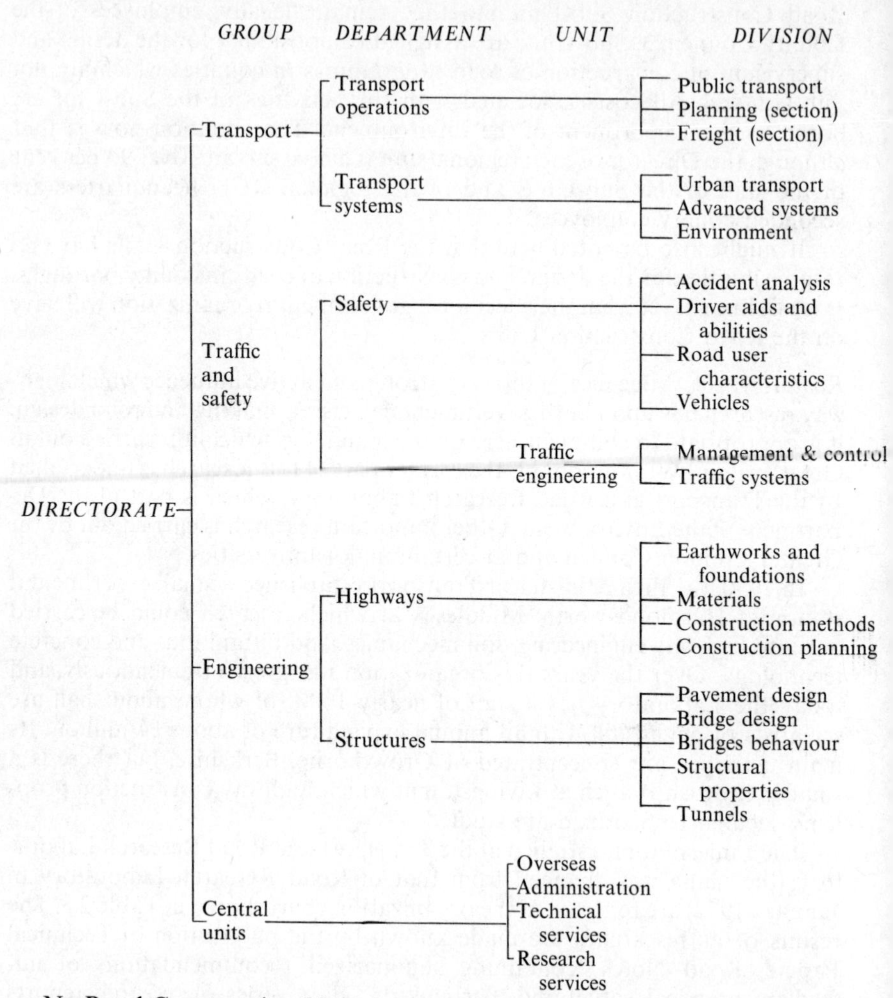

No Road Construction Units have so far been created to deal with trunk roads in either Scotland or Wales. Instead reliance is placed on the local authority organizations or, where these may have inadequate resourses, on consulting engineers.

Local authorities. The highway authorities for all non-trunk roads in England are the local authorities, i.e. normally county councils or county boroughs in Scotland and Wales, or county councils, cities or large boroughs in Scotland. In a manner similar to that for central government practice *re* trunk roads, a county council can, if it wishes, delegate its responsibilities, e.g. to non-county boroughs, urban district councils, or rural district councils in England. In this situation the responsibilities of the county council are such that, since they provide the money, they can define the policy as well as the works to be carried out.

Programming of trunk road schemes

In the mid-1960's it was noted that certain slippages occurred in the motorway programme, and this was attributed to inadequate forward planning procedures. Thus, in 1967 it was decided to change the system and to make the forward programme more flexible by providing a continuous reservoir of work. This was done by dividing the trunk road programme into two parts—a 'firm' programme and a 'preparation' pool. As is implied in the title, a scheme is included in the firm programme when a definite decision has been taken regarding its construction. Schemes considered likely to eventually get into the firm programme are first added to the preparation pool where they are very carefully assessed *re* costs and likely benefits; they are then included in the firm programme when appropriate in the light of the available resources and priority needs.

(Since 1967, also, a preparation list for principal road schemes has existed[5]. This operates in a similar way to the preparation pool for trunk roads.)

Outline of procedures. The detailed steps involved in the preparation of a trunk road scheme—they are outlined in Fig. 2.4—can be summarized as follows:[6]

A. Preliminary surveys and consultations carried out.
B. Proposals for the road line are published.
C. Objections are considered.
D. Line is fixed by Order or Scheme under the Highways Act, 1959.
E. Detailed engineering design work, including structures, is initiated.
F. Proposals for the alteration of side roads and accesses are published.
G. Objections (if any) are considered.
H. Land plans are prepared.
I. Land is acquired (includes, for practically all large schemes, the publication of draft Compulsory Purchase Orders).
J. Objections are considered.
K. Final details of engineering design are ironed out and contract documents are prepared.
L. Contractors are invited to tender.
M. Tenders are approved and contracts let.
N. Contractor begins work on site.

The primary object of the preliminary studies and consultations is to enable the Secretary of State to make an Order or a Scheme for the line of the road. An Order fixes the line of a new trunk road while a Scheme fixes the line of a motorway; the difference between the two is that the Scheme provides the authority for the exclusion of certain traffic such as pedestrians and cyclists, whereas the Order does not. There are two ways in which the line of a road can be decided.

1. It may be that a suitable line is already shown on the statutory

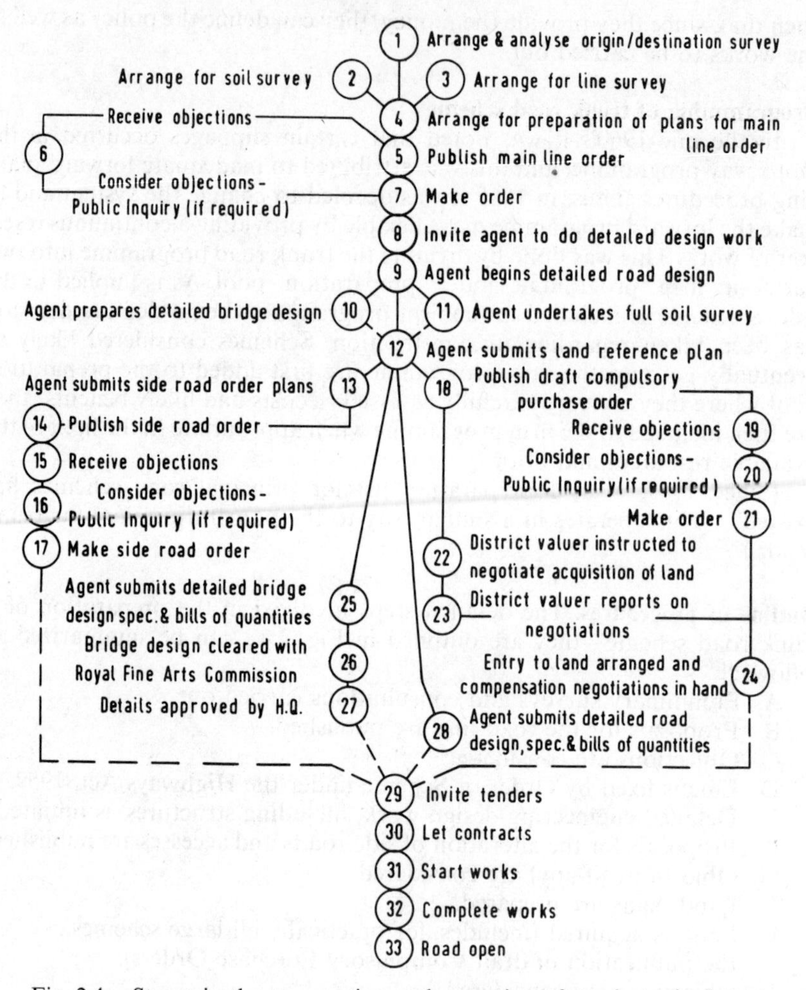

Fig. 2.4. Stages in the preparation and execution of trunk road schemes

development plan for the county or county borough. If this is so and the line is included in the approved development plan, then the line is protected against any fresh development. The value of this procedure lies in the fact that the people affected by the proposed road are given the earliest possible warning, and there is ample opportunity to assess the weight of objections and to consider alterations at a conveniently early stage.

2. When the development plan does not include a suitable line, the Secretary of State is bound by statute to take into consideration the requirements of local and national planning including the requirements of agriculture. Therefore, in addition to carrying out the preliminary surveys and field work, all interested local and national bodies have to be consulted. As soon

as the most promising provisional line has been found, then other government departments, local authorities and other interests, and public service bodies, i.e. electricity, gas and water undertakings that are likely to be affected by the proposal, are consulted informally to make sure that there are no insuperable obstacles and, where possible, to reach agreement on particular items of conflicting interests; these consultations alone may take many months. When a satisfactory line has been determined the draft Order or Scheme illustrating the line is prepared.

Whether or not the proposed line is in the development plan, advertisements must be placed in the Press, and the government departments, local authorities, and public utility undertakings notified regarding it. There is then a period of three months during which objections can be made by individuals or public bodies who disagree with the proposals. If any objections are received from individuals which cannot be resolved by discussion or which raise serious issues, a Public Inquiry *may* be held. An inquiry *must* be held if there is an unresolved objection from a local authority or, where it is proposed to build a bridge over a navigable waterway, from a navigation authority. If an inquiry is to be held, the Secretary of State appoints an independent inspector to conduct it; about three months is allowed for all interested parties to prepare their cases for this inquiry. In the light of the objections, and of the inspector's report if an inquiry is held, the Secretary may then decide to make the Scheme or Order with or without modifications, or not to make it at all. He is not obliged to accept an inquiry inspector's recommendations, but if he does not do so he must state the reason for his decision. If he decides to make substantial alterations, or abandon the initial proposal and start again, the entire procedure for the advertisement of the new line must start again from the beginning.

The next stage is the preparation of the engineering details for the layout and construction of the highway. This includes the feasibility of small changes in the route within the line established by the main Order, and the preparation of land plans. Again at this stage, other government departments, local authorities and statutory undertakers must be consulted regarding the most suitable proposals.

In a densely populated country like Britain the laying down of new roads inevitably affects, and necessitates alterations to, the existing pattern of public highways, footpaths and private accesses. Statutory authority is needed for these alterations in the same way as for the line of the new road. This again necessitates an objection period, and, in certain circumstances, a public inquiry may be held into the proposals. In the ideal situation, the publication of the line Order and the side roads Order may be combined for the one project.

Land acquisition cannot start until the design of the road is sufficiently far advanced to show the lateral limits of the land that will have to be acquired, but it may be undertaken concurrently with the side road Order procedure. When the design has reached this stage plans can be prepared showing the land needed from each of the various properties affected by the road scheme. These are accompanied by reference schedules giving the descriptions, sizes and details of ownership and occupation. The District Valuer then enters into negotiations with the owners and occupiers in order to determine the amount of compensation to be paid and to obtain agreement to enter upon the land on the required date.

If, as usually happens on all large schemes, some particular plot cannot be acquired by agreement, then compulsory powers of acquisition have to be sought. This is obtained by making a Compulsory Purchase Order, which must first be published in a draft Order describing in detail the land that is required. Notice of this draft Order is served on the owner, lessee or occupier affected, and then there is a three-week period to allow for objections. If, at the end of this period, there are unresolved objections to the Order, on any grounds other than the amount of compensation to be paid, an inquiry or hearing *must* be held. Unresolved disputes as to the amount of compensation are referred to the Land Tribunal, a procedure which does not delay the making of the Order and the use of the powers of entry given under it; often indeed the compensation cannot be finally assessed until the roadworks are completed and the actual loss or injury caused can be accurately described.

In the event of an inquiry or hearing into an objection to the draft Compulsory Purchase Order, the inspector's report is considered by the Secretary of State and he then decides whether to make the Order with or without modification, or whether to make it at all. When the Order is operative, entry to the land may be enforced within a minimum period of 14 days by service of further notices, unless certain classes of land are involved, e.g. common, open space or ecclesiastical land, when additional procedures must be followed. If land belonging to local authorities or statutory undertakers is to be acquired compulsorily in the face of unwithdrawn objections, a special Parliamentary procedure is involved before the land can be entered upon.

Thus it can be seen why, for a large highway scheme, it may take more than a year simply to obtain entry on to all the land required. Indeed, if there are serious difficulties involved in moving residents, businesses or industrial or public service undertakings on the route, an even greater length of time may be required. The time need not be entirely wasted however, since, while the land is being acquired, the detailed preparation of the scheme can be completed and contract documents prepared. The procedure for inviting tenders is now mostly by what is known as 'selective tendering'. By this method tenders are invited from selected contractors judged to have the requisite technical and financial resources and in a position to submit a realistic tender for the particular contract.

SELECTED BIBLIOGRAPHY

1. HYDE, W. S. Highway legislation, *J. Inst. Munic. Engrs*, 1961, **88,** No. 5, 169–174.
2. BRITISH ROAD FEDERATION. *Basic Road Statistics.* London, The Federation, 1972.
3. DEPARTMENT OF THE ENVIRONMENT. *Roads in England, 1971.* London, H.M.S.O., 1972.
4. FOSTER, C. D. *The Transport Problem.* Glasgow and London, Blackie, 1963.
5. MINISTRY OF TRANSPORT. *Roads in England.* London, H.M.S.O., 1968.
6. MINISTRY OF TRANSPORT. *Roads in England and Wales.* London, H.M.S.O., 1964.

3 Highway planning

PEOPLE, LAND USE AND TRANSPORT: AN OVERVIEW

Transport in general, and particularly highway transport, plays an essential role in the life of any community today. Good highway transport facilities are the result of sound planning—and more and more it is now being recognized that transport planning cannot be, and must not be, isolated from land use planning; also that when planning for the future account must be taken of the vehicles in which people wish to travel and move their goods. Before, therefore, getting involved in the detailed characteristics and factors associated with the highway planning process, it is useful to briefly consider how the ground transport systems (particularly the passenger transport ones) developed in Great Britain, their present status and situation, and how transport events likely will be shaped into the foreseeable future[1,2].

Development of 'mechanical' travel

As a general rule, travel was not *considered* to be a problem until the advent of the Industrial Revolution. Prior to then travel was, on the whole, by horse or foot—indeed, it might be noted that it was only about the middle of the 16th century that the first wheeled coach appeared in Britain. Came the Industrial Revolution, and the situation radically changed; civic life and form was changed almost beyond recognition as a great wave of migration from the countryside was initiated. The result was that villages became towns and towns became cities—and this was accompanied by the beginning of what is now known as the population explosion (viz. in 1800, the population of England and Wales was only about 9 million, two-thirds of whom could be classified as rural dwellers).

As the population increased, the need for efficient public transport facilities became evident, particularly in towns, and many far-sighted and entrepreneurial businessmen saw profits in providing these transport services (for it must be remembered that public transport was a profitable private enterprise for well over 70 years).

July 4, 1829 saw the initiation of mass public transport in towns when Mr. George Shillibeer started his 20-passenger horse 'omnibus' service in London between the village of Paddington and the Bank.

Apr. 22, 1833 saw Mr. Walter Hancock testing his 14-passenger steam omnibus service over the same route.

Aug. 30, 1860 saw the beginning of the end of the horse-bus when Mr. George Train initiated a horse-tram service in Birkenhead. It could carry more people more comfortably and more quickly than the horse-bus.

Jan. 10, 1863 saw the opening of the world's first underground railway, the first part of the Metropolitan (steam) Railway between Paddington and Farringdon.

Sept. 29, 1885 saw the first electric street tramway at Blackpool—and so initiated the death of the horse-tram. (The electric tram had a really tremendous effect on town development, for within 25 years nearly every major town in Britain had its own network of electric tramways. Typically, these radiated outward from the central area and serviced settlements clustered along these routes. Thus transport development began to be clearly associated with the now-familiar pattern whereby residential densities decline from zone to zone from the centre outwards.)

Apr. 12, 1903 saw the start of the first municipally-operated motor bus service at Eastbourne. (This is important for two reasons:

1. it marked the first occasion that a municipality assumed responsibility for providing a *public* transport system in a town;
2. the motor bus service brought the suburbs within easier reach of the town centre and, since it was not forced to stay on fixed routes, it could service housing not astride the radial arteries—and thus it helped consolidate urban sprawl.)

The advent of the motor car about the end of the 19th century had little effect on mass movement or town development for some time for the very practical reason that in those early days it could only be enjoyed by the wealthy. (Of much more importance to the general populace of that time was the 'low' bicycle.) It was not until after World War I, when the mass production methods developed in America were applied in Britain, that the price of the motor car was sufficiently reduced for it to start to become the ordinary man's carriage.

While the early horse bus, steam bus, electric tram, railway, motor bus and (following 1911) trolley bus were competing against each other for the opportunity to move passengers in built-up areas, the railway was without doubt the prime mover of people (and goods) over medium and long distances. 'Democracy in rail transport' was initiated in 1844, when an Act was passed which compelled the railways to operate a certain number of cheap trains, and effectively implemented in 1872 when the Midland Railway announced that third-class passengers would be allowed on their best and fastest trains. In 1850 the railways carried some 67 million passengers, 300 m in 1860, 600 m in 1880, well over 1,000 m in 1900, and nearly 1300 m in 1910.

The motor car and its effects

There is no doubt but that the motor car is the most marvellous instrument of *personal mobility* known to man. It is, in effect, his own private 'magic carpet' which allows him to go where he likes, when he likes, whatever the weather conditions. Above all, perhaps, it satisfies his need to make individual decisions regarding transport in his own time, at his own convenience.

The mobility provided by the motor car is well illustrated in Table 3.1. This shows not only that the effect of increasing car ownership is to increase travel from each household, but also that the increase is mostly for non-work purposes.

TABLE 3.1. *Trip generation per household per day for households in London with one employed resident per household in an urban area of average rail and bus accessibility*[3]

Purpose	Household income, £	Private transport 0-car	Private transport 1-car	Private transport Multi-car	Public transport 0-car	Public transport 1-car	Public transport Multi-car	All modes 0-car	All modes 1-car	All modes Multi-car
Work	0–1000	0·2	1·2	—	0·9	0·4	—	1·1	1·6	—
	1000–2000	0·3	1·2	1·6	1·1	0·6	0·5	1·4	1·8	2·1
	>2000	—	1·0	1·5	—	0·5	0·2	—	1·5	1·7
Non-work	0–1000	0·2	2·0	—	0·8	0·4	—	1·0	2·4	—
	1000–2000	0·3	2·7	5·7	1·4	0·7	0·3	1·7	3·4	5·5
	>2000	—	3·9	8·4	—	1·1	0·9	—	5·0	9·3

There are also, unfortunately, a number of undesirable by-products of the development of the motor vehicle. Not least among these is the number of *accidents* (see Figure 7.1, chapter on Traffic Management). Taking the figures for the past 20 years alone, it is perhaps a horrifying thought that the next 20 years could see at least 1·5 million people killed or seriously injured on the roads of this island, and a total of 6·5 to 7 million injured. Furthermore, on the basis of present experience, it is likely that some 65 per cent of the killed and seriously injured (and about three-fourths of all casualties) will result from accidents in built-up areas.

The mechanical sources of noise generally encountered in modern towns are road and rail traffic, aircraft, construction and industrial noise. Of these the most important is *road noise*[14,15]. In 1963, the Wilson Committee[4] carried out a thorough examination of the noise problem created by motor vehicles, and reported the representative existing noise climates listed in Table 3.2. To illustrate how these measured levels compare with desired levels, the Committee made the following suggestions for maximum noise levels inside living rooms and bedrooms, stating that these values should not be exceeded for more than 10 per cent of the time:

	Day	*Night*
Country areas	40dB(A)	30dB(A)
Suburban areas	45dB(A)	35dB(A)
Busy urban areas	50dB(A)	35dB(A)

Several factors are now combining to suggest that road traffic noise will get significantly worse in the future, e.g. increasing numbers of vehicles with greater power, the improving of old (or the building of new) main roads

TABLE 3.2. *Representative existing noise climates*

Group	Location	*Noise climate, dB(A) Day 8 a.m.–6 p.m.	Night 1 am.–6 a.m.
A	Arterial roads with many heavy vehicles and buses (kerbside)	80–68	75–50
B	1. Major roads with heavy traffic and buses	75–63	61–49
	2. Side roads within 13·75–18·25 m of roads in Groups A or B1 above	75–63	61–49
C	1. Main residential roads	70–60	55–44
	2. Side roads within 18·25–45·75 m of heavy traffic routes	70–60	55–44
	3. Courtyards of blocks of flats, screened from direct view of heavy traffic	70–60	55–44
D	Residential roads with local traffic only	65–56	53–45

*Noise climate is the range of noise level recorded for 80 per cent of the time. The level exceeds the higher figure for 10 per cent of the time and is less than the lower figure for 10 per cent of the time.

which bring high speed heavy traffic into residential and commercial areas throughout the day and night, the construction of elevated roads (which worsen the problem) and the construction of high buildings (which are particularly susceptible).

Today also there are few streets, residential as well as city centre streets, that are free from the *odours and fumes* associated with motor vehicles. From a health aspect, it can be said that only carbon monoxide from petrol vehicles has been shown to be a definite possible health hazard. The black smoke from diesel engines is also undesirable—not because of a health hazard but because it is unpleasant and a road accident risk, particularly in traffic congestion or in narrow or enclosed roads where it may obscure vision.

An environmental 'disamenity' which unfortunately receives very little attention is the *aesthetic deterioration* associated with vehicle intrusion, viz:

> 'the dreary, formless car parks, often absorbing large areas of towns, whose construction has involved the sacrifice of the closely knit development which has contributed so much to the character of the inner areas of our towns; the severing effects of heavy traffic flows; and the modern highway works whose great widths are violently out of scale with the more modest dimensions of the towns through which they pass'.

These words received very great attention when they were first written[5] in

1963. Today, the words are still valid, but unfortunately very many of the actions which they suggested were necessary have yet to be implemented.

There is ample evidence to the effect that, given a free choice, most people will use their *motor cars in preference to public transport* for other than long trips. This is reflected in the passenger mileage statistics shown in Table 3.3. While these data reflect only national trends, there is ample evidence elsewhere to show that there has been also a clear and continued swing away from the use of public transport in towns; that the amount of the decline has varied from town to town, and that the peak year for public transport passengers (prior to the decline) was much later in some towns than in others. At the present time it would appear that this swing with respect to passenger mileage and passenger journeys is likely to generally continue, albeit at a slower rate. On the other hand, the number of public transport vehicles per head of population may well have reached its base level and may not go much lower[7].

TABLE 3.3. *Passenger transport in Great Britain, 1959–69*[6]

Transport mode	1959 Pass-km $\times 10^9$	1959 %	1969 Pass-km $\times 10^9$	1969 %
Rail	40·8	17	34·6	9
Road				
Public service	70·6	29	57·1	15
Private transport	131·4	54	294·4	76
All	242·8	100	386·1	100

(Note also that all road vehicles travelled a total of 195×10^9 vehicle-km, including 150×10^9 car-km, in 1969: this was almost equally divided between urban and rural roads. Trunk and principal roads, which comprise about 14 per cent of the roads, accounted for about 70 per cent of the total vehicle travel.)

That the motor vehicle has a major effect upon land usage within urban areas is well recognized. For example, there is the flight of home-dwellers to the suburbs, a product probably mainly of slum clearance and higher incomes, but at least partially made possible by the mobility given by the motor car. There is also the movement of industry to sites adjacent to good road facilities. (This latter movement has been generally encouraged by town planning practice which, in the past, has generally favoured the development of the centripetal type of town—and indeed city region—of the idealized form shown in Figure 3.1.) There is also the general decrease in the number of jobs in the central areas of the big towns[8]; while it cannot be said just how much of this is a reflection of transport problems, there is no doubt but that there is a genuine fear amongst central area employers that

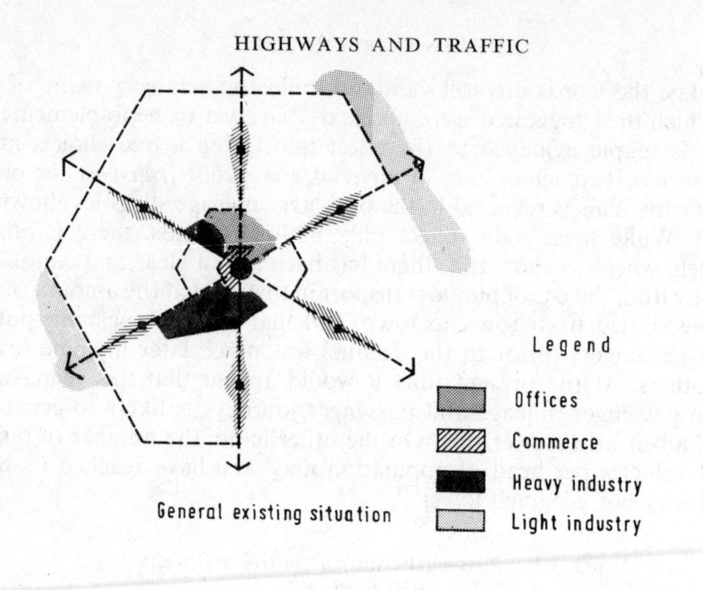

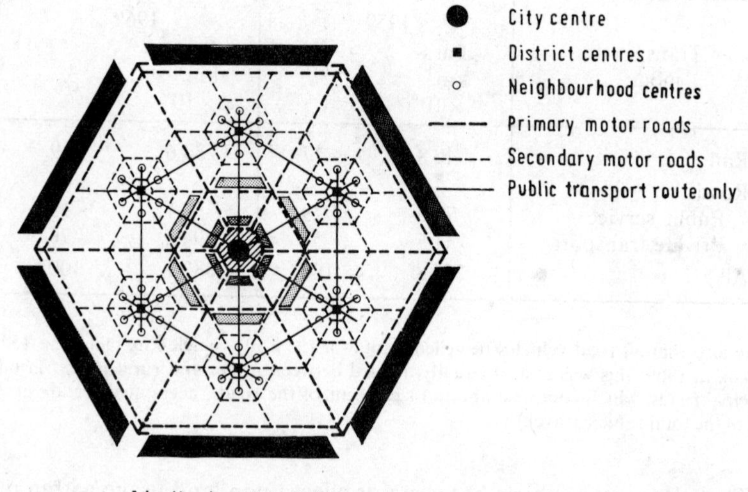

Fig. 3.1. General town planning practice with respect to land use and the transport routes

unless traffic congestion is reduced, the trend will become even more definite.

There is the general trend and continuing pressure for the establishment of shopping centres in suburbia. Of the factors listed in Table 3.4 as influencing this trend, there is little doubt but that the total travel time from the home is the main advantage of the suburban centre. This illustrates why city centre merchants are so much concerned with improving the central area-oriented transport facilities and, particularly, (as is discussed in Chapter 4, Parking) with the parking facilities for shoppers in town centres.

TABLE 3.4. *Factors influencing the decision whether to shop* in *the Central Area or in an outlying shopping facility*

	Central Area	*Suburban Centre*
For	1. Greater choice of competitive goods	1. Closer to home
	2. Opportunity of doing several different types of errands on one trip	2. Easier parking
	3. Ease of accessibility by public transport	3. More convenient shopping hours
	4. Lower prices	
Against	1. Difficulty in parking	1. Fewer types of business
	2. Very crowded	2. More limited selection of competitive goods
	3. Traffic congestion	3. Poor accessibility by public transport
		4. Prices are higher

Future trends

People. Obviously any consideration of the future relative to transport must take into account any growth in the numbers of people. While predictions of the future population are fairly notorious for their variations, they all have in common that the population is going to increase. Figure 3.2, based on 1970 data, shows that by far the greater part of the increased population will belong to the car-owning/car-driving ages.

Where will these extra millions of people live? Table 3.5 emphasizes that Britain is essentially an urban nation, and that the great majority of people live in urban areas; furthermore, that approximately one-third of the population of England and Wales lives in five conurbations. In theory, all of the future population increases could be fairly easily catered for if the extra numbers could be distributed throughout all or a great number of the urban areas. In practice, however, this is not what will take place since people cannot and should not be told where to go and what to do; what will likely happen instead is that the bigger urban areas, particularly the conurbations, will take much more than their proportionate shares of these extra numbers.

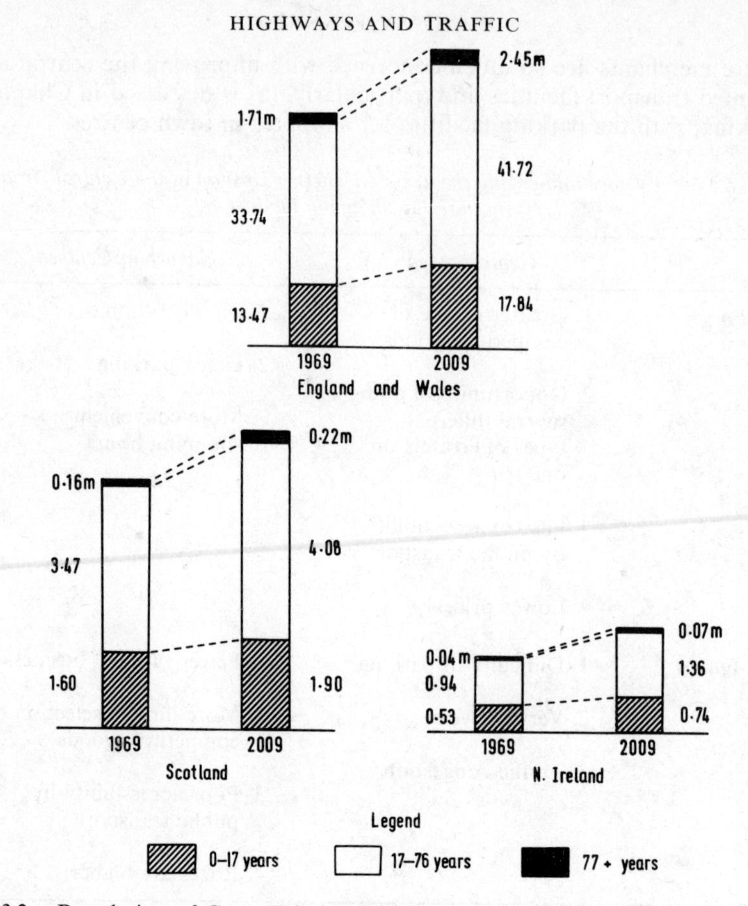

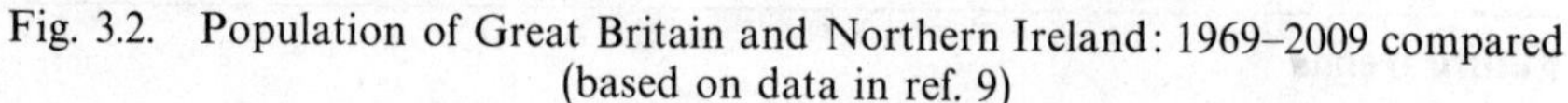

Fig. 3.2. Population of Great Britain and Northern Ireland: 1969–2009 compared (based on data in ref. 9)

It is useful here to briefly comment on conurbations for it is within these that the very major planning, transport and environmental problems arise. These conurbations (see Table 3.6) have been described[10] as consisting

'.... of groups of urban settlements more or less in a state of coalescence which have developed over a period of time, usually but not always around a central dominant city.'

London, in particular, has a structure which is strongly dominated by the central city, and this is emphasized by the form of its communication system which tends to be strongly radial and focussing on the centre. In the case of the West Midlands, South East Lancashire, Merseyside and Tyneside, the central city theme is less dominant, while the West Yorkshire conurbation consists essentially of a conglomeration of free-standing towns and cities. In all of the conurbations, the central cities are tending to lose population while the traffic regions containing them are gaining.

TABLE 3.5. *Distribution of the population in the urban areas of England and Wales, 1969* (Based on data in ref. 9.)

Population group, ×1000	No. of urban areas		Population		
	Individual	Cumulative	Aggregate, ×1000	% of total	Cumulative %
Over 5000	1*	1	7703	15·8	15·8
4000–5000	—	1	—	—	15·8
3000–4000	—	1	—	—	15·8
2000–3000	2*	3	4874	10·0	25·8
1000–2000	2*	5	3069	6·2	32·0
750–1000	1*	6	840	1·7	33·7
500– 750	1	7	529	1·1	34·8
250– 500	8	15			
150– 250	9	24			
100– 150	18	42	6001	12·3	47·1
75– 100	26	68			
50– 75	49	117	4795	9·8	56·9
25– 50	134	251			
Less than 25	504	755	10 579	21·7	78·6†

*Conurbations †Remaining 21·4% reside in Rural Districts

TABLE 3.6. *Basic characteristics of the conurbations in Great Britain*

Conurbation	Registrar General's definition			Transportation study definition			
	Area, km²	Population, in 1967	Density, persons/km²	Survey year	Area, km²	Population	Density. persons/km²
London	1595	7 880 760	4940	1962	2437	8 827 000	3623
W. Midlands	697	2 446 400	3513	1964	973	2 529 000	2598
SELNEC	984	2 451 660	2493	1965	1070	2 595 801	2428
Merseyside	388	1 368 630	3524	1966	409	1 453 333	3555
Tyneside	233	849 370	3645	1966	833	1 432 400	1718
Clydeside	777	1 764 430	2272	1964	1175	1 929 000	1640
W. Yorkshire	1255	1 730 210	1385	1966	2590	2 098 770	810

Vehicles. If it is accepted that (a) people's standards of living will continue to rise, (b) people now regard 'personal mobility' as an integral part of a high standard of living, (c) the numbers of people on this island will continue to grow, and (d) it is unlikely that any new form of personal transport will supplant the motor car within the foreseeable future, then it can reasonably be expected that the numbers of motor vehicles could increase along the lines indicated in Fig. 3.3. (Two of the more important assumptions utilized in preparing this diagram are that Britain's population will be 71·7 million in the year 2010 (a figure which now appears unlikely to be met), and that the parameters defining the elements of vehicle ownership and use, etc., during future years will be quantitatively in accordance with those applicable now.)

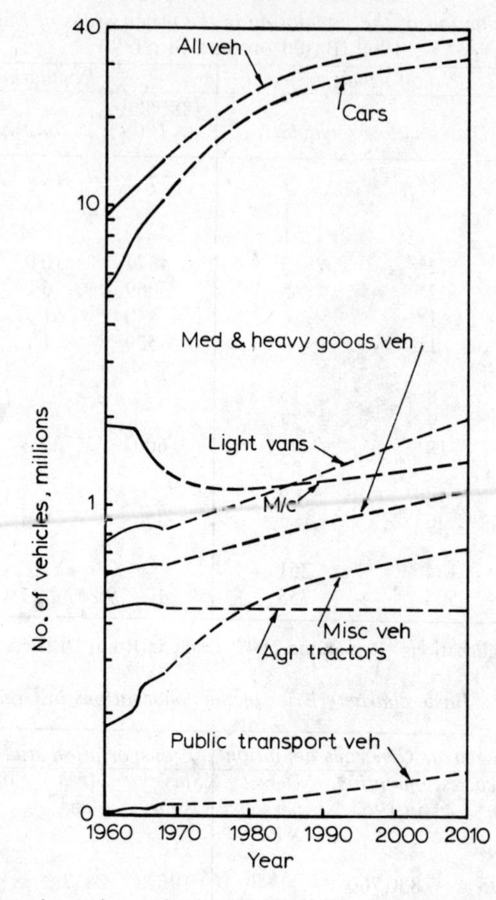

Fig. 3.3. Actual numbers of vehicles 1960–68, and forecasts 1969–2010[7]

From what has been said previously with regard to the population distribution, it is clearly of interest to have an appreciation of the manner in which these vehicle numbers, particularly the numbers of cars, will be distributed throughout the land. *Extreme* possibilities suggested[11] for various environmental locations in Great Britain for the year 2010 are as follows:

Rural counties	0·50 cars per head
Typical counties	0·45–0·50
Medium-sized towns	0·40–0·45
Conurbations	0·35–0·40
Inner London	0·30–0·35

Note that the large urban areas have relatively lower predicted car ownership values, thus reflecting a future heavier reliance on public transport in those areas.

Strategy for the long-term future

Any long-term 'solution' to the transport problem in Great Britain must inevitably be associated with significant developments in land-use planning and transport technology. Detailed summaries of developments in transport technology are available in the literature[1,2], so the following discussion will be confined to the land-use approach.

Why transport planning cannot be isolated from land-use planning is illustrated in Fig. 3.4. This figure shows that whatever the land is used for, its activities generate trips—and some activities obviously generate more trips than others (1–2); these trips, in turn, point up the need for particular types of transport facilities (2–3–4); the extent to which the transport facilities are able to cope with the trip demands determines the quality or degree of accessibility associated with the land in question (4–5); of course, the accessibility associated with the land influences its value since, logically, the land has no value if people cannot get to it (5–6); and finally, it is the land value which helps to determine the use to which the land is put (6–1).

Thus, it can be seen that the control of land-use is to a large extent the key to the control of movement. However, initiating *change* in land-use is not something that is easily done under the democratic governmental system, and hence the impact of this solution-approach may not be felt for many years to come; this is particularly so with respect to the planning of the large urban and conurbation areas where, in addition, the pressures are so great that the land-use strategies tend to get overwhelmed by the sheer needs of more immediate 'current' problems.

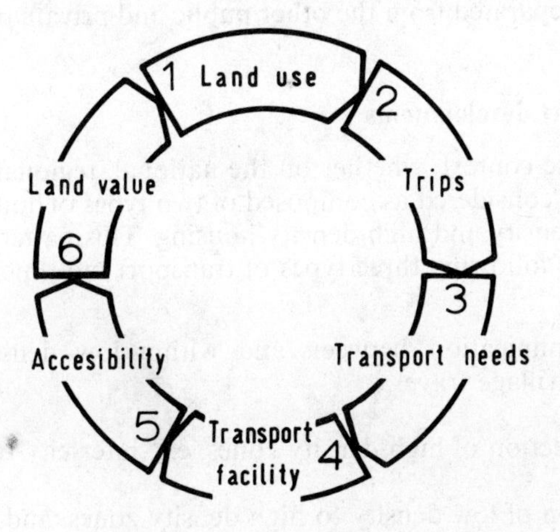

Fig. 3.4. The land-use transport cycle

The manner in which transport planning in the conurbations might develop has been outlined in concept form by Buchanan et al in their study[12] of the region linking Southampton and Portsmouth. In this study, a rational hierarchy of travelways (in the form of a 'directional grid') was proposed which would accommodate different modes of transport having scales of operation which fall naturally into a graded order. Thus, Buchanan postulated six categories of routes from '1' to '6', each successive route having a larger scale of operation in relation to distance (see Fig. 3.5). The categories proposed were such that the '1' and '2' routes correspond to paths and roads in housing areas, while '6' routes correspond to regional or national communication lines. '1' routes generally only connect with '2' routes, '2' with '3', and so on; thus at intersections, where routes cross, are concentrated the facilities which require accessibility from the two consecutive categories of urban sub-systems served by these routes, e.g. shops which need to be accessible to a residential area at one level and to wholesale distributors at another, or industry which needs to be accessible not only to its employees but also to a regional freight route. In Fig. 3.5, the main urban facilities are shown grouped on alternative 'red' and 'green' routes which Buchanan called 'spines of activity'. Thus, the 'red' routes accommodate public as well as private transport (the two systems running parallel); the 'green' routes are through-routes (possibly some used also by express transport systems) through landscaped areas—they also serve for the random movements more likely to be carried out in private transport vehicles.

In the smaller- and medium-sized towns, the land-use developments will likely continue along the lines in which they are now generally being guided i.e. toward the development of the centripetal town of the type idealized in Fig. 3.1. This type has very many advantages from a movement aspect and, practically, it would be undesirable to change at this stage. The larger of these towns may, however, also require 'rapid' public transport facilities which will be separated from the other public and private transport modes.

Current transport developments

Any land use context, whether on the national, regional or local scene, may be broadly considered as composed of two types of housing concentration, viz. low density and high density housing. This pattern of land usage gives rise to the following three types of transport movement:

1. Intercommunication between and within low density zones, e.g. village-to-village travel

2. Interconnection of high density zones, e.g. inter-city travel

3. Connection of low density to high density zones, and vice versa, e.g. suburban to central area travel.

Low density-low density travel. It does not require any special technical knowledge to appreciate that this transport task demands, and will make use of, a travel mode that has time and space flexibility—in other words, the

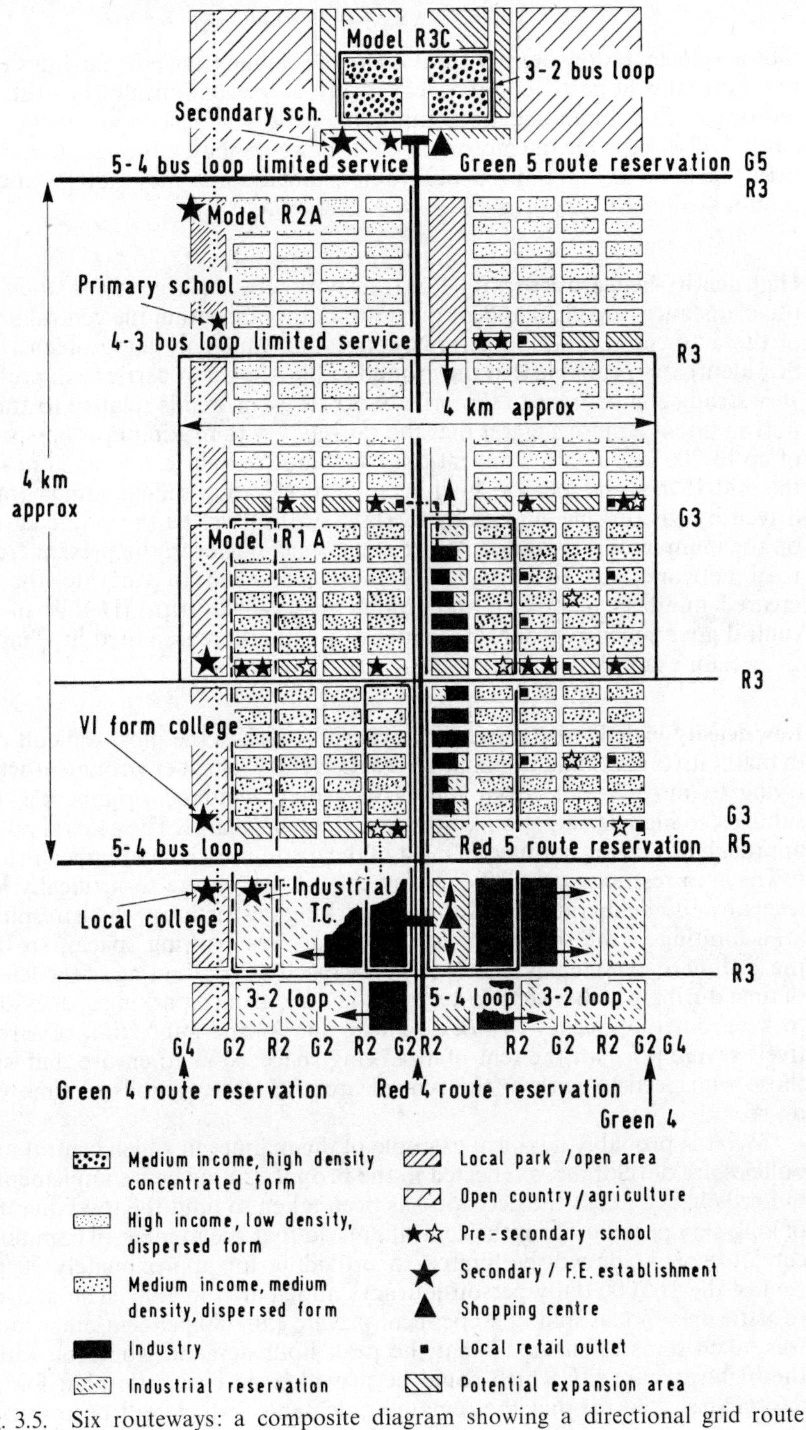

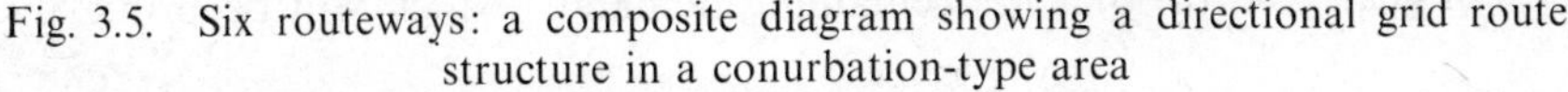

Fig. 3.5. Six routeways: a composite diagram showing a directional grid route structure in a conurbation-type area

motor vehicle. In low density rural areas the public transport facilities cannot normally hope to compete effectively and economically against the motor car. For communication within low density suburban areas, the tendency will also be for the motor car to displace the bus, although here there may be room for specialized bus routes, albeit some may not be always economically self-supporting.

High density-high density travel. For transport between large towns when the travel distances are great and/or the destinations are within the central areas of these towns (i.e. where serious congestion and parking problems are prevalent), it is probable that the ground public transport carries will present very strong competition to the private car. Present trends relative to transport in Great Britain suggest that the railway—with maximum train speeds of up to 200 km/h with 'conventional' locomotive-hauled duorail trains by the mid-1970's, and 240 km/h with rapid accelerating self-contained trains driven by gas turbine engines by the late-1970's—will be the prime carrier on the main inter-city routes. At the same time, however, the present trunk road network will be very considerably expanded to cater fo the increased numbers of private and commercial vehicle trips. (Details of the central government's proposals *re* highway expansion are noted in Chapter 2—see, for example, Fig. 2.1.)

Low density-high density travel. This transport task is the most difficult one in that it involves trying to maintain a satisfactory urban environment while trying to move people from relatively widely scattered origins, e.g. the suburbs, to high density destinations, e.g. the central area. The general policy approach now being adopted in most of the medium-sized and larger British towns is to restrict road traffic movements of this nature to artifically low levels by the deliberate imposition of parking restrictions which result in (a) a limiting of the total number of central area parking spaces, so that the feeding road system is kept within its capacity, (b) a limiting of the length of time during which a vehicle may stay in a controlled parking space, so as to keep out the journey-to-work vehicle, and (c) the imposition of a relatively severe price for the rent of a parking space, so as to ensure that only those who need to park in the central area for short periods of time will do so.

What is probably a typical example of the manner in which central area policies are developing is reflected in the proposals now being implemented in Leeds[13]: In Leeds the decision has been taken to limit the total quantity of long-stay parking about the central area so that the number of commuter car journeys made will be limited to providing for approximately 20 per cent of the 163 000 daily person-journeys anticipated in 1981. The decision to settle on a modal split of 20 per cent private car—80 per cent other mode was taken so as to ensure (i) that the peak hour demand would be within the highway capacity which could be provided in a realistic plan for the design year, and (ii) that the remaining demand—which will be primarily

(66 per cent) for bus transport—would be sufficient to support an efficient public transport service at satisfactory frequencies.

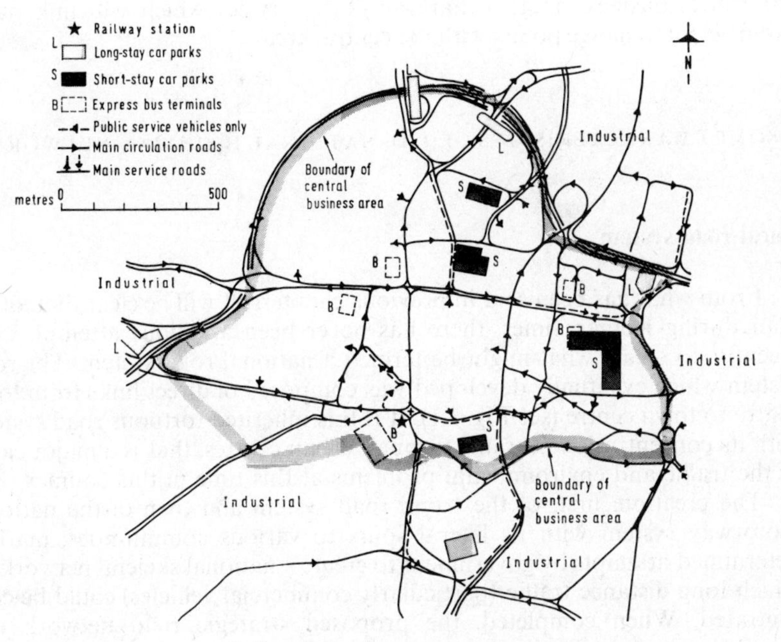

Fig. 3.6. The Leeds central area traffic plan for 1981

In addition to the above long-stay parking spaces, approximately 8000 on- and off-street car parking spaces are to be provided for short-stay shopper and business parking needs i.e. for up to 2·5 hours, within the central area of Leeds. These will be close to offices, shops and warehouses and linked with the large pedestrian precinct within the city centre.

The general outline of much of the Leeds plan is reflected in Fig. 3.6. Note, first of all, that the road system—part of which is an urban motorway—is designed to enable traffic which does not wish to enter the central area to by-pass it. The long-stay multi-storey parking facilities are all located at the fringe of the central area so as to intercept the commuter traffic; the intention is that the commuters will leave their cars at these facilities and then either travel on foot or by public transport to their work destinations. In contrast, the short-stay parking facilities and the public transport terminals are generally located well within the central area, thereby ensuring that the users of these facilities are close to their destinations.

As a complement to the above proposals, Leeds City Council is developing the following new forms of passenger road service in addition to the normal inter- and intra-district bus services: 1. Express bus services which use the primary road network and provide non-stop fast services between suburban community centres and the town centre; 2. City centre mini-bus services which provide for short (shopping and business) movements within the central business area; 3. Park-an'-Ride services which will link outer suburbia interchange points with the central area.

SOME CHARACTERISTICS OF THE NATIONAL HIGHWAY NETWORK

Rural road system

From what has been said in previous chapters, it will be clear that, other than during Roman times, there has never been a serious attempt, until recently, to create what might be termed a national road system. The road system which eventually developed was composed of direct links from town centre to town centre (see Fig. 1.1). It is this inherited tortuous road system, with its concentration of traffic through urban centres, that is a major cause of the traffic and environmental problems at this time in this country.

The creation, first, of the trunk road system and then of the national motorway system with its lateral spurs to various communities, marked determined attempts by government to create a national skeletal network on which long distance traffic (particularly commercial vehicles) could be concentrated. When completed, the proposed strategic road network (see Fig. 2.1) will undoubtedly bring great benefits to industry, as well as to the populace in general, in terms of shorter journey times, less driving stress, and significant release from, and relief to, traffic congestion in towns.

When it is considered that the lateral influence of a rural motorway-type road is spread through a width that is roughly equal to half its length, it can be seen that the strategic road system will affect traffic on nearly every highway throughout the length and breadth of Britain. Not only will it enable all unnecessary travel through towns to be avoided, but it will also have the very important 'by-product' that many overloaded rural roads will be so relieved that they can again fulfil their more local linkage functions without any extra reconstruction being required.

By-passes Notwithstanding what has been said above, it should not be thought that the answer to every town's traffic problem lies in the construction of a by-pass; it is just not that simple. As can be seen in Fig. 3.7, the need for a by-pass is primarily a function of the size of the town. In large towns there is so much local traffic, and so much of the external traffic on the approach roads is bound for town destinations, that the removal of the through traffic makes relatively little difference to the internal traffic congestion situation. It should be kept in mind however that, although in a large town the *percentage* of by-passable traffic may be small, there may be still a sufficient *volume* of through traffic to warrant a by-pass for its own sake.

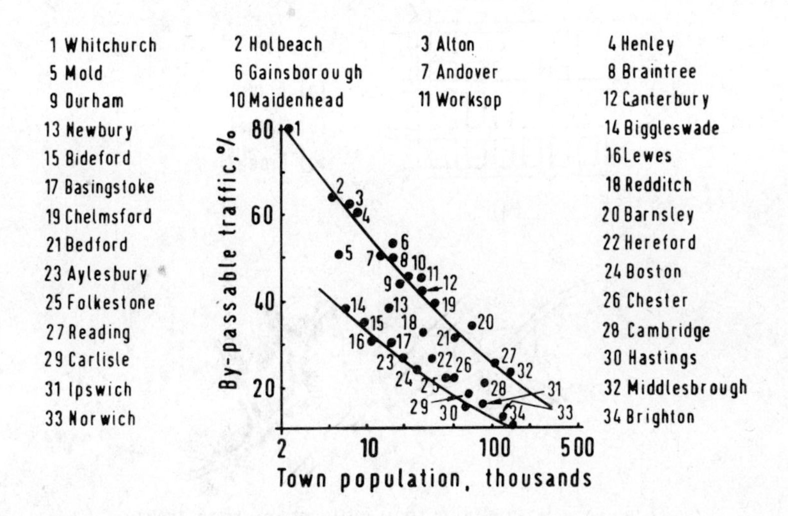

Fig. 3.7. Relationship between town size and percentage of by-passable traffic[16]

In the smaller urban areas, especially those situated on main routes between large towns, the need for by-pass facilities tends to be relatively greater. Not only do these towns act as obstacles to traffic, but the traffic itself is a major nuisance within these towns. By-passes in these cases will give much needed relief, the exact amount depending on the character and location of the town as well as its size. It is only by carrying out proper planning surveys that the exact need can be determined in any particular instance.

Urban road patterns

Broadly speaking there are three principal types of major road pattern in urban areas. As shown in Fig. 3.8, they are the radial, the gridiron and the linear patterns. Of these the radial pattern is the predominant type in this country and so it will be discussed in greater detail than the other two.

Gridiron pattern. Originally favoured by the Romans, this major road pattern is adopted extensively throughout the United States. In one sense it is by chance that this pattern came into favour throughout American urban areas. The basic reason for its acceptance is that, when the counties and townships were being laid out between a hundred and hundred and fifty years ago, it was easier for the surveyors to set them out using straight lines and rectangular co-ordinates. In time the pattern became the accepted one

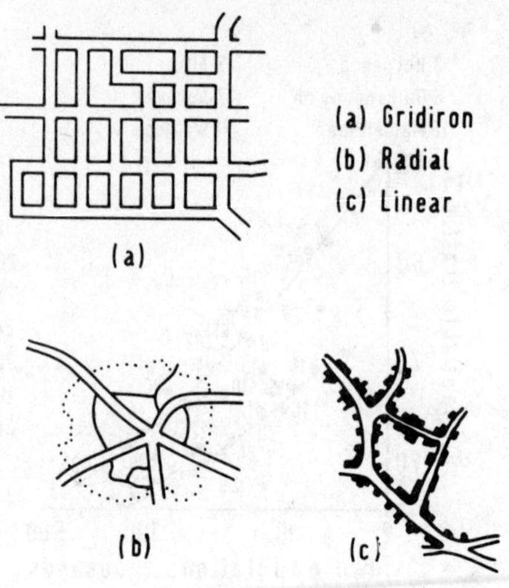

Fig. 3.8. Basic major road patterns in urban areas

so that all newer towns and town expansions tended to be set out in a similar manner. This is particularly noticeable in the mid-western and western towns in the United States, whereas parts of the older eastern cities, tend to be more unco-ordinated and 'European'.

Although the gridiron pattern can produce monotonously long streets flanked by dull blocks of buildings, it has some considerable traffic-moving advantages. It encourages an even spread of traffic over the grid, and as a consequence the impact at a particular location is reduced. It facilitates the imposition of extensive one-way street systems since alternate streets of the grid can be made one-way in opposite directions. If there is a definite central business area in the middle of the grid, it is relatively easy for through traffic to by-pass it, since there are usually alternative by-pass routes available in all four directions.

One objection to the gridiron pattern is that extra distances must be travelled when going in a diagonal direction. This has been remedied in many American cities by superimposing major diagonal routes upon the grid. While these diagonals have aided traffic movement, they also have an unfortunate effect on the architectural development of adjoining frontages, producing in many instances acute-angled plots and awkward street intersections.

There are very few examples of towns in Britain being laid out on the gridiron pattern. Individual housing estates laid out in this fashion are, however, often to be seen.

Linear pattern. Historically this type of urban road pattern developed as a result of local topographic difficulties. The most obvious example is that of a settlement which originally became established alongside a well-used track-way through a long valley. As time progressed and the settlement grew, the length of the town became too great for the then available pedestrian method of internal travel, and so the village began to grow laterally; the homes and other buildings were then located alongside feeder trackways connected to the major route. Today many examples of towns which developed in this way are found in valleys such as those of the West Riding of Yorkshire and in South Wales.

From the point of view of traffic movement, the linear pattern appears at first to have particular advantages. The most important of these is that the main traffic flow is canalized into one major roadway. In principle this concept of canalization is a good one, as evidenced by the modern usage of the motorway. In practice, however, the channelling of all traffic into the main street of a town can create more problems than it solves.

Nowadays the motor-vehicle is used both for trips within the town as well as for those with destinations outside the town. As a result the main street serves as a route for both the completely internal traffic and the internal-external traffic, due to the lack of other suitable longitudinal routes. This traffic is in turn increased by the addition of external-internal traffic and through traffic.

The net result is that many of these linear towns are now literally bisected by the heavy flows of traffic concentrated on the central major street. In addition, since the central street was never designed as a thoroughfare, and in fact usually just 'grew', it is normally easily overloaded.

Radial pattern. As discussed earlier the highway system in Britain developed in the form of a network of roads connecting town centre to town centre. Thus any given town had several roads radiating from its centre to the other towns and villages about it. Eventually as the towns grew in size they tended first to develop along the radials and then later to fill the spaces in between. Thus the towns of today have a strongly marked radial system with the business area stabilized at the centre of a more or less symmetrical cartwheel plan.

The location of the main traffic generators within the central area, the system of radial roads converging on the main source of attraction and, usually, the lack of suitable by-pass routes for through traffic all combined to produce the belief that the cause of central area congestion was traffic on the radials and that the solution might be to build ring roads to enable it to be diverted.

Usage of ring roads. By definition a ring road is a highway that is roughly circumferential about the centre of an urban area, and which permits traffic to avoid the centre of this area. In practice, three forms of ring road have come into being: an inner ring road, an outer ring road and intermediate ring roads. A town with a population of about 20 000 will tend to have but a single inner-outer ring road, whereas a city of 500 000 might have both an inner and an outer ring road. Urban areas with larger population groupings will tend to have one or more intermediate ring roads in addition to the inner and outer roads.

Perhaps the easiest way to visualize the basic urban road pattern in Britain is to liken it to a cart-wheel. The spokes of the wheel are the radial routes. The hub of the wheel is the inner ring road, its function being to serve the needs of local and local-through traffic. Its purpose is to promote the convenient use as well as the amenity of the central area by deflecting all vehicles which have no need to traverse it, while affording convenient means of entry for those to whom access is essential. Thus the location and design of the inner ring road is very closely bound up with the size, layout and usage of the central area. In practice therefore the inner ring road may be round, square, or elongated, and may be incomplete on one or more sides, depending upon the conditions prevailing.

The outer ring road can be considered as the rim of the wheel. Although it is often used by through-traffic in order to by-pass a town, its original purpose was to serve the traffic of the town itself by linking up the outer communities and acting as a distributor between the radials. To serve this purpose the outer ring road is generally located within the outer fringe of present and future urban development and not outside the limits of the town. Because of the greater availability of space at the edges of towns, the outer ring road has tended to be more formally circumferential than the inner ring road. Again, however, its completeness depends upon the needs at specific locations.

In the larger population centres, other reinforcing ring roads may be located within the outer and inner ring roads. These intermediate roads serve the needs of traffic, whether from a distance or of local origin, which desires to reach points between the outer and inner roads. They are normally located in areas already partially developed and thus, to a larger extent than outer ring roads, they incorporate many existing local roads with all their diversity of use.

Advantages of ring roads. What was originally the principal advantage of the radial system, its means of providing direct access to the town centre, is now often a liability. During peak periods especially, as the radial road is followed inward, the traffic volume builds up progressively until the central area is reached and dispersal attempts to take place. In the older towns this has meant that the greatest accumulation of vehicles occurs where road conditions are most critical and the cost of additional road space is at a premium. In addition it has been found that through trips via the radial routes not only often take longer because of central area congestion, but also they can mean greater travelling distances.

The principal advantage of a ring road lies in its ability to service a central area while circumventing it. This is suggested by the diagrams in Fig. 3.9 which show some possible routes of travel from a point 0 outside a central area enclosed by a completely circular ring road of radius r, to destinations inside the ring road. If it is assumed that the destinations are scattered at random over the central area, then the *theoretical* average distances travelled from point 0 to a destination D within are as shown in Fig. 3.9.

Although the ring route is appreciably longer than the other routes, it has the particular advantage that only 0·33*r* is travelled within the central area. The remainder of the journey, 1·57*r*, is on the edge of the area, and thus, in theory, should be free of the central congestion problems.

In practice the advantage of ring routing is not as great as this, because towns are not circular, because it is not possible to join a ring route at any point, and because, for journeys wholly within the central area, the relative advantages of the routes are different. Nevertheless, as Table 3.7 shows, the ring road concept can be very saving of road-space where it is most required.

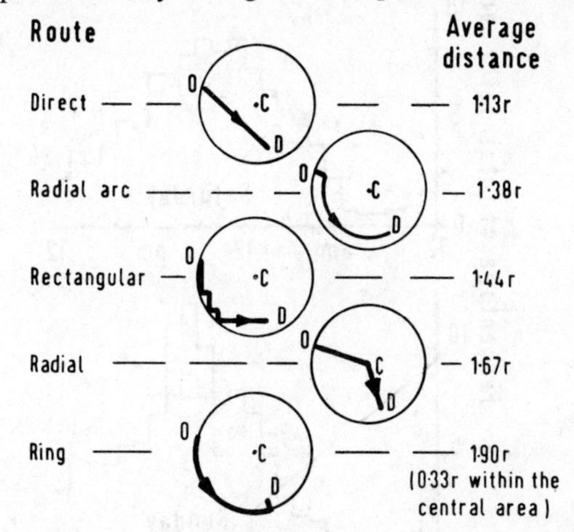

Fig. 3.9. Some possible routes for journey-to-work travel in the central area[17]

TABLE 3.7. *Effect of number of commuters upon the road space required within a central area*[17]

Commuters	*Area of road per person, m²*		
	Direct, radial-arc, or rectangular route	*Ring route*	
		Total	*Within central area*
10 000	0·74	1·02	0·28
100 000	2·60	3·44	1·02
1 000 000	9·10	12·45	3·25

(Peak period = 2 hours; 1·5 persons per car)

HIGHWAY TRAFFIC CHARACTERISTICS

It is necessary for the highway planner to be able to predict the highway volumes that he can expect on the roadway or network of roads being evaluated. Traffic volumes, however, are much heavier during certain times of the year and during certain periods of the day. Hence the engineer-planner must be familiar with the manner in which traffic behaves as a whole before he can attempt to conduct his basic traffic prediction surveys.

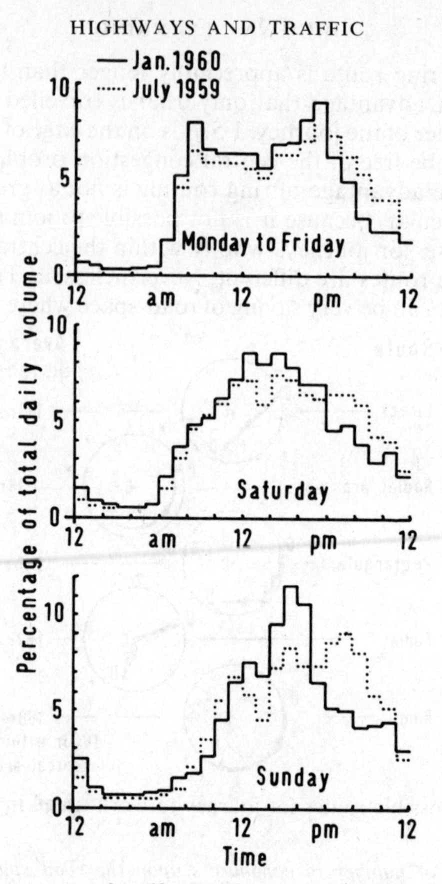

Fig. 3.10. Hourly patterns of traffic flow in Britain (Based on data from ref. 18)

There are three cyclical variations that are of particular interest to the highway planner: the manner in which traffic flow varies throughout the day and night, the day-to-day variation throughout the week, and the season-to-season variation throughout the year. In addition he must also be aware of the directional distribution of the traffic and the manner in which its composition varies. A thorough understanding of the manner in which all of these behave is a basic requirement of any highway planning programme.

Hourly patterns

Typical hourly patterns of traffic flow, based on data collected at urban and rural census points throughout Britain, are shown in Fig. 3.10. There are some characteristics of particular interest illustrated in this figure.

First and perhaps most important is that the weekday pattern is basically unchanged from Summer to Winter. In both examples shown in Fig. 3.10(a) there is a peak about 8.30 a.m. followed by a drop until the middle of the afternoon, after which there is a further rise until an evening peak is reached about 5.30 p.m., after which the traffic flow falls again to its lowest point at about 4 a.m. The most marked difference between the two patterns is that in July a higher proportion of the day's traffic is carried in the brighter evening hours after 6 p.m.

The data in Figs. 3.10(b) and 3.10(c) illustrate the fact that, although the Monday to Friday pattern is relatively consistent and is essentially as is to be expected, the patterns during the weekend can vary considerably. Not only do the Saturday and Sunday patterns differ from each other but they differ between themselves at various times of the year. The most marked variations are (*a*) the lack of a morning peak, as such, during the weekends, (*b*) a more pronounced peak on Winter Sundays, and (*c*) the significant shift and spreading out of the weekend evening peaks from Winter to Summer.

A most important point to note here is that data gathered in 1968[19] reflect the same general pattern illustrated in Fig. 3.10. In other words, although traffic volumes may grow, the relative *percentages* of traffic at the different hours of the day in, say, July are quite consistent year after year.

A further point to emphasize is that the patterns in Fig. 3.10 represent average data obtained at 49 and 39 national survey points in July and January respectively. Thus the actual pattern on any individual road may vary significantly from these data. For instance, not only are traffic volumes heavier on urban roads than on rural roads,[20] but they are also more concentrated during the peak periods of the day. Peak hour volumes are more pronounced and directional on urban radial streets. Urban ring roads on the other hand do not have such sharp peaks. The durations of peak flows also vary considerably; some roads have high sharp peaks, whereas others have peaks that may not be so high but are more critical because of their length.

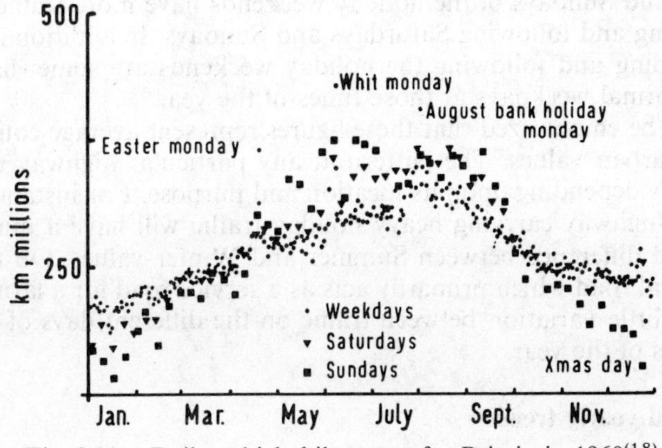

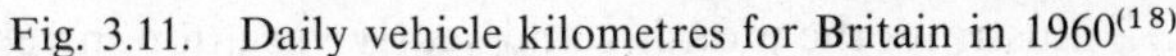
Fig. 3.11. Daily vehicle kilometres for Britain in 1960[18]

Daily patterns

Figure 3.11 gives estimates of the daily travels starting at 8 a.m., for the year 1960. The diagram illustrates clearly that the distances travelled vary considerably with the time of the year. In general, however, the weekday flows behave in a surprisingly consistent manner as do, generally, the Saturday flows. Sunday traffic on the other hand varies considerably from week to week. During the Winter the weekend traffic in general and Sunday traffic in particular is less than during weekdays; however, as the weather gets better travel increases on the weekends especially on Sundays. Again it is of interest to point out that measurements carried out in 1968[21] reflect the same general trends shown in Fig. 3.11—except, of course, that travel had increased overall by about 70 per cent.

Table 3.8 shows, over the year as a whole, a systematic but small increase of traffic from Monday to Friday. This increase is most marked in July.

TABLE 3.8. *Vehicle kilometres on different days of the week, expressed as percentages of average weekday vehicle travel*[18]

	January	*March*	*May*	*July*	*September*	*November*	*Annual average*
Monday	99·6	97·6	97·6	96·0	98·4	98·0	97·9
Tuesday	100·7	99·8	99·8	98·2	99·2	99·6	99·6
Wednesday	99·4	99·9	100·1	100·2	100·5	100·2	100·0
Thursday	99·1	100·7	100·4	100·7	100·4	100·6	100·3
Friday	101·3	102·0	102·2	104·9	101·5	101·5	102·2
Saturday	90·7	97·3	103·2	109·4	109·2	98·2	102·5
Sunday	75·1	92·8	116·8	114·8	112·8	85·8	101·5

Note also that the traffic on the holiday Mondays is in each case higher than on any other day within a week in either direction. Furthermore, the Saturdays and Sundays of the holiday weekends have more traffic than on the preceding and following Saturdays and Sundays. In addition the weekdays preceding and following the holiday weekends are somewhat higher than the normal weekdays at those times of the year.

It must be emphasized that these figures represent average countrywide rural and urban values. The pattern at any particular highway may vary considerably depending upon its location and purpose. For instance, a seaside resort highway carrying heavy holiday traffic will have a much more pronounced difference between Summer and Winter values. On the other hand, a rural road which primarily acts as a service road for a farming area may show little variation between traffic on the different days of the week and months of the year.

Monthly and yearly trends

Figure 3.12 illustrates the monthly and yearly variation in vehicle travel between 1956 and 1960. For comparison purposes the values are expressed in terms of the average monthly value for 1956. Data collected between 1961 and 1968[21] show very consistent and similar upward trends.

The average numbers of kilometres travelled in 1968 by the various classes of vehicle were 3800 for motorcycles, 13 200 for cars and taxis, 47 100 for buses, and 21 900 for goods vehicles[22]. The figure for cars reflects a gentle but continuous yearly increase (about 1 per cent per year) after remaining static at about 11 700 km between 1951 and 1958; however, it would now appear that the British figures are levelling off and no marked increase in annual vehicle usage can be expected within the foreseeable future. (In the United States, average car travel has remained essentially static at about 15 100 km over the period 1952–67.)

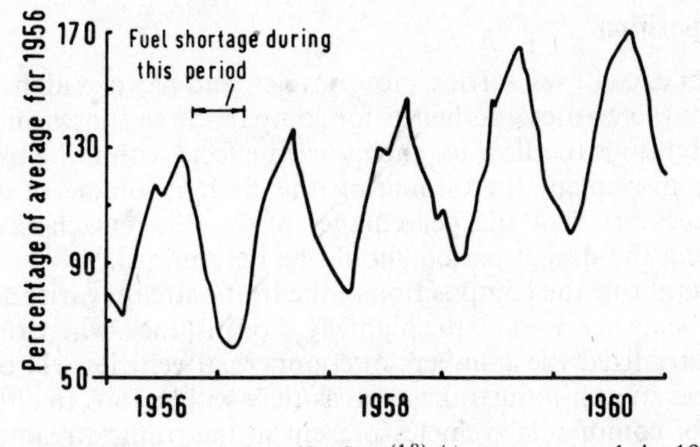

Fig. 3.12. Monthly index of traffic flow[18] (Average month in 1956 = 100)

Effect of weather

The data already shown point up partially the effect that the state of the weather has upon the flow of traffic. During the Summer months the distances travelled are high while they are down during Winter months. Just as important, however, is the fact that, at any time of the year, months that are wetter than normal tend to have less traffic than normal; the difference is statistically significant but not very large.

Directional distribution

A most important consideration in the planning of any highway is the directional distribution of traffic. The importance of this factor can be gauged by considering the case of a dual-carriageway highway designed to carry 4000 vehicles per hour. If the directional distribution of the traffic results in a 50–50 split, then two traffic lanes in each direction may well be sufficient to handle the traffic. If, on the other hand, some 70 per cent of the traffic flow is in a given direction at a given time, then three lanes in each direction may be needed to supply the equivalent level of service to the traffic on that highway.

While this example of a 70–30 distribution may seem to be extreme, it is not so, in any sense. For instance, a major radial road in Leeds has a consistent traffic distribution of 70 : 30 and 40 : 60 during the Monday to Friday morning and evening peak hours respectively.

The actual distribution to be used for design purposes can only be determined by field measurements. If an existing road is to be reconstructed, then the field studies can be carried out on it beforehand. If the highway is to be an entirely new facility, then the measurements should be made on adjacent roads from which it is expected that traffic will be diverted.

A most important feature of the directional distribution is that it is relatively stable and does not change materially from year to year. Hence, relationships established on the basis of current traffic movements are normally also applicable to future movements.

Traffic composition

Passenger cars, buses, lorries, motor cycles, and bicycles all have different operating characteristics and hence, for comparison purposes, must be given different weightings to allow for their varied effects within the traffic stream upon traffic movement. In estimating the design volume of a road it is therefore necessary that the percentages of the different classes of vehicle present during the design period should be determined.

As a general rule the composition of the traffic stream varies according to the locality being serviced by the highway. For instance, where the locality is highly industrialized the numbers of commercial vehicles will be relatively great, whereas in non-industralized areas they will be low. In either case the percentage of commercial vehicles present in the traffic stream will be less during the peak periods than during other hours of the day, since the number of passenger cars on the road is greatest at these times.

Actual values for planning and design usage again are best determined from field studies. Generally the traffic composition determined from current traffic figures is assumed stable when used for future design purposes.

HIGHWAY SPEED, DELAY AND VOLUME STUDIES

When planning a new or improved road or road system, it is necessary to know the distribution and performance of the traffic on existing roads. Not only is this of use in predicting future traffic behaviour, but it is also of value in assessing whether alterations are justified, and in deciding priorities for road improvement.

In almost all planning studies, measurements of traffic flows and speeds are needed. Often measurements are also needed of stopped times and the frequency with which these stops occur. The following discussion is concerned with the manner in which these data are obtained and analysed.

Speed studies

The term 'traffic speed' is often used very loosely when describing the rate of movement of traffic. To the highway engineer there are many different types of speed, each of which describes the rate of traffic movement under specific conditions and for a specific purpose. The vehicle speeds of most interest are spot speeds, running speeds, and journey speeds, and these are discussed here.

Spot speed. This term is used to describe the instantaneous speed of a vehicle at a specified location. Spot speeds have a variety of uses. They can be used as evidence regarding the effect of particular traffic flow constrictions such as intersections or bridges. Since spot speeds at ideal sections of highway are indicative of the speeds desired by motorists they can be used for geometric design purposes on improved or new facilities. In addition, as will be discussed in the chapter on Traffic Management, spot speeds are used in determining enforceable speed limits.

The location at which the spot speed measurements are taken depends upon the purpose for which they are to be used. Whatever the purpose of the study, it should be conducted so as to reduce to a minimum the influence of the observer and his equipment upon the values obtained. Hence the observer and equipment should be located as inconspicuously as possible. If it is not possible to measure all vehicle speeds, vehicles should be selected at random from the traffic stream in order to avoid bias in the results. Thus for instance every tenth vehicle or those whose registration numbers end in a pre-selected digit should be measured, rather than leaving it to the observer's discretion to select the test vehicles.

Methods of measurement

One method of collecting spot speed data is to observe the time required by a vehicle to cover a short distance of roadway. There are two main variations of this method. With the *Direct-timing* procedure two reference points are located on the roadway at a fixed distance apart and the observer starts and stops a stop-watch as a vehicle enters and leaves the test section. While this is a most uncomplicated way of collecting spot speed data, it has the obvious disadvantage of being subject to error because of the parallax effect. This error is of particular importance in before and after studies, where the observer may have to change his observation position.

To overcome this parallax effect, use can be made of an *Enoscope.* This instrument, also known as a Mirror-box, is an L-shaped box, open at both ends, which contains mirrors set at a 45-degree angle. One of these boxes is located at each end of the test length, and the observer takes up a position approximately midway between. As he looks into the appropriate Enoscope his line of sight is bent so as to be perpendicular to the direction of travel. Thus he can start and stop the stop-watch the instant the vehicle passes by the appropriate box, and so more accurate measurements can be obtained. Night-time measurements can also be taken by placing small lights at the reference points directly opposite the mirror-boxes. As vehicles flash by they break the beam, thus again indicating the beginning or ending of timing.

With the *Pressure-contact strip method* two contact strips, usually two pneumatic tubes, are laid across the carriageway at a fixed distance apart. When a vehicle passes over the first tube, an air impulse is sent instantly along the tube which activates a time-measuring instrument in the hands of the observer. When the second tube is passed by the same wheels of the vehicle, the timer is automatically stopped, and the reading noted either visually by the observer or by automatic data recorders.

Provision can usually be made with these devices for switching the direction of the start and stop detectors so that speeds in either direction of travel can be measured in the course of a particular study. To avoid invalid measurements due to overtaking manoeuvres occurring at the contact strips, the observer has a superimposing ready-switch which, until it is activated, does not allow any readings to be taken.

The easiest way of measuring spot speeds is to use a *Radar Speed-meter*. When this is in operation a beam of very high frequency is directed from the radar meter to the moving vehicle. The waves are bounced back from the vehicle but, because of the Doppler effect, these reflected waves have a slightly different frequency from the transmitted waves. This difference, which is directly measurable, is proportional to the speed at which the vehicle is moving. Provided that the beam of the transmitter is contained within a cone of about 20° of the line of movement of the vehicle, the spot speed can usually be measured to an accuracy of about 2 per cent with the normal type of radar meter.

The radar speedmeter is usually set up near the edge of the carriageway at a height of about 1 m above ground level. Its operating zone extends a distance of about 45 m and so it can measure all speeds within this zone, whether they be approaching or receding. A drawback associated with using this speedmeter is that individual vehicle speeds can be difficult to obtain in heavy traffic flows where vehicles are likely to overtake/mask each other. No very definite information is available on this, but it appears that it is very difficult to operate the speedmeter efficiently on 2-lane roads when the traffic is heavier than about 500 vehicles per hour. American experience suggests that positive identification of all individual speeds becomes impossible at traffic volumes greater than about 1000 vehicles per hour on multi-lane highways.

On very crowded highways, i.e. motorways or busy city streets, spot speeds can be very accurately obtained by means of photographs obtained with a *Time-lapse Camera*. This camera takes photographs at fixed intervals of time, thus obtaining a permanent record of all vehicle movements within the camera vision. After the film has been developed the speed of any type of vehicle present in the traffic stream can be obtained by comparing its carriageway position on successive exposure frames.

The main drawbacks associated with using the camera method are the relatively considerable time required and the expense involved in processing the film and analysing the data (unless sophisticated analysis equipment is available).

Analysis

Normally the spot speeds measured at a particular location will vary considerably, the degree of variation depending upon the number and type of vehicles being measured and the condition of the roadway. In analysing the data, therefore, it is necessary to consider carefully beforehand what information is required. Some of the desired values can most readily be obtained by graphical interpretation, whereas others can be easily calculated directly from the field data.

1. *Graphical analysis.* Two methods of graphically interpreting the data shown in Table 3.9 are illustrated in Fig. 3.13. Figure 3.13(a) contains examples of both a histogram and a frequency curve, each illustrating the number of occasions at which the different speeds occurred. Figure 3.13(b) is an example of a cumulative spot speed distribution curve for the same data.

The frequency curve is a very useful preliminary guide to the statistical normality of the data. If the measurements are normally distributed, then the frequency curve will be bell-shaped.

The *Modal Speed*, which is the spot speed which occurs most often, can be detected easily at the peak of the frequency curve.

The cumulative curve shown in Fig. 3.13(b) was obtained by plotting the cumulative percentage versus the upper limit of each speed group shown in Table 3.9, and then drawing a smooth S-shaped curve through the points obtained. This curve is most useful in determining the speed above or below which certain percentages of vehicles are travelling.

The *Median* speed is the middle or 50 percentile speed. It is the speed at which there are as many vehicles going faster as there are going slower.

TABLE 3.9. *Illustrative grouped spot speed data*

Speed class km/h	*Average speed km/h*	*No. of vehicles*	*Frequency %*	*Cumulative %*
10–14·9	12·5	3	1·5	1·5
15–19·9	17·5	10	5·0	6·5
20–24·9	22·5	21	10·5	17·0
25–29·9	27·5	31	15·5	32·5
30–34·9	32·5	54	27·0	59·5
35–39·9	37·5	43	21·5	81·0
40–44·9	42·5	21	10·5	91·5
45–49·9	47·5	10	5·0	96·5
50–54·9	52·5	5	2·5	99·0
55–59·9	57·5	2	1·0	100·0

The *85 Percentile* speed is the speed below which 85 per cent of the vehicles are being driven. This speed is often used as the criterion in establishing an upper speed limit for traffic management purposes. Some authorities also use the 85 percentile speed of a highway as a measure of the design speed to be selected on similar new highways.

The *15 Percentile* speed is generally considered to be the speed value which should be utilized as a minimum speed limit on major highways such as motorways. Vehicles travelling below this value on high speed roads are potential accident hazards because of their obstructive influence upon traffic flow.

2. *Mathematical analysis.* In a spot speed study, probably the most often obtained statistic is the *Arithmetic mean* or *Average* spot speed. This is the

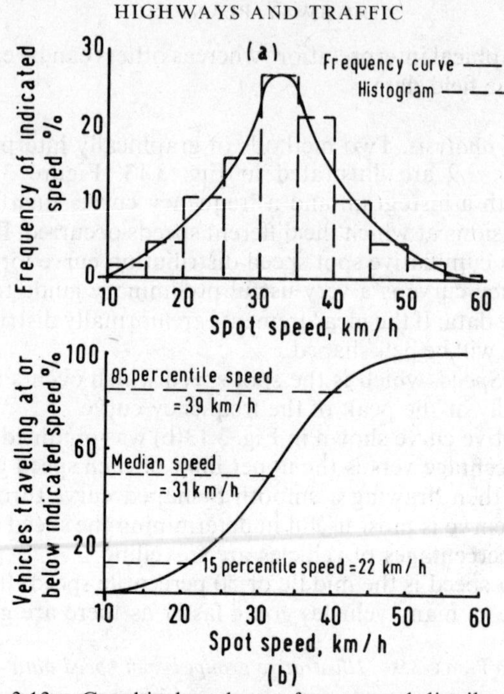

Fig. 3.13. Graphical analyses of spot speed distribution

sum of all the variable speed values divided by the number of observations. Mathematically this is expressed as follows:

$$\bar{x} = \sum X_j / n$$

where $\bar{x}$ = average spot speed,

X_j = jth spot speed,

and n = number of observations.

More usually, to avoid excessive computations at a later stage, the entire data are grouped into speed-class intervals in the manner indicated in Table 3.9. A speed-class interval of convenient size is easily obtained by using the following equation:

$$\text{C.I.} = \frac{R}{1 + 3{\cdot}322 \log n}$$

where C.I. = class interval,

R = range between largest and smallest speed values,

and n = number of observations.

For the data in Table 3.9, the fastest speed was 59 km/h, while the slowest was 12 km/h. Thus,

$$\text{C.I.} = \frac{59 - 12}{1 + 3{\cdot}322 \log 200} = 5{\cdot}5$$

It is more convenient to use integers, so a class interval of 5 was used to group the data in the table.

When the data are grouped into speed-class intervals, the average value is obtained from the following equation:

$$\bar{x} = \frac{\sum f_i x_i}{n}$$

where $\bar{x}$ = average spot speed,

x_i = average speed of the ith speed interval group,

f_i = frequency of the ith group,

and n = number of observations.

A statistical measure of the dispersion of the spot speeds is given by calculating the *Standard Deviation* of the set of observations. This is estimated by first obtaining the variance of the sample, and then taking the square root of the variance. Thus

$$s^2 = \frac{\sum (X_j - \bar{x})^2}{n} \text{ and } s = \sqrt{s^2}$$

where s = estimate of the standard deviation of the distribution,

s^2 = variance of the sample,

X_j = jth spot speed,

n = number of observations,

and $\bar{x}$ = average spot speed.

For grouped data the standard deviation is estimated as follows:

$$s = \sqrt{\frac{\sum f_i (x_i)^2 - \frac{1}{n}[\sum (f_i x_i)]^2}{n}}$$

Thus for the data in Table 3.9, $s = 8{\cdot}62$ km/h.

The usefulness of the standard deviation lies in the fact that the arithmetic mean plus and minus one standard deviation contains 68·27 per cent of the data, plus and minus two standard deviations contains 95·45 per cent, and plus or minus three standard deviations contains 99·73 per cent.

A statistic which indicates the confidence with which the arithmetic mean of the sample can be considered as the actual mean of the entire traffic speed population on that section of the highway is the *Standard Error of the Mean.* This is determined by first calculating the variance of the mean and then taking its square root. Thus

$$s_{\bar{x}}^2 = \frac{s^2}{n}$$

and

$$s_{\bar{x}} = s/\sqrt{n}$$

where s_x = standard error of the mean,

s_x^2 = variance of the mean,

s = standard deviation of the sample,

and n = number of observations.

Therefore for the data in Table 3.9,

$$s_{\bar{x}} = \frac{8{\cdot}62}{\sqrt{200}} = 0{\cdot}61 \text{ km/h}$$

This statistic enables it to be said with 95·45 per cent confidence that the actual mean of all the spot speeds at this particular location lies between 33·25±2 (0·61) km/h i.e. between 34·47 and 32·03 km/h.

The standard error of the mean is also a most useful statistic in determining whether the speed differences at different locations or in 'before' and 'after' studies are significant. This significance can be tested by calculating the *Standard Deviation of the Difference in Means*. Thus

$$\hat{s} = \sqrt{s_{\bar{x}B}^2 + s_{\bar{x}A}^2}$$

where $\hat{s}$ = standard deviation of the difference in means,

$s_{\bar{x}B}^2$ = variance of the mean of the 'before' study,

and $s_{\bar{x}A}^2$ = variance of the mean of the 'after' study.

If the difference in mean speeds is greater than twice the standard deviation of the difference in means, that is,

$$\bar{x}_A - \bar{x}_B > 2\hat{s}$$

where $\bar{x}_A$ and $\bar{x}_B$ are the mean speeds of the 'after' and 'before' studies, respectively, then it can be said with 95·45 per cent confidence that the observed difference in mean spot speeds is statistically significant. In other words, the difference in speeds is a true change reflecting the changing road conditions, and is not due to chance.

The standard error of the mean is also very useful in estimating the number of vehicles that must be sampled in order to ensure a certain degree of accuracy. For example, let it be assumed that it is desired to know how many vehicles must be sampled at the site at which the data in Table 3.9 were collected in order to ensure an error of less than 0·5 km/h in the average speed obtained. Thus,

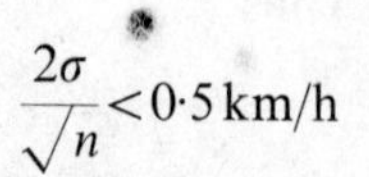

$$\frac{2\sigma}{\sqrt{n}} < 0{\cdot}5 \text{ km/h}$$

where n = desired number of observations,
and σ = standard deviation of the entire population of spot speeds. In this case it is assumed that $\sigma = s$, the previously determined estimate of the population standard deviation.

Therefore,

$$n > \frac{4\sigma^2}{(0{\cdot}5)^2}$$

$$> \frac{4(8{\cdot}62)^2}{0{\cdot}25}$$

$$> 1188{\cdot}9$$

Since n is an integer, this means that the number of vehicles sampled must be greater than 1189 to ensure the required accuracy.

In conducting before and after studies care must always be taken that similar measurement procedures are used in both stages of the study; otherwise the speed differences may not be true reflections of any change in road conditions. For instance, if the spot speeds are measured instantly by means of a radar speed-meter or pressure-contact strips placed closely together, the average speed is given by

$$\bar{x}_t = \frac{\sum X_j}{n}$$

where $\bar{x}_t$ = time mean speed, and X_j and n are as defined before. Note that the average spot speed is here called the *Time-mean* speed.

On the other hand, if a time-lapse camera is used to obtain the data, the average mean speed is obtained by

$$\bar{x}_s = \frac{\bar{l}}{t}$$

where $\bar{x}_s$ = space-mean speed,
$\bar{l}$ = average distance travelled by each vehicle,
and t = time interval between two successive exposures (a constant).

Note that the mean spot speed calculated in this way is called the *Space-mean* speed.

The basic difference between the two mean speeds is that the spot speed for each individual vehicle is calculated prior to the determination of the time-mean speed, whereas the average time (in the above illustration this is the fixed time interval t) is first determined in calculating the space-mean speed. Both methods have been examined and it has been shown[(23)] that

$$\bar{x}_t = \bar{x}_s + \frac{\sigma_s^2}{\bar{x}_s}$$

where $\bar{x}_t$ = time-mean speed,

$\bar{x}_s$ = space-mean speed,

and σ_s^2 = variance of the space distribution of speeds.

Thus the time-mean speed is always greater than the space-mean speed unless the very unlikely happening occurs that there is no variation at all amongst the speeds. The difference is usually in the order of 12 per cent.

Running and journey speeds. Although spot speeds are most useful in measuring fluctuations in speed at particular locations, they give no information regarding fluctuations in speed throughout a route as a whole. Statistics that are of value in this situation are the running and journey speeds.

Running speed is defined as the average speed maintained over a given route while the vehicle is in motion. Thus in determining the running speed, the times en route when the vehicle is at rest are not taken into account in the calculations. Normally only the average running speed and the standard deviations are the variables determined.

Knowledge of the running speed can be very useful to the highway planner. For instance the running speed is the speed value used for the purpose of the capacity determinations. It is also a speed that can be used as a measure of the level of service offered by a highway section over a long period of time. Economic studies estimating the value of highway alterations and assessing priorities for improvement may also utilize running speed values.

The motorist about to set out on a journey is primarily concerned with the time required to complete the trip. Hence the speed in which he is most interested is the average journey speed i.e. the distance travelled divided by the total time taken to complete the distance. This total time includes both the running time, when the vehicle is actually in motion, and the time when the vehicle is at rest, i.e. at traffic signals, in traffic jams, etc.

Knowledge of the journey speed and, in particular, the journey time is most useful in highway planning. For instance, it is an excellent direct measure of traffic congestion and the general adequacy or inadequacy of a road or road system. Highway economic studies utilize journey times and speeds in their analyses. In addition journey times are used as criteria on which to base decisions regarding the diversion and assignment of traffic to new and improved highway facilities.

Methods of measurement

The following three methods are those most used in obtaining the running and/or journey speeds of traffic streams:

1. *Registration number method.* With this method observers with synchronized watches are stationed at both ends of the highway test section. As each vehicle passes an observer, its registration number and time of passage are noted and recorded. At a later time the registration numbers are matched and the individual vehicle times determined. Knowing the distance

over which the vehicles have travelled, the mean running speed of traffic can then be determined.

Unless traffic is very light, it is usually necessary to employ two observers for each direction of travel; the first observer notes the registration numbers and the times while the second observer records them. Two observers can accurately record at the rate of about 300 vehicles per hour in this manner. If the traffic flow is heavier an unbiased sample of the vehicles can be obtained by noting only the licence numbers ending in certain digits.

Besides being laborious and time-consuming, the registration number method has the very obvious disadvantage that it can only be used on highway sections having minor or no intersections, since vehicles may leave the test section or stop en route unknown to the observers. It is therefore most suitable for highway sections in rural locations.

2. *Elevated observer method.* With this method two or more observers are stationed on top of a high vantage point located adjacent to the road section being evaluated. These observers then select vehicles at random from the traffic stream and observe and time their progress over the test section.

While this method has the advantage that any vehicle delays can be noted, it has the disadvantage that it is very often difficult to secure suitable vantage points. Of necessity also the road test section will usually have to be relatively short, so that the selected vehicles remain within sight of the observers. As a result, this method is most useful on relatively short sections of city streets.

3. *Moving observer method.* This method[24] of determining the journey speed can best be illustrated by considering a stream of vehicles moving along a section of road of length l in such a way that the average number of vehicles q passing through the test section per unit time is a constant. The stream can be regarded as consisting of flows q_1 moving with speed v_1, q_2 with speed v_2, etc.

Suppose an observer travels with the stream at a speed v_w and against the stream at a speed v_a. When he is travelling against the flow q_1, which is moving at a speed v_1, the rate at which vehicles pass him is greater than if he were stationary, i.e. they pass him at the rate of $(v_a+v_1)/v_1$, and the number of vehicles that pass him per unit of time is equal to $q_1(v_1+v_a)/v_1$. Similarly when he is travelling with the traffic stream the number of vehicles going in the same direction, as counted by himself, i.e. the number of overtaking vehicles minus the number of overtaken vehicles, is given by $q_1(v_1-v_w)/v_1$.

If t_w denotes the journey time l/v_w which corresponds to speed v_w; t_a denotes the journey time l/v_a which corresponds to speed v_a; x_1 denotes the number of vehicles with speed v_1 met by the observer when travelling against the stream, i.e. in time t_a; and y_1 denotes the number of vehicles which overtake the observer when travelling with the stream minus the number he overtakes, i.e. in time t_w; then

$$x_1 = q_1(t_a+t_1)$$

and

$$y_1 = q_1(t_w-t_1)$$

Similar results hold for all the other speeds, so that if q denotes the flow of all vehicles in a given stream, i.e. $q_1+q_2+q_3+\ldots$; if x denotes the total number of vehicles met in the section when travelling against the stream; if y denotes the number of vehicles overtaking the observer minus the number he overtakes when travelling with the stream; and if $\bar{t}$ denotes the mean journey time of all vehicles in the stream, i.e. $(q_1 t_1+q_2 t_2+\ldots)/q$; then summing overall gives

$$x=q(t_a+\bar{t})$$

and

$$y=q(t_w-\bar{t})$$

Therefore

$$q=\frac{(x+y)}{(t_a+t_w)}$$

and

$$\bar{t}=t_w-\frac{y}{q}.$$

The space mean speed is then given by $l/\bar{t}$. It must be remembered that the times used in the equations may (journey times) or may not (running times) include the periods during which the test vehicles are actually stopped.

In using the moving observer method to study traffic conditions on a highway section, a car (or preferably a pair of cars, one starting from each end and each carrying three observers) is started along the test section. One observer starts a stop-watch which is left continuously running and records the time history of the vehicle along the highway. A second observer counts the number of vehicles in the opposing traffic stream that are met on the journey, totals being recorded by the first observer en route at points which tend to divide the length of highway being studied into sections of reasonably homogeneous traffic conditions. Meanwhile the third observer counts the number of overtaking and overtaken vehicles, the totals being recorded at the same points as before. The values of x, y, t_a and t_w then used in the above equations are the averages of all the individual values obtained. If it is desired to know the average speed of certain types of vehicles only, the procedure is the same, except that the only vehicles taken into account in the calculations are those of the type being considered. It is generally held that 12 to 16 runs in each direction along a test section will be sufficient to give reasonably consistent speed estimates. Because of the apparent sensitivity of the method to the minute-by-minute variations in the traffic stream, however, one recent study[25] has suggested that at low traffic volumes (e.g. less than about 250 veh/h in a given direction on a two-lane road), the number of test runs required to achieve a given degree of accuracy on relatively short sections of road may be so great as to render the method impractical to use. In general, the heavier the traffic volume on a given type of road, the less the number of test runs required.

This procedure has the particular advantage that it is most economical in manpower. A small team of observers with one or two cars can collect reliable data over considerable roadway distances in a few weeks. Most important is the fact that it allows much information to be collected at the same time, i.e. journey times and speeds, delays due to intersections or parked cars, and other relevant information.

Delay studies

Delay studies can be carried out separately at particular locations or in conjunction with studies determining running and journey speeds. They are of considerable value to the highway planner since they enable him to pin-point locations where conditions are unsatisfactory as well as determining the reasons for and extent of the delays. This information can be used to indicate the urgency of need for improvement and the extent to which the improvement should be carried out. Highway economic studies automatically utilize delay times in their calculations.

Studies into the cause and extent of delays on highways must take into account the fact that there are two forms of delay. The first of these, termed *Fixed Delay*, occurs mostly at roadway intersections. It is literally the result of some fixed roadway condition, and hence it can occur irrespective of whether the highway is crowded or not. Typical highway fixtures causing this type of delay are traffic signals, railway crossings, roundabouts and stop signs.

The second type of delay is called the *Operational Delay* and this is primarily a reflection of the interacting effects of traffic on the highway or street. Operational delays can be caused by parking and unparking vehicles, by pedestrians,by crossing and turning vehicles at uncontrolled intersections, as well as by vehicles stalling in the middle of the traffic stream.

Internal frictions within the traffic streams themselves can be another cause of operational delays. For instance, vehicle volumes in excess of capacity will cause traffic congestion and result in considerable delay to traffic. Intersections adjacent to each other and carrying heavy turning movements can be the cause of a considerable amount of weaving within the traffic streams as vehicles attempt to enter and leave the main roadway; these manoeuvres inevitably reduce the mean speed of traffic and lower the general efficiency of movement.

Methods of measurement. The moving observer method is most useful in determining the cause and extent of delay encountered en route. Before the experiments are started a survey of the route is made and a journey log prepared. Then as the test vehicle moves along the route all locations and times of stopping and starting are noted, as well as ancillary data. Analysis of the journey logs easily points up the locations and extent of any delays to traffic.

While the moving observer method will determine the causes and extent of any delays it will not always provide sufficient information by itself on which remedial action can be based. For instance, at intersections it will be necessary to obtain additional detailed information such as the traffic volumes on the different approach roads, the traffic management methods in use at or adjacent to the intersection, the accident statistics, and the physical dimensions and geometric layout of the intersection, before decisive action can be taken.

Volume studies

The terms traffic flow and traffic volume are used interchangeably to define the number of vehicles that pass a given point on the highway in a given period of time. As such it is probably the statistic that is of most value to the highway planner; to attempt to design a modern high-speed highway without knowing the traffic volumes that can be expected upon it is like trying to design a structure without knowledge of the loads that it will have to carry. In addition an estimate of the relative importance of already constructed highways can be obtained on the basis of a comparison of the volumes of traffic that they carry.

The type of traffic volume data collected at any given time and location depends upon the use to which the data will be put. Similarly the method of collecting the data is dependent upon its usage. In either case it is most important to remember that, notwithstanding the timing, location and method of data collection, traffic flow studies provide only very limited information regarding existing traffic on the test roads. For instance, the 24-hour data in Fig. 3.14—the widths of the traffic flow bands in this diagram are proportional to the measured traffic on each roadway—clearly show that the traffic in Greater London is diffused over many roads at present. This can be attributed to the fact that most main roads in London provide similar moderate traffic capacities, so that motorists tend to seek indirect alternative routes, thereby distributing the demand more or less evenly on all main roads. However, other than pointing up which roads are more used than others, this figure by itself tells little; no information at all is given about the origins and destinations of vehicles and whether the volumes of traffic on each road accurately reflect its importance, i.e. whether motorists are using a particular route because it is less crowded or whether it is actually the most direct and desirable route between the various origins and destinations.

Nevertheless, despite the obvious limitations associated with the data obtained from traffic flow studies on their own, they have many and varied uses. In rural areas well served with highway facilities and in smaller urban areas, volumetric measurements can very often be used to interpret the relative importance of individual roadways and thus justify particular highway improvements or traffic management measures. When carried out in conjunction with vehicle classification counts they have a variety of uses, e.g. in capacity determinations and highway economic studies, in establishing correction factors to be applied to automatic traffic counts, and in establishing geometric design and traffic control criteria, particularly at intersections.

Figure 3.15 illustrates one example of the type of information that can be

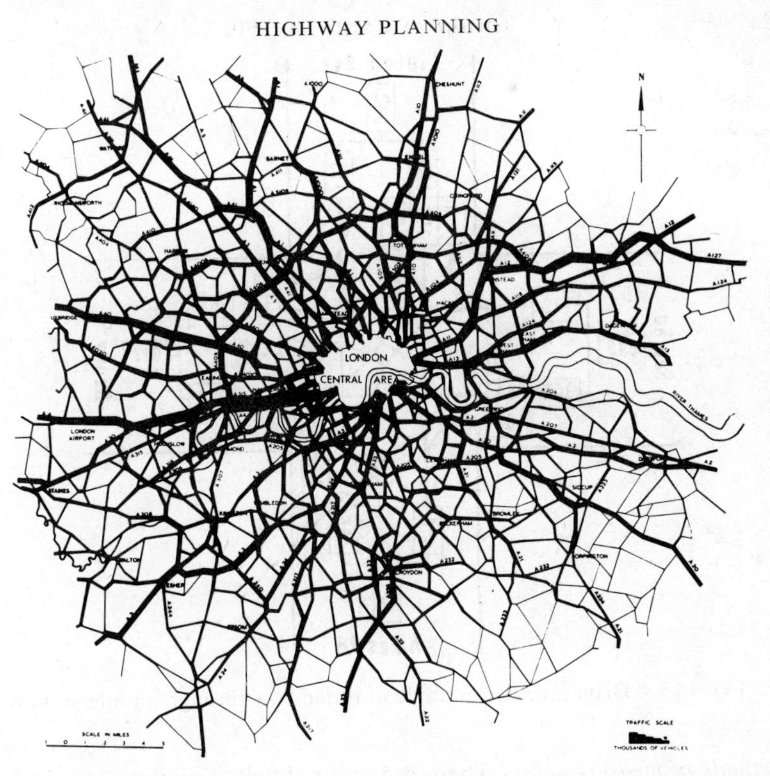

[*Crown copyright in map reserved*]

Fig. 3.14. 24-hour average weekday traffic flow pattern within the London Traffic Survey area[26] (*Courtesy of the Greater London Council*)

obtained regarding traffic flow through an intersection. This graphical method of presenting data on traffic volumes and turning movements is very informative; the relative importance of each turning movement is discernible at a glance. If a vehicle classification count is carried out at the same time then 'pie' diagrams can be drawn to show the composition of each traffic stream.

Volumetric counts conducted at cordon locations can provide valuable information regarding the accumulation of vehicles within an enclosed area. If vehicle occupancy counts and pedestrian counts are conducted as well, information is obtained regarding the accumulation of people within the cordon area, and the numbers of people using the different transportation modes.

As was discussed earlier, the manner in which traffic volumes and traffic patterns behave throughout the year and from year to year is predictable on similar highway types. Thus volumetric data collected over a relatively short period of time on individual highways can often be scaled to allow for future daily, weekly, monthly, seasonal or yearly variations, as required.

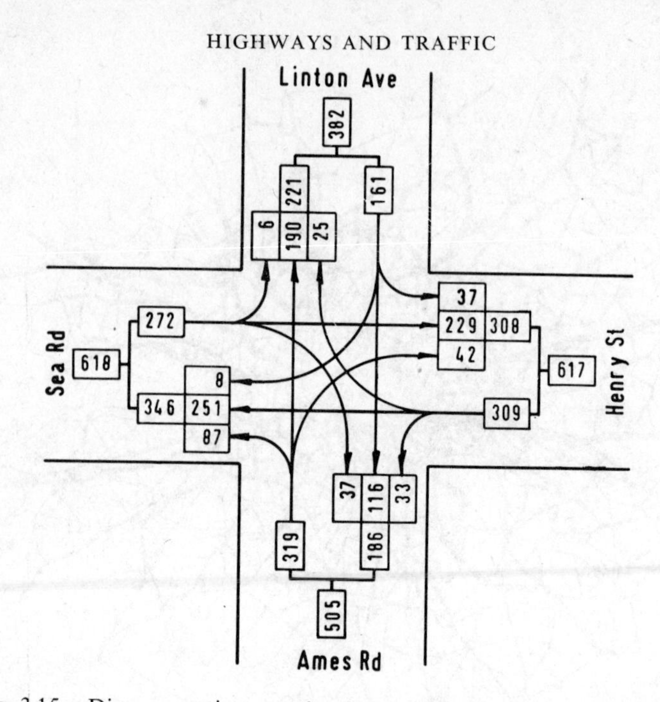

Fig. 3.15. Diagrammatic example of traffic flow through an intersection

Methods of measurement. There are many different methods of obtaining traffic volume data. These include the use of mechanical devices which can automatically count and record the data; manual counts where traffic observers do the counting and recording; a combination of mechanical and manual methods, such as a multiple pen recorder where observers actuate pens which mechanically record the data; moving observer methods; and methods involving the use of photography. Of these, the methods most widely known are the automatic methods, the manual methods and the moving observer technique. Whichever is used at any particular location depends entirely upon the facilities available and the type of information required.

The greater part of traffic counting, especially when data are required over long periods of time, is carried out using *automatic devices*. An automatic apparatus consists essentially of some device for detecting the passage of a vehicle and a counting mechanism for recording the detection pulses.

Detectors are of three main forms:

(*a*) Magnetic detectors placed on or beneath the carriageway which cause signals or impulses to be sent out as vehicles pass over them. There are two basic types of magnetic detector, i.e. the type which detects the presence of the magnetic fields associated with ferromagnetic objects in the vicinity of the detector, and the type which detects the presence of a ferromagnetic object by noting the inductance change which occurs within a wire loop when a vehicle passes over it. The inductance loop detector is now coming into fairly wide usage in Great Britain, especially at permanent counting stations on heavily travelled rural roads[27]. As it is normally inset in the top of the basecourse of the pavement, the detector is safe from traffic hazards, snow and ice, and has a very long life. Its main disadvantage is that, in urban areas particularly, the accumulation of heavy electrical installations, underground storage tanks, cables and other underground conduits can make its use very difficult, if not impossible.

(*b*) A photo-electric device which actuates a counter when a vehicle passes through a beam of light focussed across the road on a photo-cell capable of distinguishing between light and lack of light. This simple and reliable system is limited to light volume roads because of accuracy problems and its inability to distinguish between passing vehicles on individual traffic lanes.

Another similar device consists of a radar detector which is suspended over the roadway and continually transmits and receives a radio signal of known frequency. Whenever the frequency of the reflected signal differs from that of the transmitted one, a moving vehicle is detected. This is also a relatively simple and reliable detector that is very accurate and not subject to the wear and tear of traffic. However its high initial cost precludes its use in many instances.

Vehicles can also be detected by ultrasonic detectors of either the pulsed or resonant type. The pulsed type detects vehicles by measuring the time required for the echos from bursts of ultrasonic energy transmitted toward the road surface to bounce back to a detector suspended above the carriageway. When a vehicle passes beneath the unit the sound waves are reflected from its roof and so the received echo occurs earlier because of the shortened path length. In contrast, the resonant type of ultrasonic detector requires the receiver and transmitter units to be mounted opposite each other so that vehicles can pass between them and break the beam of ultrasonic energy.

(*c*) Contact strip devices laid on top of, or in a pad inserted within, the carriageway, and which actuate counters when vehicles pass over them. These can be of the electrical or pneumatic tube variety. The permanent electrical type, inserted within the carriageway pavement, consists of a steel base plate beneath a vulcanized rubber pad holding a strip of suspended spring steel. The gap between the two contacts thus formed is filled with a dry and inert gas during the assembly and sealed as a unit during the vulcanizing process. As each vehicle axle crosses the treadle pad there is a positive electrical contact closure which actuates a recorder. This detector is an entity inserted in each traffic lane and thus has the ability to count vehicles by lane.

The temporary detector most often used in Britain consists of a single length of 12·7 mm I.D./25·4 mm O.D. rubber tubing stretched across the roadway and secured to the carriageway by clips or saddles of canvas and brass. One end of the tube is sealed (except for a tiny air-hole whose purpose is to minimize the reflection of air waves in the tube) and the other end is attached to a pressure-actuated diaphragm switch. The passage of a wheel over the detector causes a pulse of air to travel along the tube so that it moves the diaphragm outwards against a contact point, thereby completing an electrical circuit which actuates a recorder housed at the roadside. This detector can be very accurate, and is especially easy to install and maintain. It has the disadvantage that it is very vulnerable to wear and tear by traffic, especially street cleaners and snow ploughs, as well as to vandalism and theft. In addition this detector, as well as all others, cannot distinguish between the number of axles on each vehicle, and thus can only count axles or pairs of axles.

There are various forms of counting units utilized in conjunction with the detecting devices just discussed. The types most useful are as follows:

(*a*) A simple accumulating counter which is read directly and as necessary by an observer. No printed record, other than the accumulated total, is available at any specified interval of time.

(*b*) A more sophisticated accumulating counter which upon actuation by a timing mechanism prints the results out on a paper tape at required intervals, say every 15 minutes or every hour.

(*c*) A counter using a circular chart recorder. This can record traffic volumes of up to a 1000 vehicles and utilizing 5, 10, 15, 20, 30 and 60 minute counting intervals, for periods of 24 hours or 7 days. The principle of this counter is a very simple one; the circular graph paper rotates at a uniform speed, and as vehicles move across the detector a recording pen moves out on the graph and records the traffic volumes.

(*d*) A punched-tape recorder which enables tape punched automatically at the roadside to be processed through a translator which, when connected to a key-punch machine, produces punched cards or tape for direct use in computers. The operation of this counter consists of an input-impulse memory, an interval timer period, and the punching and reset mechanism. The input system receives the impulses from the detector and converts these to successive positions on the pulse memory code disc. The interval timer establishes the end of each time period. The punching and reset mechanism then punches the paper tape with the accumulated total.

The normally-used automatic counter can provide only very limited information on its own regarding the numbers of vehicles that pass by the recorders. When more detailed information is required, such as the exact number of vehicles instead of the number of axles, or the turning movements at an intersection, or simply when no automatic devices are available, recourse is made to *manual counting procedures* utilizing field personnel who record the required data on previously prepared tally sheets. Manual observers are also needed to determine the percentages of the different types of vehicle present in the traffic stream: This is necessary because of the differing influences of the various types of vehicles upon the movement of traffic.

The advantages of manual counting are that the counts are more accurate, very specific information is obtained and, in general, office work is simplified. Because it is usually more expensive to obtain data in this way, the use of manual methods is normally limited to short periods of time, or to places where automatic methods cannot be used.

The *moving observer method* technique has already been described in detail with reference to obtaining running and journey times and speeds. The basic method requires that the test car or cars pass up and down the highway test section for a prescribed number of times while travelling at approximately the estimated average speed of traffic and noting the travel times and the number of vehicles overtaking and overtaken.

If it is desired to express the flow in vehicles per hour, this is given by the following formula:

$$q = \frac{60(x+y)}{t_a+t_w}$$

where q = traffic flow in one direction, vehicles/hour,
x = number of vehicles met while driving against the flow to be estimated,
y = number of vehicles overtaking minus the number of vehicles overtaken while driving with the flow to be estimated,
t_a = travel time while driving against the flow to be estimated, minutes,
t_w = travel time while driving with the flow to be estimated, minutes,
and 60 = constant, min/h.

ORIGIN AND DESTINATION SURVEYS

When new or improved traffic routes and facilities are being planned it is necessary to estimate where they should be located so as to attract or relieve most traffic, and how much traffic they will actually carry when constructed. To do this properly it is necessary to determine the pattern of the journeys that people make. The origin-destination (O-D) survey shows what amounts of travel there are between various locations. It does not normally tell much about the actual routes travelled between particular origins and destinations; instead it emphasizes the travel desires rather than the actual routes.

The scope of an origin and destination survey may be limited to one particular route in an urban or rural area, or may be extended to include any part or all of an urban or conurbation area. Thus they can range from basic motor traffic studies covering hundreds of square kilometres in rural areas[28] to being part of comprehensive urban transportation surveys of similar area magnitude.[26]

While O-D surveys can be carried out on existing facilities and within existing towns, they are obviously not practical where new towns are concerned. Instead some method of predicting the traffic-generating potential of houses, factories, commercial establishments and the multitude of other forms of land use is required. Moreover, if this information is available it will also have application to existing urban and rural areas. Another major purpose of comprehensive O-D surveys, therefore, is to determine what relationships do exist between the amount and type of travel and the traffic-generating factors.

The following discussion is a brief outline of some of the main forms of origin and destination surveys and the manner in which they are used.

Survey-area zoning

All origin and destination surveys begin with the definition of the survey area. While it would be desirable to analyse every journey individually, it is impracticable to attempt this, and hence origins and destinations are grouped by zones. All journeys with origins or destinations within a zone are then assumed to begin or end at its centroid.

O-D zones can be either large or small; what is more important is that each should include developments of a homogenous nature. Thus zones in urban areas may be smaller, closer to the central area and larger in residential suburban districts.

In practice zone boundaries are often dictated by geographical or physical limitations such as rivers, embankments, hilly topography or other such features that inhibit traffic movements and help create 'traffic-sheds'. In rural areas National Grid references have been used to delineate the boundaries of individual zones. Whichever method is used, it is most convenient to make the centre of each zone a highway or similar traffic generator. A major highway should *never* form the boundary of a zone.

Data such as were gathered in the Registrar General's population census in 1971 are invaluable in transport planning. Hence one trend in recent years has been to select zones whose boundaries are compatible with those of the enumeration districts used in the census.

O-D methods

There are many and varied procedures for carrying out origin and destination surveys. The manpower involved can vary from just a few observers to many hundreds of interviewers and analyzers. The analyzing equipment required may vary from pencils, paper and a slide rule to the most modern electronic computing equipment capable of processing extensive data accumulations.

Roadside interview method. This is a direct interview O-D method whereby motorists are stopped and questioned regarding origins and destinations and other journey data as required. Normally this interview takes place on the highway, the motorist being stopped en route to his destination. A variation of this method, however, requires the drivers to be interviewed just as they have parked their cars; this usually occurs in conjunction with central area parking studies. In either case it is confined to the drivers of the motor vehicles and normally does not provide detailed information regarding passengers.

This method is capable of supplying detailed and accurate information, since all the information is gathered directly from the motorist and there is no need to infer anything from his behaviour.

Interview sites. The location of the interview sites depends primarily on the particular problem being studied. Thus if the survey is concerned only with trip data on a single isolated minor highway, a single mid-point location may be adequate. With town studies, on the other hand, sites may be located on all main radial roads so that perhaps 95 per cent of the traffic entering or leaving the survey area can be intercepted. In the case of a survey for an isolated major highway, interview stations may also be sited on all other highways expected to be affected by the construction or reconstruction of the new facility.

In general, the interview sites should be located as near as possible to the boundaries of zones. If they are not, then apparently absurd answers can be obtained when motorists moving through survey points have both origins and destinations within the same zone.

A primary consideration in locating interview sites is the safety of both the interviewer and the motorist. Thus all sites should have a minimum of about 230 metres of unrestricted sight distance available on both sides, and should be well clear of any intersections. Signs explaining the purpose and extent of the stoppages should be located well in advance of each site in order to give the motorist time to adjust his speed accordingly.

Interview team. Each interview team must consist of at least two members, one making a classified count of all vehicles passing in the direction being studied, and the other conducting the actual interviews. More usually, however, an interview team surveying traffic on a fairly busy two-lane road and interviewing in both directions will consist of a party chief, two recorders, six interviewers and two policemen. The policemen are necessary as it is only they who have the power in law to stop and question a person on the public highway.

Sampling. It is not usually possible to stop and question all motorists, and so sampling procedures are normally used. This is necessary not only to avoid traffic delays and maintain good public relations, but also because congestion may cause local motorists to detour around the interview site, thus distorting the traffic pattern. To avoid this, one of three sampling methods is commonly used.

The first of these requires a fixed number of vehicles to be stopped. Thus three vehicles may be stopped out of every twelve and the next nine allowed to proceed, in order to give a 25 per cent sample.

The second method requires all vehicles to be stopped that arrive at the site during a pre-determined period of time. Thus all vehicles would be stopped every alternate half-hour in order to obtain a 50 per cent-by-time sample.

The third procedure is similar to the second in that no attempt is made to maintain a fixed relationship between the number of drivers interviewed and the total number of vehicles on the road. Instead a variable sampling fraction is obtained by the interviewer, after he has just completed an interview, signalling to the policeman to stop the very next but one vehicle irrespective of what it is. This has the practical advantages that the interviewer is able to work at a constant rate, he is never too busy or too slack, traffic congestion and delays are kept to a minimum, and bias is not introduced by always selecting a slow-moving vehicle leading a platoon. It has the statistical advantages that it is never left to the discretion of either the policeman or the interviewer as to which vehicles should be stopped or whether interviewing should cease because of congestion. In addition, because of the variable sampling procedure, the fraction of total traffic interviewed is greater in light traffic and lower in heavy traffic; this has advantages over the other two methods from a statistical viewpoint.

Just as it is not possible to stop all vehicles on the road at a given time, so also it is not normally possible to carry out the direct-interview survey throughout the entire year. Instead, the survey can be carried out during a period when the traffic pattern is considered to be representative of the whole year. The data are then adjusted later on the basis of knowledge obtained from continuous automatically-obtained volume data. Alternatively, the survey can be conducted during the period when the problem is most acute. It is this latter method that is most commonly used in Britain, the surveys on rural roads being often conducted during the normally busy weeks of August.

Normally O-D surveys are carried out only during part of the 24-hour day, usually from 6 a.m. to 10 p.m. Furthermore, it is not usually necessary to carry out the survey at all stations on the same day. Instead, two or three stations can be covered in one day, and another two or three on other similar days, and so on. In fact, if the number of interviewers is limited, a team can work the early shift on one day and the late shift on another similar day at the same site without causing any bias to the data.

The direct interview. The exact questions asked at the interview vary with the needs and objectives of the survey. Typical information that might be recorded for each interview is as follows; survey-station number, date, hour, vehicle type, direction of travel, number of passengers, origin and destination and purpose of trip. With respect to the O-D questions themselves a trip is considered to commence at the last place of call and end at the next place of call.

When collecting O-D data by roadside interview it is usual to exclude scheduled buses from any interview. Instead they are recorded in the classification count, and information about frequency of service and routes is obtained at a later time.

Considerable attention should be paid beforehand to the framing of brief unambiguous questions. If proper care is taken in this respect, and the interview itself is carried out quickly, courteously and efficiently, then most forms of bias can be eliminated from the questioning.

Analysis. The first stage in the analysis of most O-D survey data is the classification of the data by means of code numbers. In this way vehicles are grouped into different classes, and origins and destinations are grouped by zones. This coding is normally carried out in the office with the interviewers coding their original field sheets. At this stage it is usual to find some journeys that do not apparently make sense, and so their real meanings and the rectification of mistakes are best noted by the interviewers themselves.

It is usually impracticable to attempt to analyse the data by manual means, and so the coded information is normally transferred on to punched cards by skilled punch operators. These cards are then mechanically sorted according to station, hour, vehicle type, or any other classification that may be required.

The next step in the processing of the data is the production of expansion factors, one for each period of the survey. An expansion factor is the number of trips of each vehicle category represented by each interview during that particular period. For instance, if during a one hour sample period 13 passenger cars are interviewed out of a total of 72 cars, then the expansion factor for these journeys is 72/13. Similar expansion factors are determined for each of the categories used.

The next step is to transfer the appropriate expansion factors to the cards, and these can then be regarded as representing an actual number of trips. At this stage also an additional adjusting factor may be taken into account to allow for, say, the time of year when the surveys are carried out.

The final step in the basic mechanical analysis of O-D survey data is the sorting of the cards in order to determine the total number of trips for all classes of vehicle and between any pair of zones.

The results of an O-D survey are usually expressed in the form of *Desire-line Graphs*. One form of desire-line graph, obtained from a very limited survey involving only a vehicle count at an intersection, is that shown in Fig. 3.15. For large scale surveys, however, the form of desire-line graph used is shown in Fig. 3.16. Desire-line charts are usually drawn to represent numbers of trips between zones, the greater the band thickness the greater being the number of trips. They take no account of the routes taken by motorists but instead reflect their desires for ideal routes. Thus for instance the desire-line graph in Fig. 3.16(b) illustrates the general pattern of sector-to-sector movements across the River Thames within the internal survey area of the London traffic survey.

Desire-line charts can be presented in many forms. Separate graphs may be drawn to show desire-lines for through-trips, internal trips, and trips between internal and external survey zones. This last type is illustrated in Fig. 3.16(a). Often separate charts are drawn to show different types of trips between zones. Important zones such as central areas, or industrial districts, may require special diagrams to determine their exact influence. Again, depending upon the survey, major desire-lines may be drawn to reflect the traffic requirements of particular survey sectors, i.e. groups of zones with common characteristics.

Postcard surveys. There are two methods of carrying out origin and destination surveys using postcards. The two methods may be carried out separately or they may be used to complement each other.

The first method involves mailing return-stamped and addressed postcards to all registered motor vehicle owners within the survey area. Each

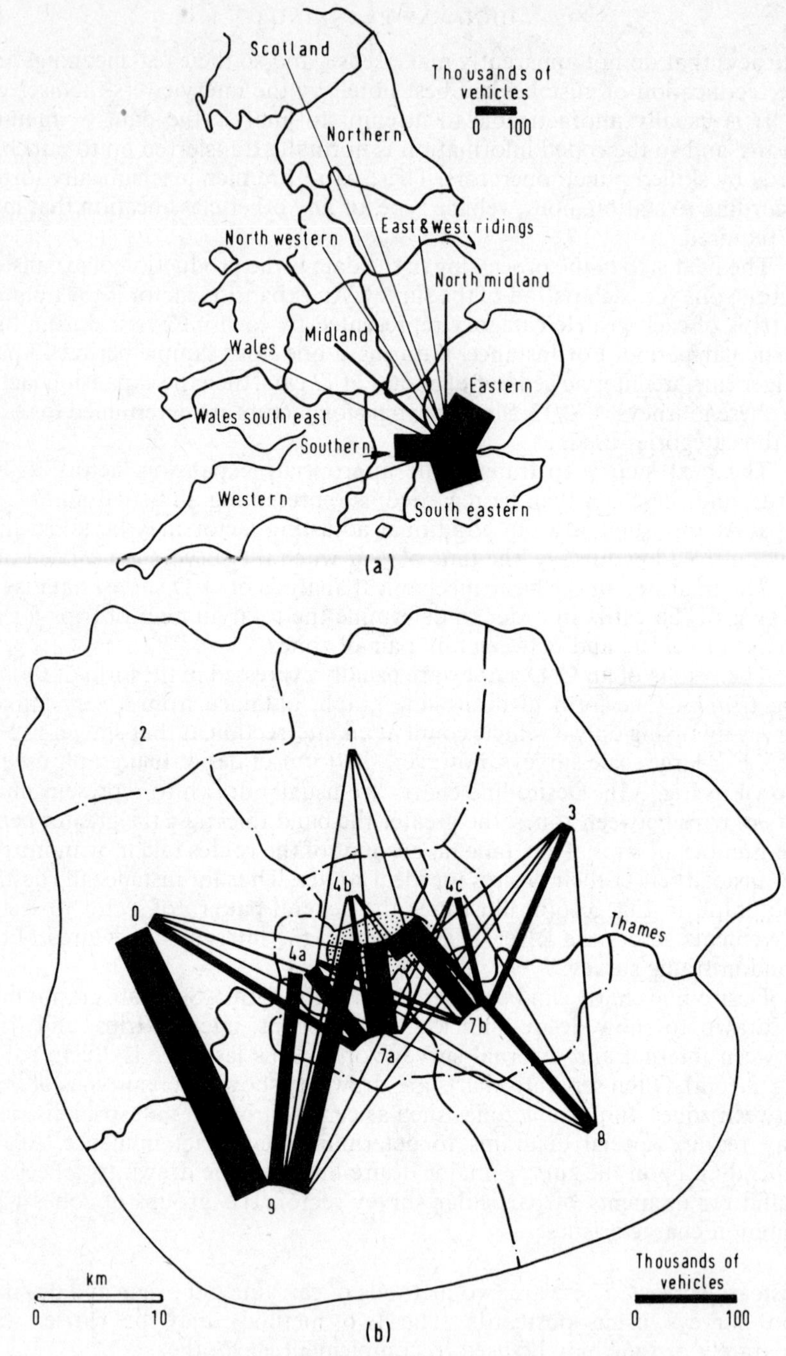

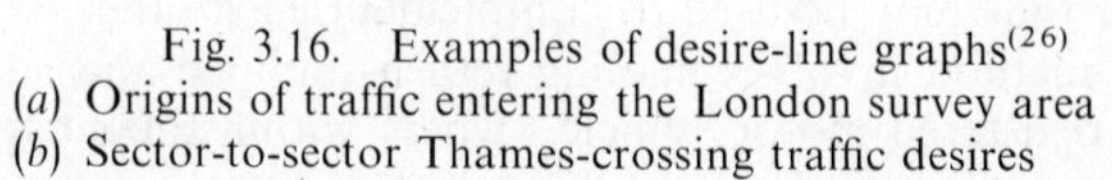

Fig. 3.16. Examples of desire-line graphs[26]
(*a*) Origins of traffic entering the London survey area
(*b*) Sector-to-sector Thames-crossing traffic desires

postcard recipient is normally asked to record all trips of his motor vehicle made on the day after the card is received; the day chosen is usually a weekday.

The second method involves handing out the postcards to motorists as they pass slowly by selected sites. If carried out in conjunction with the first type of postal survey, the sites are normally located on a cordon line about the survey area, so that information can be obtained about both external and internal trips. If carried out in isolation, the sites are normally where traffic is so heavy that vehicles cannot be stopped to enable direct interviews to take place.

In either case, the success of a postcard survey depends to a very large extent on the willingness of motorists to complete and return the postcards. This requires a very extensive publicity programme prior to the survey so that motorists will become conscious of the importance of returning the cards. Even so, the returns may only be in the order of 50 per cent. Poorly publicized surveys can be expected to have returns varying from 5 to 25 per cent.

The results obtained from postcard O-D surveys can be unreliable, primarily because of the manner in which the data are collected. Not only may the number of cards returned be small, but they may be heavily weighted in favour of facilities of particular interest to motorists in certain areas. Again, the results may be distorted because of better co-operation and returns by motorists who use writing as a regular part of their daily lives as compared to manual workers who may not bother to complete the cards. Also, it is very probable that the returns from through-trippers will be especially low and unrepresentative.

The principal advantages of the postcard survey methods are that all data are obtained at the one time for the one day of travel, and untrained personnel can be used to hand out or send the cards by post. In addition there is little interference to traffic at congested road locations.

Registration number method. Again there are two variations on this method, each of which has its particular usage. The first and more often used procedure requires observers with synchronized watches to be stationed about and/or within the survey area[29]: then as a vehicle *passes* an observer its passage is timed and its registration number is noted. (This is similar to the method described for speed and journey time studies.)

At the end of the survey the records at all the observation sites are compared, and each vehicle's trip through the survey area is traced. For the purpose of the study, the vehicle's origin is assumed to be where it is first observed, while its destination is assumed to be its last observation point. By noting the entry and exit times, the journey time of each vehicle can be estimated; when this is compared with known journey times for the same trip, the vehicle can be classified as stopping or non-stopping within the survey area.

The principal advantage of this method is that it can be used where traffic is very heavy and it is not desired to stop vehicles for questioning. In addition, since the motorist is unaware of being scrutinized, the results obtained will not be biased because of poor motorist co-operation. It has the disadvantage that, because of large manpower requirements, its use is limited to single highway facilities or to small survey areas with few exits and entrances; with this method all observation sites have to be manned for the same one day, and observations must be continuously carried out at every entrance and exit point.

The second method of conducting a registration number O-D survey is to record for a given day the registration numbers of all vehicles *parked* within the survey area. These are then compared at a later time with the motor vehicle registration lists, and the origin of each vehicle is assumed to be where it is registered while its destination is taken to be where it was parked.

The principal advantages of this method are its simplicity, low cost and freedom from bias. Its main disadvantage is that the results obtained are of limited value. Little or no information is obtained about public transport and commercial vehicle travel; all vehicles are assumed to come directly from the registered address to the parking location; no data are obtained regarding through-traffic trip purpose or time of journey; and again, its use is confined to small survey areas because of the numbers of observers required.

Tag or sticker method. This is also a moving-vehicle method that is not dependent upon the complete co-operation of the motorist. The usual procedure is for an observer to stop each vehicle briefly at the entrance to the survey area and to place a pre-coded tag or sticker, bearing the time and place of entry, under the windscreen wiper. At the same time the driver may be handed a card requesting his co-operation and explaining the nature of the survey. As the vehicle leaves the survey area, the tag is reclaimed and the time, station, direction of travel and any other possible information are recorded on it.

The main advantage of this method is that the path of a vehicle can be traced through the survey area by having intermediate observers observe the colour and/or shape of the tag on each vehicle. Thus the need for a by-pass can be determined by comparing the vehicle journey times with subsequently determined trip times over the same routes. A commonly used criterion assumes that a vehicle which takes less than 1·5 times as long as the average minimum time required to normally make the trip is a potential by-pass user.

Home-interview method. Large urban areas attract traffic from great distances outside their boundaries as well as generating intensive traffic in their own right. Within the cities, narrow streets, well-developed and highly-valued property, constrictive topography, and ever-increasing traffic volumes create obstacles which have a considerable influence on the manner in which people travel and the routes they take to their destinations. Because of the many variables involved, it is not normally possible in the larger urban areas to decide by merely observing the existing traffic flows where highway improvements should be located. Even if it were possible, it is generally not practical to obtain data within heavily trafficked areas by stopping vehicles in the streets. Not only does the possibility of costly congestion make this unrealistic, but the immense road lengths within urban areas makes it impractical to attempt to interview a representative sample of vehicles. Thus a more sophisticated method of data collection must be used to obtain the comprehensive but detailed information required by highway planners in order to be sure of the need for new or improved facilities, as well as their correct location and adequacy for the future.

The U.S. Bureau of Public Roads, with assistance from the Bureau of Census, has developed a detailed procedure for making extensive urban traffic studies that are based on a home-interview sampling technique.[30] This procedure is widely used as a model for the home-interview phase of the comprehensive transport demand studies carried out in urban areas today.

The home-interview study utilizes a sampling process similar to that used in public-opinion polls. A representative sample of homes within the survey area is selected and the inhabitants questioned regarding their travel during a particular weekday, usually the day before the interview. If the data obtained in this study are combined with those from a commercial vehicle study and from a roadside-interview study on all roads that cross the survey boundary, an almost complete record can be obtained of a typical weekday's travel within the survey area.

Sampling. As has been discussed previously, travel is an expression of individual behaviour and as such it has the fascinating characteristic of being habitual. Since it is a habit it tends to be repetitive and the repetition occurs in a definite pattern. Thus it is not necessary to obtain travel information from all residents of an urban area nor is it necessary to interview individuals over long periods of time in order to ascertain their travel habits or need for travel facilities.

Because of this habitual pattern, therefore, statistical sampling methods can be used with confidence in determining travel patterns within urban areas. A sample will be representative if the persons included in it are distributed geographically throughout the survey area in the same proportion as the whole population is distributed. Consequently, sampling based on the place of residence is the most practical means of assuring that the proper geographical distribution is obtained. In Great Britain, the residences are normally selected either from the *Register of Electors*, which contains a list of the names and addresses of all qualified electors, or from the *Valuation List*, which is a register of all rated units as held by each local authority.

The size of the sample selected is variable and depends upon the population of the survey area, the degree of accuracy required and, to a certain extent, the density of population within the survey boundaries. The adequacy of various sample sizes has been tested, and from the results of these studies the values given in Table 3.10 were compiled as a guide in determining the size of sample needed in a particular urban area.

LONDON TRAFFIC SURVEY – 1962

HOUSEHOLD INTERVIEW REPORT PART I — HOUSEHOLD INFORMATION

Page Number........ of........

Card Number ... ...

Serial Number ...

Sampling District ...

Zone Number ... ...

Report Classification ... ...

L.T.E. Sample 1-17

ADMINISTRATIVE RECORD

....
Interviewer's Name

CALLS

1. .... Date .... Time

2. .... Date .... Time

3. .... Date .... Time

4. .... Date .... Time

Telephone No. ....

REPORT CLASSIFICATION

SUPERVISOR'S COMMENTS

OTHER COMMENTS

Report Completed ....
Office Check ....
Field Check ....
Approved for Coding ....
Coding Approved ....

London Traffic Survey
Freeman, Fox & Partners
118, Westminster Bridge Road,
London, S.E.1. Tel.: WAT 2512.

A. Interview address ....

B. Name of owner of sample vehicle ....

C. Vehicle registration mark .... Vehicle make ....

D. Does any person living at this address have regular use of a car? .... Sample owner? ....

E. Day of travel (yesterday) .... Date of travel ....

F. How many persons live in this household? ....

G. How many persons living in this household are employed? ....

H. How many cars are owned? .... Company or Government cars available yesterday ....

I. Is there off-street parking space available here for these cars? ....

J. How many motorcycles are owned? .... How many other vehicles are owned? ....

K. Relative income level of total household ....

L. Total number of journeys by mode 18-33

PERSON INFORMATION (for persons 5 years of age or over)

M		N				O	P	Q	R	No. Journeys by Mode Group (Office use only)		
Per-son No.	Person Identi-fication	Jrny's Made		Inter-viewed		Sex and Age	Main Mode of Travel to Work or School	Place of Work or School (Address)				
		Yes	No	Yes	No			Occupation—Employment Status	Industrial Classification	C.D.	P.T.	Total
34												49
						50						64
						65						79
34												49
						50						64
						65						79

Card Number

HOUSEHOLD INTERVIEW REPORT PART II — JOURNEY INFORMATION

Page......of......

S	T	U	V	W	X		Y	Z	Public Transport Journeys			
Person No.	Journey No.	Origin of Journey (Address and Land Use)	Destination of Journey (Address and Land Use)	Purpose of Journey	Time of Journey		Mode of Travel	Persons in Car	Bus Route	Ticket Type	Cost of Journey	Special Notes
					Start	End						
18		Address Land Use	Address Land Use	0. Work 0. 1. Employer's Business 1. 2. Personal Business 2. 3. Entertainment 3. 4. Sport 4. 5. Social 5. 6. Shopping (conv.) 6. 7. Shopping (goods) 7. 8. School 8. 9. Home 9. X. Serve Passenger X. Y. Change Travel Mode Y.	A.M. P.M.	A.M. P.M.	0. Car Driver 1. Car Passenger 2. Cycle Driver 3. Cycle Passenger 4. Public Transport Bus 5. L.T.E. Rail 6. British Railways 7. Taxi Passenger 8. Coach 9. Goods Vehicle Passenger and Other	1 2 3 4 5 6 7 8 9	0 1 2 3	0 1 2 3 4 5 6 7 8 9 X	45	
46		Address Land Use	Address Land Use	0. Work 0. 1. Employer's Business 1. 2. Personal Business 2. 3. Entertainment 3. 4. Sport 4. 5. Social 5. 6. Shopping (conv.) 6. 7. Shopping (goods) 7. 8. School 8. 9. Home 9. X. Serve Passenger X. Y. Change Travel Mode Y.	A.M. P.M.	A.M. P.M.	0. Car Driver 1. Car Passenger 2. Cycle Driver 3. Cycle Passenger 4. Public Transport Bus 5. L.T.E. Rail 6. British Railways 7. Taxi Passenger 8. Coach 9. Goods Vehicle Passenger and Other	1 2 3 4 5 6 7 8 9	0 1 2 3	0 1 2 3 4 5 6 7 8 9 X	73	
18		Address Land Use	Address Land Use	0. Work 0. 1. Employer's Business 1. 2. Personal Business 2. 3. Entertainment 3. 4. Sport 4. 5. Social 5. 6. Shopping (conv.) 6. 7. Shopping (goods) 7. 8. School 8. 9. Home 9. X. Serve Passenger X. Y. Change Travel Mode Y.	A.M. P.M.	A.M. P.M.	0. Car Driver 1. Car Passenger 2. Cycle Driver 3. Cycle Passenger 4. Public Transport Bus 5. L.T.E. Rail 6. British Railways 7. Taxi Passenger 8. Coach 9. Goods Vehicle Passenger and Other	1 2 3 4 5 6 7 8 9	0 1 2 3	0 1 2 3 4 5 6 7 8 9 X	45	
46		Address Land Use	Address Land Use	0. Work 0. 1. Employer's Business 1. 2. Personal Business 2. 3. Entertainment 3. 4. Sport 4. 5. Social 5. 6. Shopping (conv.) 6. 7. Shopping (goods) 7. 8. School 8. 9. Home 9. X. Serve Passenger X. Y. Change Travel Mode Y.	A.M. P.M.	A.M. P.M.	0. Car Driver 1. Car Passenger 2. Cycle Driver 3. Cycle Passenger 4. Public Transport Bus 5. L.T.E. Rail 6. British Railways 7. Taxi Passenger 8. Coach 9. Goods Vehicle Passenger and Other	1 2 3 4 5 6 7 8 9	0 1 2 3	0 1 2 3 4 5 6 7 8 9 X	73	
18		Address Land Use	Address Land Use	0. Work 0. 1. Employer's Business 1. 2. Personal Business 2. 3. Entertainment 3. 4. Sport 4. 5. Social 5. 6. Shopping (conv.) 6. 7. Shopping (goods) 7. 8. School 8. 9. Home 9. X. Serve Passenger X. Y. Change Travel Mode Y.	A.M. P.M.	A.M. P.M.	0. Car Driver 1. Car Passenger 2. Cycle Driver 3. Cycle Passenger 4. Public Transport Bus 5. L.T.E. Rail 6. British Railways 7. Taxi Passenger 8. Coach 9. Goods Vehicle Passenger and Other	1 2 3 4 5 6 7 8 9	0 1 2 3	0 1 2 3 4 5 6 7 8 9 X	45	
46		Address Land Use	Address Land Use	0. Work 0. 1. Employer's Business 1. 2. Personal Business 2. 3. Entertainment 3. 4. Sport 4. 5. Social 5. 6. Shopping (conv.) 6. 7. Shopping (goods) 7. 8. School 8. 9. Home 9. X. Serve Passenger X. Y. Change Travel Mode Y.	A.M. P.M.	A.M. P.M.	0. Car Driver 1. Car Passenger 2. Cycle Driver 3. Cycle Passenger 4. Public Transport Bus 5. L.T.E. Rail 6. British Railways 7. Taxi Passenger 8. Coach 9. Goods Vehicle Passenger and Other	1 2 3 4 5 6 7 8 9	0 1 2 3	0 1 2 3 4 5 6 7 8 9 X	73	

Fig. 3.17. Typical home-interview field sheet[30]
(*Courtesy of the Greater London Council*)

The home interview. The actual interview itself is the most important part of the home-interview study. The interviewer provides the only first-hand link between the survey and the general public, thereby determining whether or not the public response is to be co-operative. The completeness of the data, and thus the validity of the decisions based upon them, depends upon the willingness of the members of each household to provide travel data.

The interviews are conducted within the selected domiciles and questions normally refer to travel by each member of the household above the age of five on the day prior to the interview. Since the accuracy of the survey depends to a large extent on obtaining the desired information from pre-selected dwellings, most interviews are conducted in the evening because of the greater chance of finding people at home. In order to maintain the unbias of the sample, 'call-backs' must be made rather than the alternative of interviewing the next-door neighbour when there is nobody at home at a pre-selected residence. Each interview usually requires roughly one hour; this time includes an allowance for travel time and call-backs. The type of information collected normally is indicated in Fig. 3.17.

TABLE 3.10. *Recommended sample sizes for conducting home-interview O-D studies*[30]

Urban population	*Sample size*
<50 000	1 in 5 dwelling units
50 000–150 000	1 in 8
150 000–300 000	1 in 10
300 000–500 000	1 in 15
500 000–1 000 000	1 in 20
>1 000 000	1 in 25

Commercial vehicles. Normally surveys of commercial vehicle travel are carried out in conjunction with the home-interview study. If the complete pattern of travel within the survey areas is desired, this phase of the survey must be carried out with the same care as, and in a manner similar to, the home-interview study. The samples of vehicles are usually selected at random from numerical or alphabetical registration lists; customarily the sample rate for lorries and similar heavy vehicles is twice that in the home-interview survey. Desirably the taxi sample should amount to at least 20 per cent of the total number of taxis registered within the survey area; at the very least the percentage of taxis should be as great as that used for the goods vehicle study. In either case the desired information is gathered by visiting the company offices where the vehicles are registered and obtaining the vehicle records for the previous day.

Information needed concerning public transport vehicles can be obtained from the records of the companies involved. The locations of the existing routes, schedules of operation, and the numbers of passengers carried on an average weekday in the course of the survey are examples of typical public transport data collected.

External-cordon study. A roadside interview study is usually carried out in conjunction with the home-interview study in order to obtain information about trips with origins and/or destinations outside the internal survey area. In carrying out this phase of the study a cordon-line is drawn about the survey area and motorists are stopped and interviewed as they cross the cordon-line.

In order to determine the travel patterns of the non-residents of the survey area, and to obtain corroboratory information regarding the travel patterns of the survey area residents, it is usually practice to sample the traffic streams in both directions of travel at each interview station on the cordon. It is evident that some duplication of information will occur at this stage. This duplication should be eliminated by removing the appropriate trip information from the internal survey data before the analysis is undertaken.

Evaluation of survey accuracy

It is most important that the home-interview data should be checked for completeness and accuracy. Three methods are most often used to check the reliability of the data; they involve comparisons with data obtained at screen-lines, at control-points, and/or at the external cordon. If these comparisons reveal considerable differences, then the survey data are usually adjusted accordingly.

Screen-line comparisons provide the best means of checking the reliability of the vehicle trip data collected in both the internal and external phases of the study. A natural barrier to traffic, such as a river or railroad track, that passes through the internal area makes an ideal screen-line. The purpose of a screen-line is to divide the internal area so that the number of vehicles moving from one part to another can be determined by direct count at all screen-line crossings. The value obtained at the screen-line count is then compared with the number indicated by the home-interview study.

In order to function properly the screen-line must extend across the survey area from cordon-line to cordon-line. In addition the screen-line should coincide with zone-lines in order to permit the cards for trips with origins on one side of the screen-line and destinations on the other to be separated for comparison purposes.

The results of one screen-line comparison study are shown in Fig. 3.18. It is normally considered that the main internal and external interview studies are sufficiently complete and reliable if the expanded interview results account for 85 per cent or more of the total volume as determined by the direct counts at the screen-line. In the case illustrated in Fig. 3.18 the results obtained were considered to give excellent correlation and no special adjustments to the original expansion factors for the home-interview and commercial-vehicle interview data were judged necessary.

Control-point comparisons are used only when the routes of trips are obtained in the course of the home interviews. Two or three points, well known to motorists and preferably in different quadrants of the internal survey area, are selected as control points before the survey is begun.

Viaducts, large bridges, underpasses, and other points through which large volumes of traffic are funnelled are excellent for this purpose. As with the screen-line count, a comparison of the traffic volumes obtained from the external and internal surveys and the actual traffic volumes at these control sites gives a measure of the accuracy and completeness of the data.

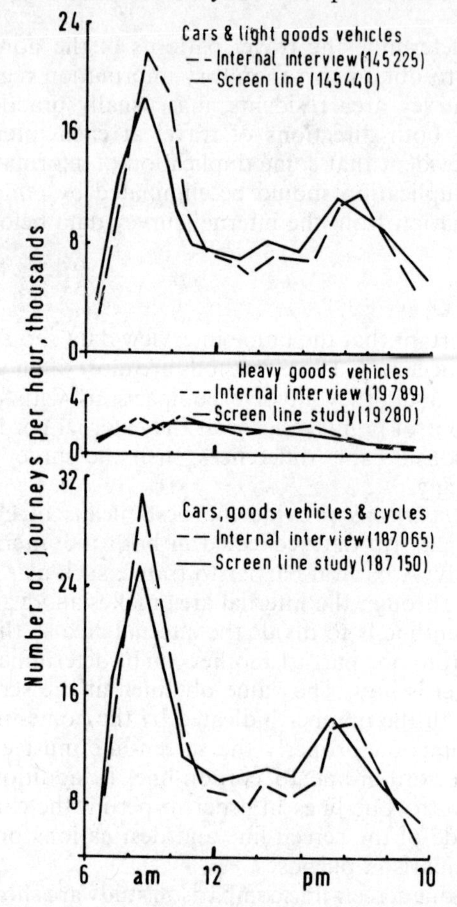

Fig. 3.18. Screen-line traffic volume comparison[26]

Control-points should never be adjacent to the external cordon-line. Although both external and internal travel data would be collected, the primary purpose of these checks is to measure internal travel reliability, and it is very probable that the amount of traffic near the outskirts of the survey area would be too small to provide a satisfactory check. The control points should not be so close to the central area, however, that the reliability of the data may be in question because of the passing and repassing of vehicles looking for places to park.

If it is not possible to use a screen-line, and if a few good control-points are available, a control-point comparison can be a most useful method of checking home-interview data reliability. The main drawback associated with this method is that it entails extra questioning during the internal and external studies.

Cordon-line comparisons are the easiest type of comparison since they do not require any extra data. In order to obtain data for comparison purposes it is first of all necessary to derive from the internal survey the passenger car and commercial vehicle trips that cross the external cordon-line drawn about the survey area. These volumes are then compared with similar trips across the cordon-line made by internal survey residents as determined from the external-cordon survey.

The principal advantage of the cordon-line method is that the comparison is made from data gathered as an integral part of the regular survey. However, the external-cordon check obviously cannot check on the reliability of all data, and hence it is normally considered as complementary to the screen-line or control-point comparisons and not as a substitute for them i.e. the trips recorded in the internal survey which pass through the cordon-line are not normally representative of internal travel within the survey area as a whole.

Selection of O-D survey method

The choice of which origin and destination method should be used in a given situation is primarily a function of what information is required. Unfortunately also, however, the type of survey utilized is only too often controlled by the available funds. This is to be deplored since, as has been determined in so many instances, data obtained on the basis of inadequate surveys can often be more misleading than if the surveys had never been carried out at all.

The registration number and tag-on-car methods are normally easy to organize and carry out. However if they are not to become unwieldy in analysis they are restricted to localized situations such as determining the manner in which vehicles move through a complicated intersection or in determining the need for a by-pass about a small town.

If more detailed information is required, such as for the location of a new river crossing, then a basic direct-interview survey on existing bridges, tunnels and ferries will adequately decide the most suitable location. Similarly the location of a new bus terminal can be determined by direct interviews at the other existing terminals or on the buses en route.

In urban areas and on major rural roads, more complete evaluations of drivers' needs are required. In particular detailed origin and destination surveys are required in order to properly estimate highway needs in urban areas. The following five systems of O-D surveys may be considered appropriate for collecting travel data in urban areas.

External-cordon surveys. An external-cordon O-D survey on its own is usually adequate in built-up areas with populations of less than about 5000. In urban areas of this size it is the traffic from outside which usually exerts the major influence on the traffic patterns. Furthermore, any traffic movements not recorded during the course of this survey are normally not sufficiently stable to have any significant impact on long-range highway plans.

This type of survey can also be used in urban areas with populations between 5000 and 75 000 when the predominant flow of traffic are on through routes and where problems associated with public transport or the movement of traffic on the overall street system are not important considerations.

The external cordon itself is an imaginary line set up about the survey area at which motorists are stopped and asked questions about their trips. The cordon-line should be situated sufficiently far beyond the built-up areas as to intersect a minimum number of roads, yet not so far out that it takes in much rural area. Ideally it should encompass the general area of daily commutings and at the same time allow for future urban growth.

External-internal-cordon surveys. This type of survey is most useful in urban areas containing populations of between 5000 and 75 000 where most of the traffic is oriented towards the central area and there are no major deficiencies in the street system outside the central area.

In this type of survey, roadside interviews are conducted at two cordon-lines. The external cordon is located outside the edge of the built-up area, while an internal cordon is placed about the central area. The set-up of the interview stations on both cordon-lines is essentially the same as for the external cordon alone and the interviews are geared to obtain the same basic type of information.

Information given by this type of survey can be used to establish the pattern of passenger car and commercial vehicle movement fairly comprehensively in compact motorized communities. However, urban areas which rely heavily on public transport or have public transport-routing problems, may require special additional studies. For instance, direct interviews with bus passengers or postcard questionnaire studies may very usefully be employed to obtain any additional information required.

External cordon-parking survey. This type of O-D survey has been shown to be most useful in urban areas with populations between 5000 and 75 000 where the principal traffic destinations are within the central area, and where a parking problem also exists within the central area. Thus the survey is composed of an external-cordon study, as described before, and a comprehensive parking study within the central area of the city.

In essence the information obtained is similar to that obtained with the external–internal-cordon survey, with the additional bonus that data are obtained about the parking needs of the central area itself. Its principal disadvantage is that no information is obtained about trips that are completely internal and which do not terminate within the town centre. Thus if it is expected that a substantial amount of non-stopping traffic may pass through the central area, or if one of the major purposes of the survey is to determine the need for an inner ring road, this method of origin and destination survey should not be utilized.

External cordon-controlled postcard survey. A survey of this type can be utilized in urban areas of up to about 200 000 population. The postcard method of obtaining information is used for the internal survey phase of the study while the external-cordon study takes care of the trips originating externally.

The results obtained should be checked for accuracy by means of a screen-line comparison.

External cordon–home interview survey. This type of survey can be used successfully in cities of any size. However, because of the very large cost of conducting the survey and analyzing the data, it is primarily applicable to the larger urban areas and is rarely justified in cities with less than 75 000 population.

As described before, the home-interview data are supplemented with data on trips originating outside the survey area, obtained from roadside interviews conducted at the external cordon-line. Internal commercial vehicle movements are determined by sample interviews of vehicle operators and/or drivers. The combined data, when expanded, should give a composite of practically all trips made within the survey area on a typical weekday during the period of the survey.

CONSTITUENTS OF THE HIGHWAY DESIGN VOLUME

The final step in the traffic analysis leading to the design of a highway or highway system is the determination of the volumes of traffic which will have to be handled. This can be a most complicated process, particularly in urban areas. Why this should be is most easily illustrated by considering the basic ingredients which make up the volume finally selected for highway design purposes.

Design year

One of the first problems facing the highway planner is the decision as to what year in the future should be used for design purposes. At the present time it is generally accepted that 20 years hence is suitable as a design period for rural highways and motorways. Prediction beyond this period is limited by the possibility of changes in land use and population, or regional economy or even in the methods of transport.

In urban areas the problem of choosing a design year is much more complex. Once a major highway is constructed within a city, extensive building development can be expected alongside, and so there is much less scope for changing the geometric design in the future. The evidence for this statement can be seen in any crowded city or suburban centre today. Since modern buildings are normally considered to have an economic life of at least 30–50 years, it is not uncommon to think *strategically* in terms of this length of time when planning a new major highway system, particularly in the larger urban areas. Practically, however, it is now becoming more common to design-plan in terms of a more immediate future, say 10 to 15 years, while ensuring that the design-plan does not contradict long-term strategic developments. In other words, rather than planning for the long-term future on the basis of a 'one-off' transport study (with all of its consequent risks), current thought is turning toward transport planning in a 'rolling programme' context, i.e. where one designs for a relatively short (and predictable) period ahead, and then constantly revises the transport and land use plan as time progresses, further data become available, and policies become clearer.

Traffic prediction components

Design traffic volumes for some future date are derived from knowledge of current traffic and estimates of future traffic. The basic factors which constitute the design volume for an individual highway are illustrated graphically in Fig. 3.19. From this figure it is evident that there are two values of current traffic and three forms of future traffic that have to be taken into account.

Current traffic

By current traffic is meant the number of vehicles that would use the new or improved facility if it were open to traffic at the time the current measurements were taken. If an already existing highway is being reconstructed and improved, then the current traffic is composed of vehicles already using the old highway plus existing traffic attracted from other adjacent roadways. In the case of a completely new highway the current traffic is composed entirely of existing attracted traffic.

Depending upon the type and location of the highway, current traffic can be determined from traffic counts on existing roadways that are likely to influence traffic flow on the new or improved highway and/or from roadside-interview origin and destination studies. In large urban areas more comprehensive transport demand studies will normally be required.

On low-volume highways in rural areas use of classified traffic count data alone is usually sufficient to evaluate the current traffic volumes. In this case the amount of attracted traffic can be estimated by a highway engineer having a thorough knowledge of local conditions. The importance of local knowledge must not be underestimated in this case, since the amount of attracted traffic is dependent upon the attractiveness of the new roadway as compared with other existing and adjacent roadways. For instance when the existing roads are heavily travelled the percentage of attracted traffic will obviously be greater than when there is little congestion on the roads. The extent of this changeover can best be estimated by a local engineer familiar with the area rather than a highway engineer from 'outside'. It must be emphasized here that this method of estimating the attracted traffic volumes is of course very crude and should be limited to usage on low-cost low-

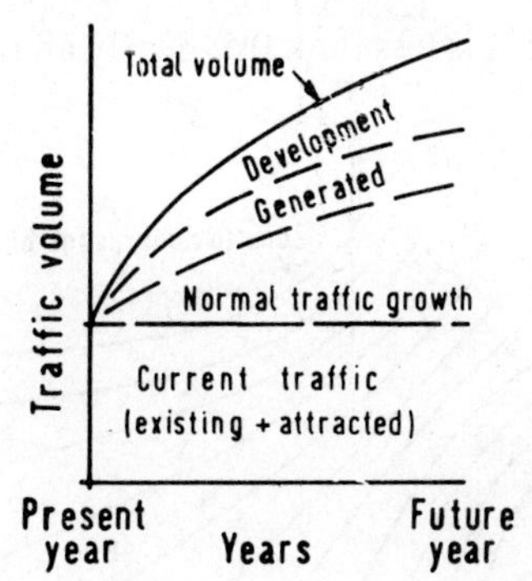

Fig. 3.19. Constituents of a design traffic volume

volume roadways and where the cost of any underestimation or overestimation is minimal.

On high-volume rural roads or roads through the smaller urban areas, a combination of classified traffic counts and roadside interviews is required in order to obtain the data on which to base estimates of the current traffic volume. This was the method used in obtaining the current traffic volumes for the M1 motorway;(28) in this case twenty-three roadside-interview stations were established on the major highways within the zone of influence of the line of the new motorway. For major highways within large urban areas, the required information can best be obtained from comprehensive transport demand studies.

The number of vehicles attracted to a new or improved high-volume highway is most difficult to evaluate. Certainly it is evident that intuitive practical recommendations based on local knowledge are inadequate here. More sophisticated methods must be used in order to justify the large economic investments in the new facilities. These methods of estimating the attracted volumes of traffic are known as *traffic assignment methods.*

The factors generally considered to have the greatest influence on the routes taken by motorists are the comparative travel times and distances, and it is these which form the basis of most diversion studies. Their importance is reflected in the diagrams in Fig. 3.20. The assignment curves in these diagrams are actually composites derived from a series of separate studies in the United States.

The chart in Fig. 3.20(a), which was developed to suit California traffic conditions, attempts to express the amount of attracted traffic as a function of both journey times and distances. In so doing it illustrates that many motorists are willing to travel further and take longer in order to use motorway facilities, whereas others will do exactly the same in order not to use the motorway.

In Fig. 3.20(b) the emphasis is placed entirely upon the time factor. The upper curve in this graph assigns traffic to a motorway with complete control of access, whereas the lower curve is for a major arterial but non-motorway highway in an urban area. For both types of highway the percentage of attracted traffic approaches zero when the travel-time ratio is greater than 1·5. When it is less than 0·5, nearly all the traffic tends to use the new facility.

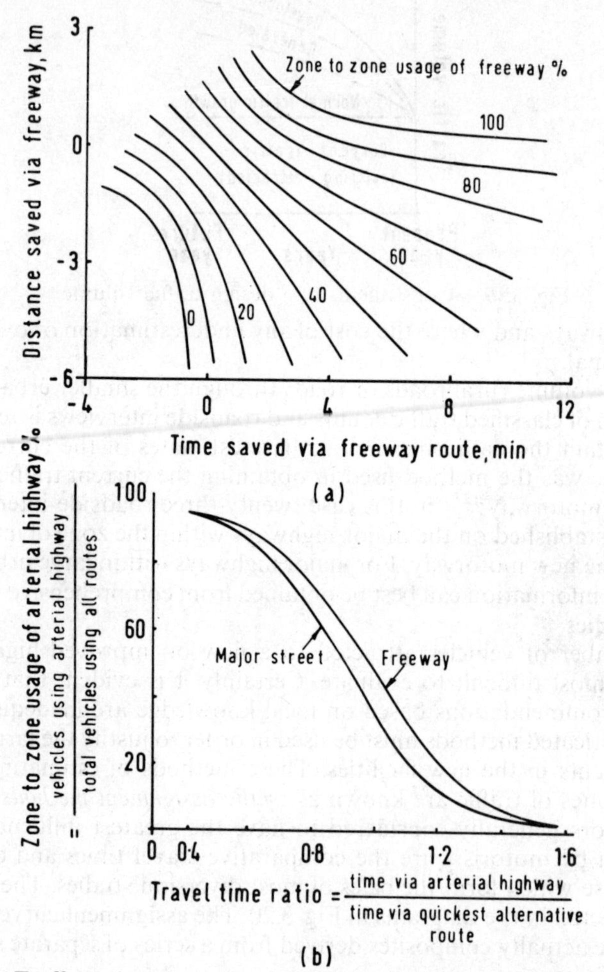

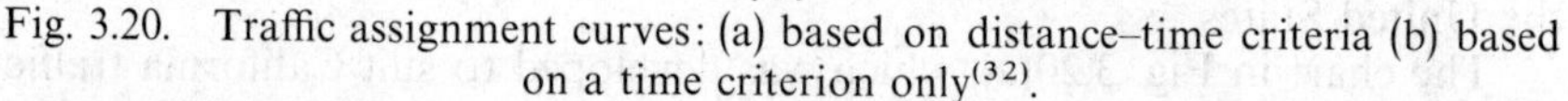

Fig. 3.20. Traffic assignment curves: (a) based on distance–time criteria (b) based on a time criterion only[32].

Future Traffic

As Fig. 3.19 illustrates, there are three components which make up the future constituents of the design traffic volume. These are the increases due to the normal traffic growth, the traffic generated by the new roadway itself, and traffic developed because of changes in land use alongside the new or improved facility. Each of these components can be considered separately.

1. *Normal traffic growth.* This is the increase in traffic volume due to the general normal increase in the numbers and usage of motor vehicles. The very evident desire of people for the mobility and flexibility offered by

the motor vehicle, coupled with the fact that the motor-vehicle industry has assumed such economic importance in the country as a whole, makes it inevitable that further substantial increases in motor-vehicle ownership can be expected in Britain (see Fig. 3.3). Care should be taken, however, in utilizing the national projection figures for particular design purposes, as they may not reflect the growth rate in the area under consideration. Data have been published (in 1966) which suggest the likely number of vehicles in various licensing authorities in Great Britain in different future years[11].

A further important factor which must also be considered is the possible growth in motor-vehicle usage. With a continually rising standard of living, coupled with the present trend towards shorter working weeks and longer leisure periods, it is not unlikely that the average car usage could increase in future years.

2. *Generated traffic.* This component refers to future motor vehicle trips, other than those made by public transport, that would not occur at all were it not for the new highway. Generated traffic has itself got three constituents, each of which should, ideally, be evaluated separately.

The first type of generated traffic is the *Induced Traffic.* This is traffic which did not previously exist in any form and which results entirely from the introduction of new or considerably improved highway facilities. For example, a new motorway connecting two adjacent cities such as Leeds and, say, Harrogate might generate a considerable amount of induced traffic because of the increase in accessibility to both cities.

A measure of the amount of induced traffic that will occur can be obtained by examining the estimates for attracted traffic. It can be postulated that where the maximum attraction occurs the maximum amount of induced traffic will also occur. On the other hand, where adequate capacity is already available or where the travel time ratio (i.e. time via the new highway divided by the time via the quickest alternative route) is high or where terminal facilities are poor, then the amount of induced traffic can be expected to be low.

The second category of generated traffic can be termed *Converted Traffic.* This traffic is created as a result of changes in the usual method of travel. Thus the building of a highway facility such as a motorway may make a route so attractive that traffic which formerly made the same trip by bus or railway, or even airway, may now do so by passenger car or lorry.

The amount of converted traffic is dependent upon relative journey times, and comparisons of convenience and economy. These in turn depend upon the dynamics of the various transport technologies, their management, the extent of governmental regulation and investment, and operating and maintenance costs. In practice, estimates of converted traffic are sometimes obtained by consultation with public transport companies about their records of similar effects in other localities.

The final category of generated traffic may be called the *Shifted Traffic.* This traffic consists of trips previously made to entirely different destinations, but which change due to the attractiveness of the new highway. Thus a new highway which now provides easy access to a previously unpatronized shopping centre may result in changing the shopping patterns of connected residential areas.

In summary, generated traffic as a whole can be attributed to the convenience and attractiveness of, and the better accessibility provided by, the new roadway. Generally the greater the length of the highway and the greater the degree of improvement, the greater is the amount of generated traffic. New highways leading to destinations with adequate parking facilities generate more new trips than ones leading to poor parking facilities. Areas with poor public transport facilities also produce more generated traffic than normal if provided with an improved highway. Controlled-access highways generate more traffic than ones subject to frequent intersections with resultant interferences and dangers.

Few definitive data have been isolated *re* the extent of generated traffic in given situations. American experience suggests that generated traffic on urban motorways may amount to as much as 20 to 30 per cent of current traffic volumes,[32] while on rural motorways it may vary from 5 to 25 per cent of the current traffic.[33] A most important characteristic of this traffic is that it occurs within a relatively short period of time, usually one to two years.

3. *Development traffic*. This is the portion of the future traffic volume, due to improvements on land adjacent to the highway, that is over and above that which would have taken place had the new or improved highway not been constructed. Increased traffic due to normal development of the adjacent land is a part of normal traffic growth and so is not considered a part of this development traffic. Experience with highly improved roadways indicates that lands adjacent to them tend to be developed at a more rapid rate than normal, and it is the extra traffic generated by this that is considered to be development traffic.

In a country like Britain which is predominantly urbanized and industrialized, development traffic may contribute substantially to the long-term component of the future traffic growth along new major highways. For instance, there will be tremendous pressure for land alongside major routes leading to the proposed Channel Tunnel which (when built) will connect this country to the continent. Industry in particular will want to be close to the direct routes, and thus these could generate considerable amounts of development traffic. The exact amounts generated will, of course, depend upon the extent to which regional and local authorities allow this to take place.

For this reason estimates of this type of traffic require consulting the planning authorities involved and determining the proposed land usage alongside the new roadway. The numbers of zone-to-zone trips expected to be produced by each type of development can then be estimated on the basis of previously obtained travel habit data.

TRANSPORT DEMAND STUDIES IN URBAN AREAS

As noted previously, the long-term planning for highways within an urban area presents a most difficult proposition. As well as being complex, it is also a new proposition, i.e. it is only since the mid-1950's that attempts

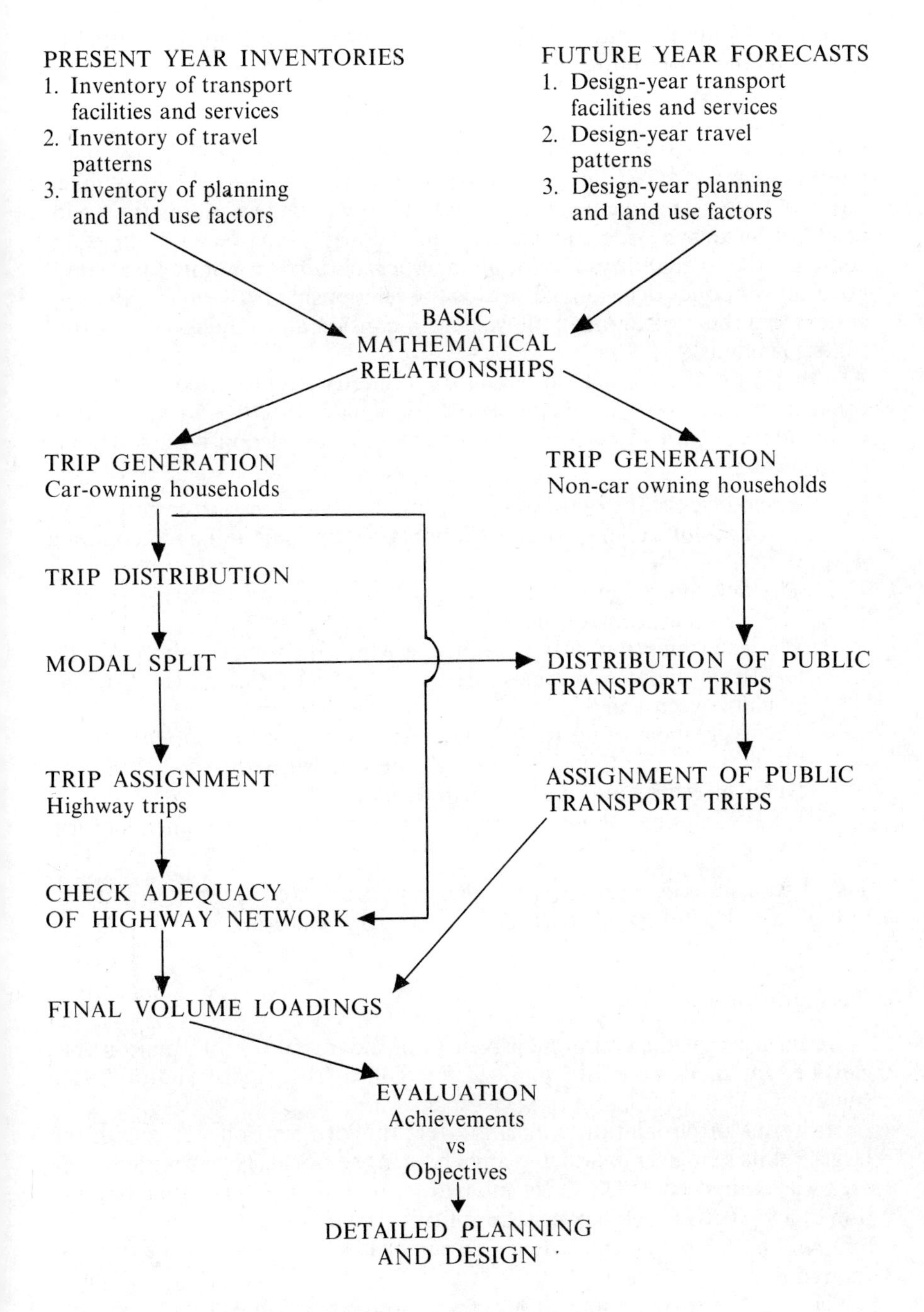

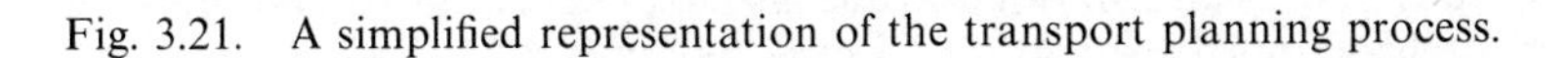
Fig. 3.21. A simplified representation of the transport planning process.

have been made to relate the design requirements of transport facilities to the social and economic status of the area being studied.[34]

At this stage, it can be said that two points are very evident regarding urban traffic prediction, planning, and design. Firstly, planning for highways within a large urban area cannot be considered in isolation from the planning of other forms of transport. Secondly, there is little doubt that complete reliance upon past and current traffic trends only when forecasting the future needs of an urban area can be very hazardous; factors such as land-use controls, the availability of parking facilities, natural economic forces, and government policy decisions regarding the relationship between the highway system and the environment, all exert very considerable influences on future traffic demands.

Obviously, the manner in which travel needs are predicted in different urban areas will vary according to the particular objectives in mind. Generally, however, the more complex studies can be divided into the following main phases:

1. Inventories of the main traffic facilities, public transport services, present and future land uses, and appropriate population and economic data.
2. The determination of the existing interzonal travel patterns, and the factors which control them.
3. The determination of the manner in which the travel growth characteristics of individual zones interact and affect future travel distributions between zones.
4. The assignment of future travel to alternative transport modes, i.e. the modal split, and then to different routes between zones.
5. Detailed evaluation of the transport plan.
6. Detailed system planning and design to meet the requirements of the future.

The relationships between these various steps is reflected in the flow diagram in Fig. 3.21. The following discussion is primarily concerned with phases 1–4 as noted above.

Basic inventories

A measure of the comprehensiveness of the inventory information that can be gathered may be obtained by considering the London traffic survey.[26]

In terms of population and land area, this project ranks as one of the largest of its kind ever conducted anywhere in the world. The internal survey area was composed of 933 zones and had an area of 2437 km^2 with a population of 8 827 000. Ninety-six per cent of this population resided in 2 959 000 households; hospitals and other communal establishments accounted for the remaining 4 per cent.

During the survey some 50 000 homes were visited in order to carry out home interviews, while about 100 000 drivers were stopped and questioned at 91 roadside interview stations. Traffic volume counts were conducted in order to gross-up the road-interview data to total average daily traffic levels. The interviews were designed not only to give information on the origins and destinations of traffic but also data on journey purposes, modes of travel, land uses served and times of travel. Goods vehicle movements and taxi journeys were also studied. Travel within the survey area by commercial vehicles based within the external cordon was determined by direct contact with the owners of over 37 000 representative vehicles. Nearly 3200 km of roadway were surveyed for the volumes of traffic carried and the average journey speeds.

In conjunction with the O-D studies an extensive inventory was made of the transport facilities and services within the survey area. This inventory included a traffic volume census, measurements of roadway widths and other existing features, as well as the journey times on the different routes. Journey-time studies were made for two purposes. Firstly, they reflected the quality of the traffic flow and helped to identify those sections of the main road system that were most in need of improvement. Secondly, the journey time data were later used in evaluating the basic travel habits and in traffic forecasting.

The third basic type of study carried out was an inventory of the planning factors known to influence the generation and attraction of journeys. Factors studied included population and household trends, and the distribution of employment and car ownership. The importance of each zone with respect to retail, wholesale, recreational and educational activities was also determined.

The fundamental assumption underlying all methods of trip forecasting is that there is some order in human travel behaviour in urban areas that can be measured and described, and that the present-day inventory data form the basis for formulating these descriptions. The process by which the mathematical descriptions are used to predict future journeys involves the preparation of models for trip generation, trip distribution, modal split, and traffic assignment.

Trip generation

Trip-end generation models are concerned with the estimation of the number of trips into and out of various traffic zones. They are based on the principle that land use generates trips, and that, for example, both the numbers and types of 'from home' trips are influenced by (*a*) socio-economic variables such as car ownership or availability, household income and size, occupational status, household composition (e.g. number of workers per household), (*b*) location variables such as population or residential density, rateable value and distance of household from the town centre, and (*c*) public transport accessibility variables. (This last variable is of particular importance if the modal split is to be included as part of the trip generation model; in the following discussion the modal split will be treated separately from the generation stage.)

As may be gathered from the above, the number of variables considered to exert a casual effect on trip generation can be very great. The purpose of the trip-end modelling process is to identify those which are meaningful determinants of trip-making behaviour, and express their effects in a mathematical way so that they can be used with confidence as a predictive tool. In general, the traffic analyst has the choice of three main methods of developing trip-end models: 1. Expansion Factor method. 2. Regression Analysis method. 3. Category Analysis method.

The *Expansion Factor* method is mentioned here primarily to emphasize that its use is now generally out-dated in relation to trip generation in fast-changing complex urban areas. Methods of this nature, which simply place reliance upon past growth rates as a means of predicting future trips, should be confined to short-term forecasting in rural areas.

The *Multiple Regression Analysis* method is probably that which has been used in the majority of transport demand studies in the past. The extensive use of the 'best fit' least squares method, which involves developing a linear equation of the form

$$Y = b_0 + b_1 X_1 + b_2 X_2 + \ldots\ldots. b_k X_k$$

is primarily due to the increasing accessibility of sophisticated computer programs which, automatically and speedily, can carry out the extensive calculations required by the process. The following example from Cardiff, illustrates the applicability of the method:

$$Y_w = 0{\cdot}097X_1 - 351X_2 + 0{\cdot}773X_3 + 0{\cdot}504X_4 - 43{\cdot}6$$

where Y_w = zonal work trips by all modes
X_1 = population of zone
X_2 = number of households in zone
X_3 = number of employed residents in zone
X_4 = number of cars owned in zone
and $-43{\cdot}6$ = constant.

Similar equations for other types of trips were also developed so that when the various Y-values were combined, ΣY gave the total number of trips associated with each zone for the year of the study. On the assumption that the relationships expressed by the various Y-equations remain stable, the future numbers of zonal trips were estimated by substituting appropriate future estimates of the independent variables.

It may be noted that the use of regression equations derived on a zonal basis has been rigorously criticized[35] on the grounds that the models are not stable from one area to another and, thus, are unlikely to be stable over time into the future. Instead, it is recommended that regression equations should more properly be derived initially on a household rather than on a zonal basis; that household regression models provide a more meaningful description of the factors underlying trip-making behaviour, and yet can be easily expanded to provide zonal trip and estimates as required.

In an attempt to free the traffic analyst from many of the assumptions and problems inherent in the use of multiple regression techniques, another trip-end modelling method—*Category Analysis*—has been developed[36] which considers the household as the fundamental analysis unit, but assumes in this instance that the number of trips from a household is a stable function of three main parameters: household income, car ownership or availability, and family structure. In all, the following 108 different categories of household have been defined, and each associated with a trip rate:

Income classes, £/annum (as reported in 1967)	*Ownership, cars/ household*	*Family structure, persons/ household*
⩽500	0	0 workers + 1 non-employed adult
501–1000	1	0 workers + ⩾2 non-employed adults
1001–1500	2+	1 worker + ⩽1 non-employed adult
1501–2000		1 worker + ⩾2 non-employed adults
2001–2500		⩾2 workers + ⩽1 non-employed adult
>2500		⩾2 workers + ⩾2 non-employed adults

Category analysis has some appeal as a method of predicting trip generation and/or attraction. One obvious reason for this is that it is intuitively appealing to categorize households in a given zone, both now and in the future, and associate each with an expected trip-making behaviour. More important is the fact that its usage cuts down very considerably on the amount of current home interview data which need be gathered and analysed, since the household information required for any given urban area can be obtained directly from the 1971 National Census, while trip rates are available from previous studies (and only require sample checks in the area being considered).

Trip distribution

Trip distribution is the procedure utilized to distribute generated and attracted trips from each zone to any other zone. Three main models have been developed to distribute these trips: 1. Growth Factor methods. 2. Gravity method. 3. Opportunity methods. All of the models assume that travel between any two zones is a function of both the attractiveness of the activities in the zones, and the resistance to travel (expressed in travel distance, time, or cost) between the zones.

Growth factor methods. Four different types of growth factor model have been developed: these are the Uniform Factor, Average Factor, Fratar, and Detroit methods.

The *Uniform Factor* method involves the determination of a single growth factor for the entire survey area; all existing interzonal trips are then multiplied by this factor in order to get the future interzonal flows. The expansion factor used is the ratio of the total number of future trip ends to the existing total.

The *Average Factor* method represents an attempt to take into account the fact that rates of development in zones are normally different from the rate for the urban area as a whole, and this can be expected to be reflected in different interzonal trip growth rates. Thus it utilizes a different growth factor for each interzonal movement: this is composed of the average of the growths expected at each pair of origin and destination zones. As might be expected, this method, as described, will not give total future flows originating and terminating in a given zone which agree with the figures derived from the trip generation model. To overcome this an iterative process, based on revised interzonal growth factors, must be used until an acceptable balance is achieved.

The *Fratar* method of successive approximations[37] is perhaps the most widely known iterative process. More complicated than the Average Factor method, it gives more rational answers, albeit at the expense of a considerable amount of computer time. A variation of the Fratar method which requires considerably less computer time is known as the *Detroit* method, (see also ref. 38). This latter model is easiest described mathematically, viz:

$$T_{ij} = t_{ij}\frac{F_i F_j}{F}$$

where T_{ij} = future trips from zone i to zone j

$$F_i = \frac{T_i}{t_i} = \frac{\text{future total trips originating at zone i}}{\text{existing total trips originating at zone i}}$$

$$F_j = \frac{T_j}{t_j} = \frac{\text{future total trips ending at zone j}}{\text{existing total trips ending at zone j}}$$

$$F = \frac{T}{t} = \frac{\text{future total trips in the entire survey area}}{\text{existing total trips in the entire survey area}}$$

and t_{ij} = existing trips from zone i to zone j $= t_i\left(\frac{t_j}{t}\right)$

Comment. Although relatively easy to understand, the Growth Factor methods are now rarely used. Major weaknesses of all of the methods are that they completely underestimate growths in existing trips which are at or near zero, (even though land use may change significantly as, for example, in a growing town), while they assume that the resistance to travel between zones remains a constant into the future (even though, for example, a new motorway might be opened).

Gravity model. The Gravity model approach to trip distribution came to transport planning after having been developed in other fields, notably the telephone industry and, to a lesser extent, sociology. Its utilization is based on the premise that all trips starting from a point are pulled by various traffic generators or land uses, and that the degree of pull of each 'magnet' varies directly with the form of the generator and inversely with the distance (or travel time) over which the attraction is generated.

The general form of the gravity model as used today is:

$$T_{ij} = \frac{G_i A_j F_{ij} K_{ij}}{\Sigma_{j=1}^{n} A_j F_{ij} K_{ij}}$$

where T_{ij} = future number of trips from zone i to zone j
G_i = total number of trips generated in zone i
A_j = total number of trips attracted to zone j
F_{ij} = empirically derived travel time factor, calculated on an area-wide basis
and K_{ij} = empirically derived zone-to-zone adjustment factor which takes account of other socio-economic influences not included in the model (not usually necessary in towns with less than 100 000 population).

Existing-year data are used to calibrate the model, i.e. to determine appropriate values for the F- and K-factors.

Comment. The Gravity model is by far the most widely used trip distribution process. As well as recognising the gravitational 'pull' of different land uses, and changes in travel time, it can also take into account the influences of differing trip purposes. What used to be its main disadvantage, the relative complexity of the iterative calibration process, is now largely overcome as a result of the development of well-established standard routines.

Opportunity model. A major criticism that can be levelled at the Gravity model is that it does not take direct account of individual behaviour patterns. An opportunity model, which is derived on the basis of probability theory, assumes that as trips which are generated in zone i move away from that zone, they incur increasing opportunities for their purposes to be satisfied at any zone j, and therefore there is an increasing probability that they will not proceed beyond zone j. Two forms of opportunity model have been developed, the *Intervening Opportunity* model and the *Competing Opportunity* model. The difference between them relates to the manner in which the probability function utilized in the distribution equation is calculated, viz. the Intervening model requires trips to remain as short as possible and they only become longer if suitable destinations cannot be found closer to zone i, whereas the Competing model assumes that the probability of a trip from zone i stopping in zone j is dependent upon the ratio of the trip destination opportunities within a given time boundary.

The basic Intervening model is expressed as

$$T_{ij} = G_i[e^{-LD} - e^{-L(D+D_j)}]$$

where T_{ij} = future number of trips from zone i to zone j
G_i = total number of trips generated in zone i
D = total number of destination opportunities closer in time to zone i than are those in zone j
D_j = number of destination opportunities in zone j
and L = probability that any given destination opportunity will be selected.

Comment. The opportunity model approach has been rigorously tested in three studies in the United States, and it has been shown that in general the intervening opportunity model can give results of a reliability equal to that of the gravity model. As of this time, the opportunity approach has not been reported in the technical literature as having been used in Britain.

Modal split

The purpose of the modal split analysis is generally considered as being to determine the proportion of total trips that can be expected to use the private car as against public transport (i.e. walking trips are normally omitted in city-wide studies). So far, two basic analysis approaches have been developed, viz. Trip-end and Trip-interchange models. The main difference between these two is that trip-end models are developed before the trip distribution phase of the study, usually as part of the trip generation process, whereas trip-interchange models allocate person-trips to different transport modes after the trip distribution process (as shown in Fig. 3.21).

Trip-end models assume that the principal factors which affect modal split are the characteristics of the journey origin and of the traveller, e.g. income and car availability, family size and composition, residential density, zone location relative to the central area of the town, and zone accessibility. It will be noted that all of these factors, except accessibility, have been discussed in relation to trip generation. The term 'accessibility' is used to describe the ease with which people with origins in a given zone may use particular forms of transport, e.g. a zone with high accessibility is one well serviced with public transport facilities. Regression or category analysis procedures are then used to predict the number of trips by the various modes, in a manner similar to that described for trip generation studies. Thus, for example, in the London study previously described, the 108 household classifications were increased to 108 × 6 by assigning each zone a low, medium or high accessibility index which reflected its ease of access to bus or rail facilities, viz.

$$I_{\mathrm{Bus}} = \frac{\Sigma_{\mathrm{i}}(N_{\mathrm{ij}})^{1/2}}{(0{\cdot}39A_{\mathrm{j}})^{1/2}} \text{ and } I_{\mathrm{Rail}} = \frac{\Sigma_{\mathrm{i}}(N'_{\mathrm{ij}})^{1/2}}{0{\cdot}39A_{\mathrm{j}}}$$

where N_{ij} = off-peak frequency of buses on route i and passing through zone j

N'_{ij} = number of trains during the off-peak period stopping at station i in zone j

and A_{j} = area of zone j, km^2.

Trip-interchange models assume that the levels of service provided by the transport systems are the main factors affecting modal split. Proxies for level of service which have been used in various studies include relative travel times, excess travel times, and travel costs.

A typical travel time ratio is

$$R_{TT} = \frac{t_1+t_2+t_3+t_4+t_5}{t_6+t_7+t_8}$$

where t_1 is the time spent walking to, say, the bus stop at the trip origin; t_2 = waiting time at bus stop; t_3 = time on bus; t_4 = time spent changing from one bus to another; t_5 = time spent walking from bus stop to destination; t_6 = time spent driving car; t_7 = time spent parking car at destination; t_8 = time spent walking from parked car to destination.

A typical excess travel time ratio is

$$R_{ETT} = \frac{t_1 + t_2 + t_4 + t_5}{t_7 + t_8}$$

where t_1, t_2, t_4, t_5, t_7 and t_8 are as defined above.

A representative travel cost ratio is given by

$$R_{TC} = \frac{nf}{c_1 + c_2}$$

where n = average car occupancy; f = fare by public transport; c_1 = cost of fuel (petrol and oil) for car trip; and c_2 = parking cost. (Note that road tax, insurance, or maintenance costs are not included in the R_{TC} formulation as they are not taken into account by a motorist when deciding about the mode for a particular trip.)

Whichever of the above ratios (or other combinations) is selected in a particular study, the end result is the preparation of a series of multiple regression equations or diversion curves which incorporate the relevant variables and are used to predict the number of public transport trips between each pair of zones. The most complex modal split model[(38)] so far developed involves progressing through a series of diversion curves incorporating relative travel times, relative travel costs, relative excess travel times, income of the traveller, and trip purpose, in order to determine the modal split. Another important procedure[(39)] extended regression techniques to separate groups of trip-makers into different mode users by utilizing probability techniques based on a combination of variables which related to both the journey characteristics and those of the originating household.

Comment. The main advantage of the 'pure' trip-end type of model is its simplicity; its weakness is its insensitivity to the effect on the future modal split of, for example, improving specific inter-zonal travel corridors. In contrast, the trip-interchange model is much more sensitive to adjustments in the inter-zonal network, but at the cost of losing sensitivity to the effects of the trip generation characteristics. It is for these reasons that recent models have tended to try and incorporate both sets of variables in their formulations.

Another major criticism of present-day modal split models is that they are based on current levels of service, and these have a built-in bias in favour of the private car. Thus, future estimates of public transport usage could err significantly on the low side should, for example, the bus service be significantly improved by the provision of reserved trackways in combination with a stringent parking or road pricing policy.

Traffic assignment

The next stage in the transport planning process is the assignment of the various modal trips to the different highway (or rail) routes within the survey area. A typical urban network assignment process might first involve an initial assignment by means of an *All-or-Nothing* procedure (i.e. whereby the total trips between any two zones are assigned to the path, or 'tree' with the minimum travel resistance—usually specified in terms of travel time), or by a proportional diversion procedure (such as is illustrated in Fig. 3.20)—and then use one of a number of iterative *Capacity Restraint* procedures (e.g. see refs. 40 and 41) to re-distribute the traffic volumes until the situation is reached where the travel speed assumed in the calculation for each link of the network is relatively unaffected by the assignment process.

The general approach is easiest illustrated by considering the iterative procedure utilized in the Detroit Area Traffic Study. In this study, traffic was first assigned to the various links of the proposed network using the all-or-nothing approach; the speed assumed for initial assignment purposes was the free (or unrestrained) flow speed. Obviously, traffic on a network will not always operate under free flow conditions, and the speed on each link is affected by the flow; hence, each link's travel time was modified as follows:

$$T_A = T_0 e^{(v/c-1)}$$

where T_A = adjusted travel time,
T_o = original assigned travel time (a function of the desired operating speed), or the travel time on a link when $v = c$
v = assigned volume
and c = computed capacity.

The second iteration was accomplished using the new travel times to determine a new series of minimum paths or trees. The volumes so determined were then added to the results of the previous iteration, and the average link load determined. Successive iterations recalculated the T_A-value based on the model using the average link volume for the v-value until a balanced network was obtained.

Comment. Realism in traffic assignment has been an ideal long sought after by highway and traffic engineers. As with other aspects of the transport planning process, this ideal became more practical with the advent of digital computers of sufficient size to manipulate the large arrays of data required by all assignment techniques. The Detroit Transportation Study in 1958, noted above, is generally considered to be the first technique to depart from the traditional procedure of using only diversion curves of the form illustrated in Fig 3.20. Since then a number of different techniques have been developed, each of which has some particular advantage over another. Even so, the results or answers from a traffic assignment program still require considerable examination and judgment before being applied.

In real life, a motorist selects his journey route for a variety of reasons, e.g. time, distance, monetary cost, convenience, comfort and safety. In theory, there exists in each situation an optimum route which offers the driver the best combination. Whilst much progress has been made, it cannot yet be stated with surety that the procedures available at this time are able to fully interpret the complexities of human behaviour and the dynamic ever-changing events that take place in an urban area.

SELECTED BIBLIOGRAPHY

1. O'FLAHERTY, C. A. *Passenger Transport: Present and Future.* Leeds, Leeds University Press, 1969.
2. O'FLAHERTY, C. A. People, transport systems and the urban scene: An overview, *International Journal of Environmental Studies*, 1972, **3**, 265–285 and 307–320.
3. SMITH, W. S. Urban transport co-ordination. *Traffic Engineering and Control*, 1966, **8**, No. 5, 304–306.
4. REPORT OF THE COMMITTEE ON THE PROBLEM OF NOISE. Cmnd. 2056, London, H.M.S.O., 1963.
5. BUCHANAN, C. D., *et al. Traffic in Towns.* London, H.M.S.O., 1963.
6. MINISTRY OF TRANSPORT. *Passenger Transport in Great Britain.* London, H.M.S.O., 1971.
7. TULPULE, A. H. Forecasts of Vehicles and Traffic in Great Britain: 1969. *RRL Report LR288*, Crowthorne, Berks., The Road Research Laboratory, 1969.
8. HALL, P. Transportation. *Urban Studies*, 1969, **6**, No. 3, 408–435.
9. REGISTRAR GENERAL. *Annual Estimates of the Population of England and Wales and of Local Authority Areas, 1969.* London, H.M.S.O., 1970.
10. BUCHANAN, C. A. AND PARTNERS. *The Conurbations.* London, The British Road Federation, 1969.
11. HERRMAN, P. G. Forecasts of Vehicle Ownership in Counties and County Boroughs in Great Britain. *RRL Report LR200*, Crowthorne, Berks., The Road Research Laboratory, 1968.
12. BUCHANAN, C. D. AND PARTNERS. *South Hampshire Study: Report on the Feasibility of Major Urban Growth.* London, H.M.S.O., 1967.
13. LEEDS CITY COUNCIL *et al. Planning and Transport—The Leeds Approach.* London, H.M.S.O., 1969.
14. MILLARD, R. S., *et al.* A Review of Road Traffic Noise. *RRL Report LR357*, Crowthorne, Berks., The Road Research Laboratory, 1970.
15. PRICE, B. T. *et al. Urban Traffic Noise.* Paris, O.E.C.D., 1971.
16. REYNOLDS, D. J. The assessment of priority for road improvements, *Road Research Technical Paper* No. 48. London, H.M.S.O., 1960.
17. SMEED, R. J. The traffic problem in towns, *Town Planning Review*, 1959, **35**, No. 2, 133–158.
18. TANNER, J. C. and J. R. SCOTT. 50-point traffic census—The first 5 years, *Road Research Technical Paper* No. 63. London, H.M.S.O., 1962.
19. DUNN, J. B. and HUTCHINGS, I. J. The Distribution of Traffic in Great Britain Through the 24 Hours of the Day in 1968. *RRL Report LR295*, Crowthorne, Berks., The Road Research Laboratory, 1969.
20. TANNER, J. C., H. D. JOHNSON and J. R. SCOTT. Sample survey of roads and traffic of Great Britain, *Road Research Technical Paper* No. 62. London, H.M.S.O., 1962.

21. DUNN, J. B. 50-Point Traffic Census Results for 1968. *RRL Report LR302*, Crowthorne, Berks., The Road Research Laboratory, 1970.
22. TULPULE, A. H. Recent Trends in Vehicle Use in the United States and in Britain. *RRL Report LR354*, Crowthorne, Berks., The Road Research Laboratory, 1970.
23. WARDROP, J. G. Some theoretical aspects of road traffic research, *Proc. Inst. Civ. Engrs*, Part II, 1952, **1**, 325–378.
24. WARDROP, J. G. and G. CHARLESWORTH. A method of estimating speed and flow of traffic from a moving vehicle, *Proc. Inst. Civ. Engrs*, Part II, 1954, **3**, 158–171.
25. O'FLAHERTY, C. A. and SIMONS, F. An Evaluation of the moving observer method of measuring traffic speeds and flows, *Proceedings of the Australian Road Research Board*, 1971, **5**, Pt. 3, 40–54.
26. FREEMAN, FOX & PARTNERS, *et al. London Traffic Survey: Existing traffic and travel characteristics in Greater London*, Vol. 1. London County Council, 1964.
27. JONES, D. R. Inductance loop detectors for automatic vehicle traffic counting: Trials in Gloucestershire, *Traffic Engineering and Control*, 1970, **12**, No. 5, 236–239, 243.
28. COBURN, T. M., M. E. BEESLEY and D. J. REYNOLDS. The London-Birmingham motorway, *Road Research Technical Paper* No. 46. London, H.M.S.O., 1960.
29. CLARKE, R. H. and DAVIES, D. H. Origin and destination survey by registration number method, *Traffic Engineering and Control*, 1970, **11**, No. 12, 600–603.
30. BUREAU OF PUBLIC ROADS. Conducting a home-interview origin and destination survey, *Procedure Manual* 2B. U.S. Public Administration Service, 1954.
31. CALIFORNIA DIVISION OF HIGHWAYS. *Planning Manual*. Sacramento, California, 1962.
32. A.A.S.H.O. *A Policy on Arterial Highways in Urban Areas*. Washington, D.C., American Association of State Highway Officials, 1957.
33. A.A.S.H.O. *A Policy on the Geometric Design of Rural Highways*. Washington, D.C., American Association of State Highway Officials, 1965.
34. LAMB, G. M. Introduction to transportation planning, (Pts. 1–6), *Traffic Engineering and Control*, 1970, **11**, Nos. 9–12, and **12**, Nos. 1–2.
35. DOUGLAS, A. A. and LEWIS, R. J. Trip generation techniques, *Traffic Engineering and Control*, 1970/1971. **12**, Nos. 7–10.
36. WOOTTON, H. J. and PICK, G. W. Trips generated by households, *Journal of Transport Economics and Policy*, 1967, **1**, No. 2, 137–153.
37. FRATER, T. J. Vehicular trip distribution by successive approximations, *Traffic Qly*, Jan. 1954.
38. HILL, D. H. and VON CUBE, H. G. Development of a model for forecasting travel mode choice in urban areas, *Highway Research Record* No. 38, 78–96, 1963.
39. QUARMBY, D. A. Choice of travel mode for the journey-to-work: Some findings, *Journal of Transport Economics and Policy*, 1967, **1**, No. 3, 273–314.
40. HUBER, M. J., BOUTWELL, H. B., and WITHEFORD, D. K. Comparative Analysis of Traffic Assignment Techniques with Actual Highway Use. *National Co-operative Highway Research Program Report 58*, Washington, D.C., The Highway Research Board, 1968.
41. CHU, C. A review of the development and theoretical concepts of traffic assignment techniques and their practical applications to an urban road network, *Traffic Engineering and Control*, 1971, **4**, No. 13, 136–141.

4 Parking

There are few towns in today's highly developed countries—and in many newly developing ones as well—which do not experience traffic congestion and car parking problems. Parking, in particular, is such a sensitive issue that parking policies in many communities often appear to be both conflicting and indecisive. Nevertheless, there is no doubt but that the general public is now becoming reconciled to the need to grapple firmly with the parking problem whilst civic leaders, in turn, are coming to appreciate the need to develop parking policies in the context of land use and transport policies for the urban area as a whole.

The importance of this latter point cannot be over-emphasized. *Parking policy must always be considered in the light of its effects on land use and transport policy. Parking control is, in many towns, the key to proper traffic control.*

When discussing the parking problem, and means of tackling it, one must inevitably touch upon certain 'political' aspects of the problem, as well as the technical ones. Thus, the approach taken in this chapter is first to describe some of the detriments known to be associated with the parked vehicle. The development of the town centre parking plan is discussed next, after which the technical features influencing the location, design and management of on- and off-street parking facilities are described.

TRAFFIC AND PARKING: SOME ASSOCIATED EFFECTS

As noted previously, a typical British town has a centripetal type of layout, such that most of the main roads radiate outward from the central area. These towns, which 'naturally' evolved during the days of horse-drawn and pedestrian traffic are now having to cater for large volumes of vehicles which never existed when the street systems were developed. It is useful, therefore, to consider some resultant effects since these influence the extent to which the motor car is provided for in the town and, ultimately, the manner in which this 'sensitive' parking problem is tackled.

Environment

An environmental 'dis-amenity' which receives very little attention is the aesthetic deterioration often associated with the parking of vehicles. The simplest way of describing this detriment is to quote from the 'Traffic in

Towns' report[1] regarding the visual consequences which can be associated with the parked motor vehicle in the town:

> '. . . the crowding out of every available square yard of space with vehicles, either moving or stationary, so that buildings seem to rise from a plinth of cars; the destruction of architectural and historical scenes; the intrusion into parks and squares; the garaging, servicing and maintenance of cars in residential streets which creates hazards for children, trapping the garbage and the litter and greatly hindering snow clearance; and the indirect effect of oil-stains which render dark black the only suitable colour for surfaces, and which quickly foul all the odd corners and minor spaces round new buildings as motor cycles and scooters take possession. . . . the dreary, formless car parks, often absorbing large areas of towns, whose construction has involved the sacrifice of the closely knit development which has contributed so much to the character of the inner areas of our town . . .'

Business

Commercial interests consider that they are directly affected by the parking situation and, even in this day, many merchants and business people in city centres tend to regard packed kerbs as necessary visual evidence of trade prosperity. Hence the reason why regulations to control parking in towns are so often viewed with suspicion by Chambers of Commerce.

Attempts have been made to relate the prosperity of a town centre to the availability of parking spaces (see, for example, ref. 2), but no conclusive results have been derived to this effect. Although inadequate parking is widely believed to be one of the main drawbacks of the central area of a large town, analysis suggests[3] that, when other factors are taken into account, it does not necessarily greatly affect the shopping orientation of persons. At the same time, however, it must be noted that another survey[4] reported that between one-tenth and one-fifth of the respondents in the households queried (belonging to Automobile Association members in Britain) stated that they avoided going to town centres wherever possible because of parking difficulties.

Accessibility and congestion

Traffic congestion is an important consideration affecting the viability of any shopping centre. Anything that can be done to reduce congestion and allow people to travel to the town centre in a shorter time, will make the central area more accessible—and, thus, will help people to decide to shop there as against in the suburbs or out of town.

One way of reducing traffic congestion is to eliminate/control parking at the side of the road, i.e. parked vehicles cause congestion simply by occupying road space that could be used by moving vehicles, thus reducing capacity. (This is especially noticeable if the parked vehicles are at, or close to, an intersection.) Table 4.1 shows that a small number of parked cars can have quite a significant effect on capacity, albeit the relative effect diminishes as the parking intensity increases. If the volume of traffic is fixed, then parked vehicles will also reduce traffic speed and, hence, increase journey time which, in turn, reduces town centre accessibility.

TABLE 4.1. *Effect of parked vehicles on road capacity.* (*Based on data in ref.* 5)

No. of parked vehicles per km (both sides added together)	*Effective loss of carriageway width, metres*	*Loss of capacity at 24 km/h, p.c.u./h*
3	0·9	200
6	1·2	275
30	2·1	475
60	2·6	575
125	3·0	675
310	3·7	800

Accidents

Although vehicles parked at, or manoeuvring into or out of, kerb parking spaces can be an important cause of accidents, the number of studies which have been carried out in order to produce an understandable and accurate description of the role of parking in accident occurrence are relatively few. What is probably the prototype study of this nature—it was carried out in the United States in 1946[6]—found that 17 per cent of city accidents and 10 per cent of rural road accidents involved vehicles which were parked, manoeuvring into or out of a parking position, or stopped in traffic. In 1950 and 1951 about 80 children, mostly under the age of 4 years, were killed each year in Great Britain when stationary vehicles moved away[7]. In 1958, about 7 per cent of the cyclist casualties in this country were attributed to the careless opening of the doors of parked cars, and vehicles parked at the kerb formed at least 4 per cent of the total number involved in accidents in Britain.

Apart from other general statistics of the type given above, (e.g. 17 per cent of the urban areas accidents in the American State of Connecticut were estimated[8] to involve vehicles which were parked, manoeuvring into or out of a parked position, or stopped in traffic) there is only one major study reported in the literature which has attempted to look deeper into the relationship between kerb parking and traffic accidents. This study[9] which was carried out in 1965–66 in the United States, examined in detail (including statistical analysis) some 11,620 road accidents on 152·36 km of arterial and collector streets in 32 cities, representing 17 states and the District of Columbia. The main results of this most comprehensive study—which while confirming some previous concepts of traffic engineers *re* this problem, also raised some questions regarding others—are as follows:

1. An average of 18·3 per cent of all accidents studied involved parking either directly or indirectly.

2. Almost 90 per cent of the accidents involving parking were a direct result of parking activity—either a parked vehicle or a vehicle entering or leaving a parked position—while in slightly more than 10 per cent, parking was a factor only.

3. There was no significant difference in parking accident experience between street segments on which parking was prohibited and those on which parking was restricted to less than 2 hours. A rather significant increase was found in parking accidents where the parking operation was unrestricted.

4. Parking accidents have a higher rate in residential areas than in either commercial or industrial areas.

5. Accidents involving parking have a tendency to decrease slightly as the roadway width increases.

6. Parking accidents in town centres have a somewhat higher rate than in intermediate or outlying areas, the latter two showing rates which were practically identical.

7. In 92 per cent of all parking accidents, cars only were participants. This compared with about 86 per cent of the total vehicle-km operated by such passenger vehicles.

8. Almost 40 per cent of the total number of vehicles involved in accidents during the parking operation were attempting to drive forward into the kerb. In slightly over 20 per cent of the cases, vehicles were backing into the kerb whereas, in about 24 per cent, they were leaving the kerb in a forward direction.

9. 46 per cent of the vehicles in motion were attempting to drive straight ahead on the street in question and collided with a vehicle which was parked or was involved in a parking or unparking operation. Vehicles attempting to park were involved in a little over one-third of the total parking accidents.

10. In almost 94 per cent of the accidents involving kerb parking the vehicles were reported as legally parked.

11. In slightly over 77 per cent of the parking accidents, male drivers were the participants. This is slightly higher than the percentage of drivers who were men.

12. Almost 13 per cent of drivers involved in parking accidents were under 20 years of age, as compared with approximately 8·5 per cent of the total national driving done by persons in the same age group.

13. The month of December was significantly higher than any other month in both the number and the percentage of parking accidents experienced.

14. Almost as many parking accidents occurred during hours when the road surface was either wet or covered with ice or snow as compared with the periods when the pavement was dry—even though it may be reasonable to assume that the carriageway surface was normal during the greater percentage of the total time.

15. Night-time was not a major factor in accidents involving parking operations.

16. Of the almost 2180 parking accidents examined in these urban areas, only one (this involved a pedestrian stepping from behind a parked vehicle)—resulted in a person being killed. Very few injuries were recorded and, in most cases, these also involved pedestrians who stepped from behind parked vehicles.

PARKING SURVEYS

Parking surveys are carried out in order to obtain the information necessary to provide an assessment of the parking problem in the area(s) being studied. The objective of any such study is not simply to gather 'interesting data', but rather to determine facts which will provide the logical point of departure in relation to indicating parking needs. Common to all types of parking studies, whatever their scale, are parking supply and parking usage surveys.

Parking supply survey

Parking supply surveys are concerned with obtaining detailed information regarding those on-street and off-street features which influence the provision of parking space, the existing situation with regard to parking space, and how it is controlled. A typical survey would require an inventory of the on-street accommodation, and of all off-street car parks and parking garages serving the traffic area being studied. Whether the survey area involves the central area of a town (which may be broadly defined as the area in which all streets have business frontage), or a suburban shopping area, or a hospital or university, it should include the surrounding fringe area where vehicles are parked by persons with destinations within the survey area.

A parking supply survey can be considered as being composed of three main parts, viz. an on-street space inventory, a street regulation inventory, and an off-street space inventory. Data usually collected in each of these survey phases are listed in Table 4.2; they are obtained by simple inspection of the survey area. For summary purposes, these data may be marked on a map with a scale of, say, no more than 1:2500, (more usually 1:1250), using figures to indicate the number of spaces available in various sub-area locations, and letters, symbols and colours to differentiate between parking classifications. Tabulations are usually also prepared showing, for each sub-zone within the survey area, the total number of street and off-street parking spaces available by classification, together with the space available for the loading of lorries and vans, or reserved for buses, taxis, etc.

The information obtained from the parking supply survey is an invaluable reference in connection with many of the routine decisions which have to be made in connection with the development and implementation of a parking plan. When questions arise as to what are the kerb or off-street conditions at any particular location, the answers are immediately available. The supply study also provides the basis for evaluating the available space-hours of parking as well as the parking turnover at any particular location.

TABLE 4.2. *Typical data recorded during a parking supply survey*

Street (incl. alley) inventory	*Street regulation inventory*	*Off-street inventory**
1. Pavement crossings, access to premises 2. Loading bays 3. Bus stops 4. Taxi stands 5. Pedestrian crossings 6. Visibility splays at junctions 7. One-way streets 8. Private streets 9. Service and rear access alleys 10. Vacant or unused land suitable for temporary or permanent parking space 11. Carriageway widths 12. Other local factors, e.g. areas of special amenity	1. Controlled parking a. by regulation (including discs, with each type separately classified) b. by meters (classified by type of control) 2. Parking prohibited a. always b. during peak hours 3. Controlled loading and unloading 4. Uncontrolled parking (the remainder)	1. Type a. ground level only b. multi-storey or upper level only c. underground only 2. Ownership and use a. publicly owned, for public use b. privately owned, for public use c. private use only 3. Commercial vehicles only 4. Payment a. fee-charging (subdivided by rates of charge) b. free 5. Time-limit a. up to 2h only b. up to 4h only c. over 4h or no limit 6. No. of spaces provided 7. Size of parking area 8. No. (and location on map) of entrances and exits

Parking usage survey

The parking phenomenon is very much based on the law of supply and demand, whereby supply is the total number of spaces available within a designated area, and demand is the desire to park based solely on the location of the trip destination. However, unlike the true supply and demand situation, there is a third variable—*usage*—associated with parking which reflects the desire to park close to the destination, *but within the limitations imposed by the available supply*, as well as the desire to park at a reasonable cost. In other words, demand is a constant, reflecting the desire to park at the trip destination, whereas usage is a variable that depends on the conditions at the terminal area, the characteristics of the trip, as well as of the tripmaker.

If the parking supply is in excess of the parking demand, then the *true* demand can be determined by means of a parking concentration survey—usually carried out in conjunction with a parking duration survey. If the parking supply is less than the demand, then an *indication* of the demand may be obtained by means of a direct interview parking survey; however, the *true* demand which might include 'suppressed' vehicles not able or willing to find parking within the survey area, can only be properly estimated from comprehensive land use/transport surveys of the nature noted previously in this text.

Concentration study. The purpose of a concentration survey is to determine not only where vehicles do park, but also the actual number parked at any given instant at all locations (on- and off-street) within the survey area. The street phase of the survey is usually carried out by dividing the survey area into 'beats', each short enough to be toured on foot or by car within a predetermined time interval. Information regarding the numbers of vehicles parked (legally and illegally) at each street and alleyway location are then noted on prepared forms or spoken into a tape recorder for transcribing at the end of the day. Each 'beat' normally forms a closed circuit so that no time is lost by the observer(s) in returning to the starting point for successive inspections.

The selection of a suitable length of beat—or, more properly, the 'trip interval'—depends upon the accuracy required, and upon the amount of time, money and labour available for the survey. A long trip interval, whilst cheaper in total survey time, cost and analysis, can overlook a significant proportion of short-term parkers, thus completely invalidating the survey. For this reason, it is usually most desirable to first carry out continuous observation on a sample of streets on each beat in order to determine an acceptable trip interval, as well as to estimate the correction factors which will need to be applied to the data subsequently obtained. Whilst experience suggests that trip intervals of 0·5 h and 1 h may be satisfactory for many on- and off-street parking surveys, respectively, this may not always be so in the parts of town centres where parking durations are short and where sharp fluctuations in demand are likely to exist.

Where manpower is limited, it is generally better to carry out the on-street survey on a sample basis using short trip intervals, rather than attempt to cover the whole area using longer intervals. If sampling is used, care must be taken, however, to ensure that it is representative, covers all types of parking, i.e. commuters, visitors, shoppers and residents, and that the results can be scaled up to give a picture of the whole area.

The off-street phase of the concentration survey can be carried out in the same manner as the on-street survey, i.e. by counting the number of vehicles parked at regular intervals in each facility. However, if traffic counters are available, it is simpler to use one of these at each entrance and exit in order to monitor the traffic—in essence, to carry out a cordon count of the vehicles entering and leaving the car park, from which the accumulation of vehicles can be calculated.

The selection of time and season for a concentration survey (as for a duration survey, described next) is primarily dependent upon the characteristics of the town, and on the extent of the variation in parking usage likely to be experienced throughout the week. In most towns in Great Britain, seasonal demand is at an 'average' level in September/October, and a normal weekday then (from, say 7 a.m. to 7 p.m.) may well be convenient for the survey. Days immediately preceding or following holidays, special shopping days, or days when shopping hours differ from the usual, should not be generally considered as suitable survey days. On the other hand, there may well be a case in a particular town for carrying out a special study on, say, the market day or, in the case of holiday resorts, at special times of the year.

Where the survey area includes fringe residential areas, there should be a special survey so as to isolate the residents' parking usage.

Duration survey. As the term implies, the primary purpose of a duration survey is determine the lengths of time that vehicles are stored within the survey area. In so doing, data normally collected during a concentration survey can also be collected (in essentially the same way, in the case of the on-street survey) during the duration survey.

As with the concentration survey, the most accurate way of carrying out a duration survey is to continuously observe parkers at all possible locations. For practical reasons, this is not usually possible; instead, the duration information is obtained in a more economical fashion by an observer noting the first three numbers or four letters/numbers of each vehicle's registration number as he patrols the beat (or possibly from parking tickets in the case of some off-street facilities). As with the concentration survey, the trip interval utilized during the duration survey is again critical, e.g. if a 0·5 h trip interval is utilized, a car which is observed but once during the survey may possibly have been parked for as little as 1 minute or for as long as 59 minutes while a significant number of cars may have been parked for periods up to 29 minutes without being observed at all.

Since they also obviously measure the degree to which the existing parking regulations are observed, duration surveys should always be carried out as inconspicuously as possible. The accuracy of any survey's results could be seriously biased if parkers thought that the check being made was one which might be used for law enforcement purposes.

Parker interview survey. This, the most expensive—and comprehensive—of the parking surveys, normally involves interviewing motorists at their places of parking and questioning them regarding the origins of the trips just completed, the primary destinations whilst parked, and the purposes of the trip.

TABLE 4.3. *Categories of parking: based on parking durations*[10]

Category of parking	*Approx. duration*	*Examples*
Stop-and-go	0–5 min	Drop-off of shoppers and others; pick-up of persons and packages; parcel drop-off and pick-up; taxi loading and unloading; bus loading. (Note: Most of this category of short-term parking is at the kerb. The driver remains in the vehicle, but a parking space is occupied)
Errand	0–15 min	Brief shopping and business errands or pick-up; bank deposits and withdrawals; drop-off of dry-cleaning; buying cigarettes or newspapers; paying bills, taxis awaiting fares; dashing in for a cup of coffee. (Note: Short duration but averaging longer than above. Driver typically leaves the vehicle. Mostly but not entirely kerb parking)
Convenience	0–30 min	Purchasing of convenience goods other than large orders; purchasing or ordering equipment for business or consumer use; deliveries. (Note: Split between kerb and off-street parking)
Services	0–1 h	Trips to doctor, dentist, lawyer, travel agents, etc
Basic	0–4 h	True 'shopping' for major purchases or group of purchases; multiple-purpose trips; entertainment (movies, etc.); dining-out; cultural purposes; salesman parking; loading and unloading of large trucks; parking by repair and service vehicles; tourists
Employee	0–8 h	Parking by bosses, employees, professional men, some business visitors, hotel guests, tourists, convention
Night-time	0–15 h	Resident parking; hotel guests; tourists; entertainment parkers; work vehicles; out-of-service vehicles; buses and taxis. (Note: These are long-term, off-hour parkers)

Particulars regarding parking duration and concentration (of the type already noted) may also be gathered during this survey. If required, information may also be obtained regarding the number, sex and age of each vehicle's occupants.

With this type of survey, the survey area is usually divided so that each sub-area can be covered for the desired period(s) of the day by the interviewing team. Each individual interviewer is then given a specific section and is responsible for interviewing each driver and recording each parking incident in that area; while each section will obviously vary in size according to the parking concentration and durations, typically it is unlikely to exceed 75–100 m of kerbside space in central areas. At off-street facilities, the duration phase of the survey is simply carried out by interviewers stationed at entrances or exits.

When the survey manpower available is limited, and particularly in areas subject to long parking durations, information normally obtained by a direct interview survey may instead be obtained by means of a parking postcard survey. In this case, postcard questionnaires are either handed to the motorists as they park or else placed under the windscreen wipers of all vehicles parked at the kerb or off-street within the survey area, with the request that each driver answer the questions posed and then return the card by post. The problem with this type of survey is that expansion of the data obtained is statistically difficult; whilst a good survey may result in, say, one-third of the cards being returned, little is known about the extent to which these returns are representative of the population being sampled, e.g. 'satisfied' drivers are less likely to return the questionnaires than those having difficulty in obtaining parking places, and parking habits such as duration differ according to trip purpose and type of driver (see Table 4.3). The result is that the data obtained from the postcard survey, unless treated with considerable care, may be such as to result in entirely incorrect conclusions being deduced.

Information provided. Parking usage surveys can provide very useful data regarding existing parking characteristics within the survey area. In the first place the variation in the concentration of parked vehicles with time of day may be obtained. Figure 4.1a shows, for example, the variation in the number of vehicles parked on the streets in the central area of Edinburgh[11]. Note that the parking peak in this city occurs at about 11.30 a.m. and that a second major peak occurs at 3.30 p.m.

If p is the number of vehicles parked in a street, or group of streets, and P is the number of kerb spaces available (obtained from the parking supply survey), then a parking index defined by

$$P.I. = \frac{100p}{P}\ \%$$

may be used to assess the relative amounts of parking at particular locations.

It is usual to show these *P.I.* as numbers on a map of the survey area. Alternatively, the supply may be compared with the concentration in order

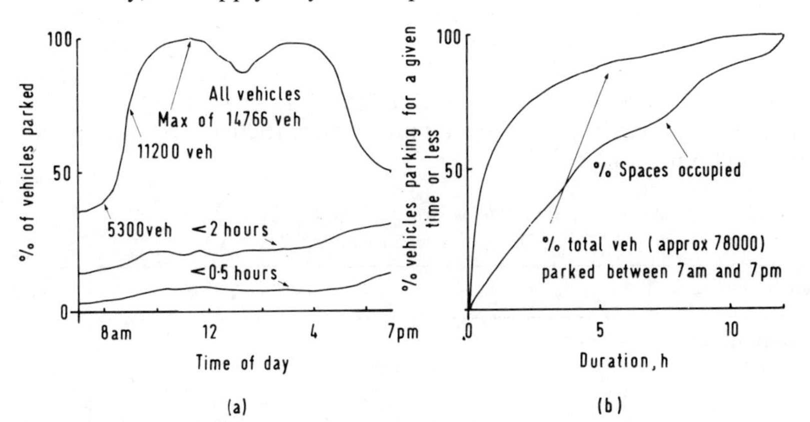

Fig. 4.1. Parking characteristics in the central area of the city of Edinburgh

to determine to what extent the occupancy values exceed the supply values; as well as indicating where extra capacity is required, this also shows the areas where illegal parking, e.g. double parking, etc. takes place, thereby implying the need for enforcement procedures at these locations.

The duration survey provides very useful information regarding the lengths of time that people park at particular locations. Fig. 4.1b shows, for example, that of the 77 970 vehicles parked on the streets of the central area of Edinburgh at the time of the survey, approximately 71 per cent parked for 2 h or less—but they only occupied 28 per cent of the 21 856 parking space usages that day. Fig. 4.1a also shows the percentage of short-term parkers in the central area of Edinburgh at any particular period of the day.

The peak hour data may be further subdivided as shown in Table 4.4 to enable a judgement to be made of the effect which might result from restricting parking to some particular maximum length—say, 2 h.

TABLE 4.4 *Some data relating to the peak use of street parking spaces in central London*[5]

No. of times each vehicle was seen	1	2	3–4	5–8	9–12	13–20	21+	*Totals*
Approx. duration of parking, hrs	0·5	1	1·5–2	2·5–4	4·5–6	6·5–10	10+	
Total no. of vehicles	2910	2358	5651	6848	7952	12 560	1344	39 362
Percentages	7·4	5·7	14·3	17·3	20·1	31·9	3·3	100

Another way of presenting daily data is shown in Table 4.5. Note that in this instance, the emphasis is placed on the distribution of the total vehicle-hours of occupation i.e. the total turnover, thus enabling an estimate to be made of the revenue which might be expected to accrue from a system of paid parking. Alternatively, the average turnover per parking space i.e. the number of times a space is used by different vehicles during a specified time, may be calculated by dividing the total numbers of vehicle-hours for the period being considered by the numbers of particular types of parking spaces laid out.

TABLE 4.5 *Some data relating to the turnover of parking spaces in Central London*[5]

No. of times seen	1	2	3–4	5–8	9–12	13–20	21+	*Totals*
Approx. duration of parking, hrs	0·5	1	1·5–2	2·5–4	4·5–6	6·5–10	10+	
Total no. of vehicles	50 768	25 872	26 416	19 616	10 048	12 560	1344	146 624
% vehicles	34·6	17·7	18·0	13·4	6·8	8·6	0·9	100
Vehicle-hours of occupation	25 384	25 872	46 228	63 752	52 752	103 620	13 440	331 048
Percentages	7·7	7·8	14·0	19·3	15·9	31·2	4·1	100

In relation to turnover, it may be noted that a 10-hour turnover of ten vehicles per space for kerb parking, and two vehicles per space for off-street car parks, is considered a good rule-of-thumb standard for maximum comfortable usage in the average central business district in urban areas in the United States.[12] Although no data are presented here, experience shows that the average turnover rates for all parking facilities combined in and about central areas, increase as the size of the urban area increases. Furthermore, American experience suggests that kerb parking spaces average turnover rates tend to be three to four times higher than off-street spaces, and surface car park spaces average higher turnover rates than garages.[13]

MANAGEMENT OF ON-STREET PARKING

The question of when and where to impose some form of kerb parking management is inextricably involved with a town's parking policy. Nevertheless, if one temporarily leaves aside policy as a controlling feature, it is possible to consider objectively the general situations where direct kerbside parking management is desirable, and how it should be carried out.

Prohibited parking

In many small towns, particularly towns which have been developed over the past century, the number of 'natural' on- and off-street parking spaces in and adjacent to the central area is often sufficient to meet the parking demand. In such towns the only parking management measures which may have to be initiated are:

1. *Intersections.* Cars should never be allowed to park within about 50 m of a major junction. (Desirably, the no-parking zone should extend even farther back from the junction.) While this prohibition can be justified on road capacity considerations, even more important is its possible effect on safety, i.e. cars and pedestrians at junctions must have adequate sight distances, while large commercial vehicles must be given sufficient space to negotiate left-hand turns (in Britain).

2. *Narrow streets.* Very often it will be necessary to initiate kerb management measures because of the relative narrowness of streets in relation to the needs of moving vehicles. Guides to when this is necessary are indirectly given in the published figures for the capacities of urban streets.[5] Parking should never be permitted on two-way carriageways in central areas which are less than about 5·75 m wide, and on one-way streets which are less than about 4 m wide; there is just not sufficient room for safe movement *and* parking on these streets.

3. *Driveways.* On no account should parking be permitted in front of driveways from houses and other buildings. Other vehicles must be allowed to enter and leave buildings.

4. *Pedestrian crossings.* For safety reasons, parking should be prohibited on or adjacent to (within, say, 8 m) pedestrian crossings. Pedestrians should be able to step off the kerb without having their view obstructed by parked cars. Similarly, the driver of a moving vehicle must immediately be able to see any pedestrian leaving the kerb.

5. *Curvature and grade conditions.* Parking should be prohibited if a study shows grade and/or curvature conditions where the removal of parking will improve the safe and efficient movement of traffic. American experience in this respect suggests[14] that parking should be prohibited (*a*) on the inside of any horizontal curve with a centreline radius less than 91 m, and a carriageway width of less than 11 m; and (*b*) on one side of any streets with a crest-type vertical curve which allows less than 49 m of sight distance when the roadway width is less than 11 m.

6. *Road bridges and tunnels.* Although there is no reason why parking should be prohibited on bridges or in tunnels simply because bridges are bridges or tunnels are tunnels, it will be found more often than not that these structures are narrower than the roadway in general and so parking may have to be prohibited.

7. *Pedestrian concentrations.* For safety reasons, it is desirable that parking be prohibited at places of heavy pedestrian concentration, e.g. at exits from schools or hospitals, etc.

8. *Priority locations.* Parking should never be permitted at kerb locations where priority must, of necessity, be given to public services. Thus, for example, parking should be forbidden at or adjacent to fire hydrants, bus stops, etc.

Time limit parking

In many town centres the need to impose parking management measures is indicated by the double parking of vehicles, by cruising vehicles awaiting the opportunity to park, or by low turnover and high occupancy of kerb space. In such instances, the measure which suggests itself is that of *rationing* the kerbside spaces so that parking preference is given to the people who are normally the life-blood of the town centre, i.e. the shoppers and visiting business people. This normally means the initiation of a time-limit parking scheme to control the lengths of time that vehicles may spend at given kerbside parking spaces.

There are three types of time-limit parking schemes now in general use:

a. Limited waiting schemes under police or warden control
b. Parking meter schemes
c. Parking disc schemes.

Before discussing these different approaches, however, it is useful to consider the basic features which should be built into any time-limit parking scheme.

Scheme design features. Initially, at any rate, the introduction of a time-limit parking scheme into a community is usually viewed with suspicion by both the motorist and the businessman, principally because of the regulating function of the scheme and inherent doubts as to 'how will it affect me'. To overcome these suspicions, and to justify its introduction to those who suspect its motive and impact, any parking management scheme should contain the following favourable features:

1. *Within the area outlined by the scheme it should result in a reduction in parking durations and consequent higher parking capacity*

A properly conducted 'before' study will indicate whether there is a real shortage of space for short-time parkers, or whether the problem is simply one of available existing space being occupied by long-time parkers. It will also suggest what time limits might be used in the scheme. An 'after' study should reveal whether the management measure is effective in driving away from the kerb the long-stay, journey-to-work parker; it should also tell whether the time limits are adequate, or whether they should be changed entirely or at certain locations only.

While, ideally—as is suggested above—the time limits prescribed for any given kerb location should be determined in the first instance on the basis of the results from a parking duration study, realistically the decision as to what should be the exact limit is very often an administrative one. The result is that, on the national scene, there is relatively little variability with regard to the time limits actually employed in town centres, e.g. the time limit which is undoubtedly most commonly used is 2 h maximum, with 4 h maximum on the edges of the central area and, in the 'core' part of large towns, a time limit of 40 min maximum is often used. (Not commonly used in this country, unfortunately, are 15 min zones near 'errand' type establishments such as banks, post offices, chemist shops, public-utility offices, commercial establishments that prepare telephone orders for immediate collection, etc.)

2. *Within the designated area a motorist should be able to find an empty parking space quickly within a reasonable walking distance of his destination*

The optimum utilization of parking spaces is generally considered to be about 85 per cent, leaving about 15 per cent available at any one time during peak periods.[15] The degree of utilization in any instance is, however, regulated by the allowable parking durations and, in the case of parking meters, by the charges imposed. As with the first design feature noted above, the adequacy of particular charges and durations can only be determined from before-and-after studies. (An objective method of evaluating this adequacy is discussed in the literature.[16])

The distances from the time-limit zone to principal traffic generators, e.g. large department stores, hospitals, etc., should obviously be taken into account whenever a time limit is being established. In locations where there are no distinctive parking generators, i.e. in a 'well-balanced' central area, the distances that particular types of parkers are willing to walk from particular time limit zones will need to be taken into account. Unfortunately, no definitive data are available in the technical literature as to what are acceptable walking distances from particular time limit zones—although, of course, measurements have been made in particular towns in order to determine actual walking distances in those towns (e.g. see reference 17).

3. *The kerbside parking spaces should be arranged so as to make the most efficient use of the road surface with the minimum inconvenience to moving traffic*

In most British towns, this criterion usually means the utilization of parallel parking at kerbs. It does not necessarily mean that the parking spaces are always marked on the carriageway; however, marking with white paint should always be carried out when high turnover is experienced and/or parking meters are used, and/or parking is not parallel to the kerb and/or where non-marking might result in the inefficient use of space.

If parallel parking is utilized (say, in a meter scheme), and if 6·1 m is taken as the space necessary to manoeuvre a car into a parking position, then two possible marking arrangements are shown in Fig. 4.2. Of these, the arrangement composed of pairs of 4·87 m bays with a 1·22 m manoeuvring space between each pair is generally to be preferred, i.e. it reduces the kerb length required for two vehicles by 10 per cent, still allows each vehicle 6·1 m of manoeuvring space, and yet permits adequate parking space for the rare 'long' car.

Where the carriageway is very wide, consideration may be given to using angle parking, since this provides more spaces for the same length of kerb. (If angle parking is utilized, the bays must be marked out with continuous white lines, discontinuous lines, or T-marks.) Another advantage of angle parking is that parking or unparking vehicles cause relatively little delay to

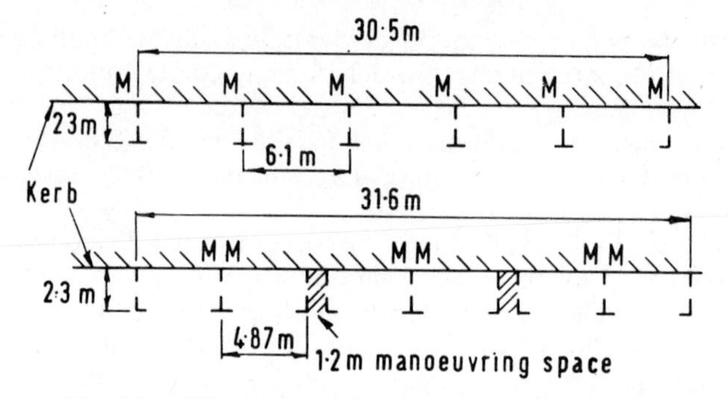

Fig. 4.2. Alternative parallel parking arrangements

moving traffic, e.g. it is reported[18] that the average driver takes 12 sec to back out of an angle bay and proceed ahead in the traffic stream, (driving *into* the angle bay results in no lost motion and the minimal interference to traffic), whereas for parallel parking the average driver requires 32 sec to back into a bay and clear the adjacent traffic lane; this indicates that the vehicle engaged in the parallel parking manoeuvre causes the greater hindrance to passing traffic. Another consideration favouring the angle stall is that it results in less delay and hazard to the traffic stream, i.e. an unparking vehicle can choose to enter the traffic stream when the flow is light, whereas the entering parallel-parking vehicle must carry out the manoeuvre immediately it arrives at the parking space.

Notwithstanding the above, considerable care should be exercised when deciding when/when not to use angle-parking. In particular, the accident potential at the location should be queried as cars are particularly liable to accident involvement when backing out of an angled space.[19] In general, therefore, angled parking at the kerb should not be permitted unless the street is exceptionally wide and traffic is light.

4. *Within the designated area the motorist should have no doubt whatsoever as to where, when, and for how long a vehicle may be parked. Within the controlled area, the time limit parking should be properly enforced*

The first of these criteria can be ensured if the controlled area is clearly defined by distinctive road and kerb markings combined with proper signing. The most efficient way of enforcing the parking restrictions in controlled zones is to use patrolling traffic wardens—this applies irrespective of whether parking is free or for payment.

Limited waiting schemes. These are schemes where the authorized time-limits are displayed on signs which also show the extent of the permitted (free) parking. Enforcement is carried out by patrolling police or traffic wardens.

This form of time-limit parking can be relatively effective at locations where the total number of parking spaces available (on- and off-street) is known to be sufficient to meet the parking demand, so that the parker has little difficulty in leaving his car close to his destination for roughly the time desired—which, under the conditions specified, is likely to be less than the posted time limit. In other words, there is little incentive to exceed the specified time limit, and so no great effort is required in relation to enforcement.

Unfortunately, the conditions described above are generally found only in the central areas of quite small towns. In larger towns, this type of scheme requires a very stringent enforcement effort, and will likely fail.

Parking meters. The parking meter was primarily designed to assist in the enforcement of parking regulations, and to increase parking turnover. The meter is so effective at meeting these objectives that it is now governmental policy to encourage the use of parking meters at town centre locations where time limits need to be strictly enforced.

There are two main types of parking meter: the 'manual' meter and the 'automatic' meter. The manual meter, which was the first type developed, is operated by the motorist inserting an appropriate coinage and then turning a knob or handle to activate the timing mechanism within the meter. The second type (which is the only meter used in Great Britain) is automatic in that the insertion of the coinage starts a pre-wound clockwork timing mechanism.

Both types of meter can be set to allow parking for a single fixed period of time for a fixed charge or, dependent upon the coinage inserted, for varying lengths of time up to a predetermined maximum. In either case, as soon as the motorist has used up the allotted parking time, a signal is displayed on the meter which is easily seen by a patrolling warden or policeman. With certain of these meters, a yellow flag bearing the words 'Excess Period' appears in the dial when the time bought has expired. This means that, until a red flag is displayed, the parker owes the local authority a specified but much higher fee for the parking space. The excess-charge meter has the advantages that (a) if the charge is set sufficiently high, the motorist will be disinclined to use the time, thus making the kerbside space available for other short-time parkers, and (b) it has the practical effect of an 'instant' fine, thus reducing the work of the Courts.

Mention might also be made here of another type of meter which has a built-in facility whereby the timeclock always starts from zero when additional money is inserted. This meter, which is in use in some cities on the Continent, ensures that every parker pays for his own parking time.

Parking discs. The most unpopular feature of the parking meter is that it levies a charge on the motorist. Thus, although it is government policy to use meters in time-limit schemes in Great Britain, a number of towns have opted for fee-free disc control instead as a means of tackling their parking problem. With the disc system a driver parking at the kerb displays on the vehicle's windscreen a cardboard disc[20] with two apertures: one aperture shows the approximate time of arrival, the other, automatically, the precise time by which the vehicle must leave.

Meters vs discs. On the basis of experience gained in this country and abroad, the following are the principal *advantages of a disc scheme over a meter scheme:*

1. *Within a disc scheme, parking is generally free to motorists*

The discs are usually obtained free of charge at police stations, garages, newspaper and tobacconists' shops, department stores, etc.

2. *There are no expensive meters to install and maintain*
3. *Parking spaces in disc schemes do not have to be delineated on the carriageway and, hence, more cars of various sizes can be parked along the kerb*

This third advantage is challenged, however, in one British study[21] which has shown that the average length of kerb per parking space under a disc scheme is 5·40 m—which does not show a great economy over the 5·48 m per car space required for a double-headed meter arrangement.

4. *Discs can be easily designed to allow variable waiting periods in the same parking space at different times of the day*
5. *There are no 'unsightly' parking meters along the footway to obstruct the view and/or pedestrian movement*

This fifth advantage may be considered particularly important in an historical town centre or in the vicinity of buildings of architectural merit.

6. *Vandalism is not encouraged*

A positive way of combating meter vandalism in vandal-prone locales, however, is to install meters with substantial malleable iron vaults to protect the coin-box.[22, 23]

The main *disadvantages of disc systems as compared with meter ones* are as follows:

1. *There is no surplus revenue available to help obtain needed off-street parking spaces, or to sustain complementary public transport facilities*

The surplus revenue data shown in Fig. 4.3 suggest that this can be a very beneficial incentive in some municipalities.

2. *The cost of administering a disc-controlled zone is borne by the ratepayers as a whole and not by the users, i.e. the motorists*
3. *The disc scheme offers greater opportunities for infringement*

It is generally considered that greater numbers of enforcement personnel are required by disc schemes as compared with metered ones. For example, patrolling wardens can experience difficulty in reading discs when inclement weather conditions prevail. Also, fraud is easier as the responsibility is placed upon the motorist to set and display the disc correctly.

4. *The disc-scheme is not as effective a traffic control measure as a meter scheme*

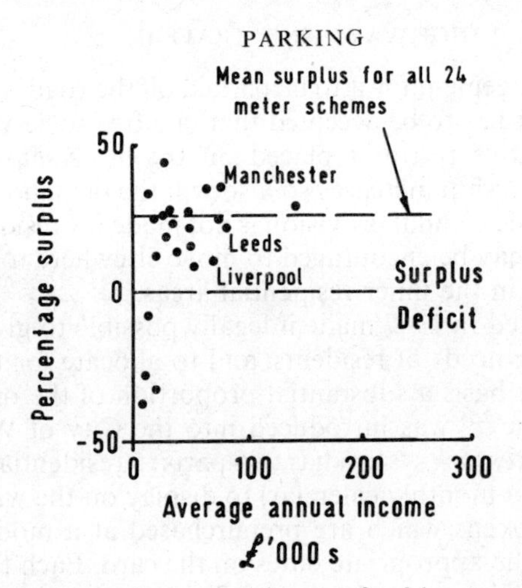

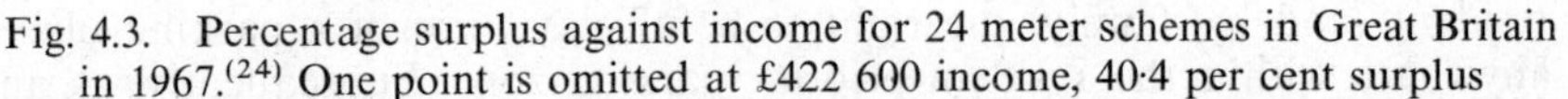

Fig. 4.3. Percentage surplus against income for 24 meter schemes in Great Britain in 1967.[24] One point is omitted at £422 600 income, 40·4 per cent surplus

The additional 'deterrent' of the parking charge is missing from the parking plan.

5. *The visiting motorist can have trouble obtaining a disc*
6. *Fee-free on-street disc parking does not encourage private enterprise to construct fee-charging off-street parking facilities*

The logic here is irrefutable. Payment for on-street parking promotes the use of off-street car parks, and the higher the on-street meter charge the more likely the financial success of the off-street facility.

Overall, it is probably true to say that the disc system is most applicable to smaller towns where the demand for parking spaces does not greatly exceed the number of spaces available, and especially where (as in a country market town) the bulk of the demand arises in a limited area of the town. In these smaller towns, there is little fear of off-street parking space development being retarded through the introduction of a disc scheme; in addition, the cost of supervising any such scheme will be relatively small as the scheme itself must, of necessity, be small. The primary function of the scheme can be then viewed as bringing order from disorder and, in this respect, a disc scheme can be every bit as effective as a meter-controlled one.

'Resident' parking

An oft forgotten part of a time-limit parking scheme is its effects on the on-street parking needs of householders living on residential streets, e.g. typically in and about the central area of a large town or adjacent to large traffic generators such as hospitals, universities, etc. It can, of course, be argued that 'garaging' a car on the public highway is anti-social in that it destroys the amenities and pleasant living conditions of a neighbourhood; therefore, that it should not be permitted, and that everyone owning a car should make an arrangement for it to be parked off the road when it is not in use. As against this, it has to be accepted that in a free society there is likely to be no administrative restraint placed on the *ownership* of a car and, therefore, as car ownership increases so also will the demand for residential on-street parking space—and if provision is not made for residents' cars then car-owning families may be encouraged to move elsewhere to the detriment of the social balance in the inner residential areas.

Legislation initiated in 1967 made it legally possible to give special consideration to parking needs of residents, and to allocate for their exclusive use on a prepayment basis a substantial proportion of the on-street space. The first such scheme[25] was introduced into the City of Westminster in January, 1968. This *Respark* system has two parts: a residential parking card (on which is printed a monthly calendar) to display on the windscreen, and 'stick-on' parking tokens which are prepurchased at a moderate cost for superimposing over the appropriate dates on the card. Each token gives the resident (and his family) the right to a full day's use of a parking place anywhere within the resident-priority parking zone. Subsequently, as an improvement on the token system, season tickets for residents parking have been made available as an alternative to the parking card.

The Respark system only applies to parking during the day, and no attempt has been made to exercise control at night.

Facilities for commercial vehicles

Commercial vehicle service is vital to the prosperity of a city's central area. Lorries and vans must be able to load/unload merchandise at or close to (say, within 30 m) of business premises. Buses must be able to pick up and drop passengers at convenient locations.

Freight facilities. All loading/unloading of store goods should preferably be handled at off-street facilities such as loading docks or alleyways. Kerb-loading should only be allowed when these facilities are lacking, and even then only to the minimum degree necessary during the normal working day.

Attempts to tackle this aspect of the parking problem have tended to incorporate the following features:

1. Adoption of regulations requiring new or substantially altered commercial buildings to provide adequate off-street parking facilities.
2. Provision of limited kerb-loading zones where justified, each street being studied separately for this purpose.
3. Prohibition of freight-vehicle parking on busy streets during peak periods.
4. Prohibition of freight-vehicle parking on critical streets during the working day.
5. Continuous review of existing kerb-loading zone needs and uses.

Time-limit parking for lorries and vans is usually difficult to enforce because of opposition by business interests. The easiest restriction to enforce, and the one that is usually of most value in reducing congestion, is the prohibition of parking during the peak traffic hours. If traffic conditions are so chaotic that prohibition must be enforced throughout the day, it may be necessary for deliveries and pick-ups to be made outside the usual commercial hours.

It is likely that future years will see the development of special parking areas for freight vehicles at locations near the edges of towns (*a*) where environmental damage will be minimized, (*b*) within easy reach of motorway, trunk, and principal roads, (*c*) near areas of industrial activity, (*d*) where there are adequate access roads, and (*e*) on level, well drained land. The aim will be to encourage heavy freight vehicles to break their loads at these lorry parks, the goods then to be carried into town for distribution by a larger number of smaller vehicles. The central government is now attempting to build up a national network of these lorry parks.[26, 27]

Bus facilities. Kerb-parking spaces must be provided for buses. However, if a bus cannot pull up at a lay-by inset into the kerb, it will block a lane intended for traffic as it picks up and discharges passengers. If buses do not stop frequently at desirable locations, then passengers will be encouraged to use their own motorcars, thereby increasing traffic congestion—and the demand for road and parking facilities. Undesirable though it may be from a traffic aspect, very many bus stops are located not far from intersections as this is most convenient for passengers. Each stop should have a clearly defined and marked zone and, where appropriate, a lay-by of sufficient length to contain all buses that can be reasonably expected to stop at a given time.

LOCATING TOWN CENTRE OFF-STREET PARKING FACILITIES

The correct siting of a central area car park is critical to its success in attracting motorists. *If the parking facility is to thrive, then it must be sited conveniently for the customers it is intended to serve.*

The above principle is the governing one influencing the location of a large parking facility. It holds true irrespective of whether the car park is offered as an inducement (i.e. if it is free or only a nominal fee is charged) in an area where adequate parking space already exists, or whether its function is to meet a demand at economic charges in an area deficient in parking supply. It cannot be over-emphasized that there is little point in taking a wedge of land, simply because it is lying there 'wasted', and putting a car park on it, if that piece of land happens to be in the wrong place.

Obviously there are other factors also which influence the decision as to where a public car park should be located, viz. the locations of the parking generators, the origins of the travellers, the adequacy of the access street(s), the nature of the topography, and the intended usage of the car park. Another important factor—albeit one which may not necessarily be taken into account on all occasions—is the effect of car park siting upon individual businesses.

Location and business

An indication of the importance of location in relation to an individual business may be gathered from the results of a study[28] carried out in Leeds (pop. 511 000). This showed that a department store (or complex of shops) which is situated on a main pedestrian route between a car park and the main shopping area of a town will be the first destination of a considerably greater proportion of shoppers from the parking facility than if the store (or complex) is at an equal or greater distance, but in an alternate direction from the car park. In other words, a shopping facility which is close to a car park, and is between it and the centre of attraction of a central area, is most likely to benefit from the car park. It is equally obvious that shopping facilities which are not in such favourable positions may lose a certain amount of trade.

Nevertheless, it is not always practical for the planner/designer of a public parking facility to attempt to take its effect upon individual businesses into account. Indeed, for him to do so could be considered wrong (if he is a governmental employee), since he must primarily think in terms of its effects in relation to the community as a whole rather than upon any one particular member.

Parking generators

The location of the parking generators (i.e. the destinations of the travellers after parking) is probably the major criterion influencing the siting of a car park. Cost factors also prevail here in that the part of the town where land is low cost is very often in run-down or slum areas at a distance from the main generators. The result is that, given a choice, parkers may not use the parking facility if they have to walk large distances to their destinations, particularly if their walk is through distressed areas.

No definitive statement can be made regarding how far people are willing to walk from an off-street car park. The controlling factors are, primarily, the extent to which there are other closer parking spaces, the size of the town, the degree of parking enforcement, the attractiveness of the destination(s), and the cost of parking in the facility in question as compared with alternate ones. As a general principle, however, it can be stated that *the larger the town the farther parkers are willing to walk, and worker parkers will accept greater walking distances than shopping parkers.*

Origins of travellers

In many urban areas it is usually found that great numbers of particular types of parkers, e.g. shoppers, workers, etc., come from certain areas of town. In such instances, the car park(s) intended for these users should be situated on the side of the central area which is toward their origins. Not only will these locations be more attractive to their potential patrons, but they will also have the added effect of causing less traffic congestion within the town centre, i.e. congestion caused by car trips through the central area and a walk back to the destination.

Access streets

A car park should always be considered as forming an integral part of the traffic plan for a town. This means, amongst other things, ensuring that the location is selected with a mind to the capability of the street system to safely and efficiently handle the entering and leaving vehicles, e.g. the sudden disgorging of a car park at peak times may well create chaos on either the immediate access street(s) or at important nearby junctions as a result of, for example, heavy movements.

The road access problem is one of the main reasons why it is often recommended (e.g. ref. 5) that off-street parking facilities should be limited to capacities no greater than, say, 400–500 cars, especially if they are to be used by commuters. Whilst some locations in large towns can, and do, justify larger car parks, it is generally held to be more desirable to have several car parks of lower capacity strategically located throughout the area, rather than a few car parks of high capacity. Alternatively, larger car parks may be used in certain areas if they are located so that their entrances and exits are dispersed on to lightly trafficked (preferably one-way) streets leading to high capacity routes.

Car parks should preferably be located at mid-block locations as compared with corner ones. No car park exit or entrance should be within at least 50 m of a street junction, due to the possibility of interference with or from the intersection control. Indeed, if the intersection is an important one, it may well be that the car park entrance or exit will be blocked by back-up traffic. Alternatively, if the car park is small, and the demand for parking is high, queueing cars may extend back onto the access road and into the junction.

Very many towns either have, or are developing, ring roads about their town centres, the aim being (*a*) to allow through traffic to by-pass the central area, and (*b*) to minimize the intrusion by vehicles which must enter the town centre. In such instances, radial route car parks may logically be placed on the left-hand inbound side—this minimizes hold-ups in the morning—connecting the ring road to the core of the town centre. Indeed, if the ring road and/or radial road is a high quality route, it may be possible to incorporate the car park in its design, e.g. under an elevated intersection structure.

Entrances and exits. The inter-action between the car park and its immediate access road has been examined in some detail[29] and it is now possible to estimate the extent to which, in particular, exit traffic may be compatible with traffic on the service road. This is best described by means of an example problem.

EXAMPLE. Assume that a 900-space multi-storey car park is to be located so that its exits discharge into a 7 m wide one-way street carrying a peak flow of 200 veh/h. The car park, which will stand on a rectangular site about 30 m wide by 120 m long, will be primarily used by commuters. The peak hour departure demand, during the evening, is estimated to be 800 car/h. An early decision was that exit gates of the lifting-barrier and pay-from-car type would be used. Determine the compatibility of the entrance and exit system with the service roads.

Step 1 Calculate the number of exit gates required

Exit gates of the type specified can handle about 200 veh/h. Thus, four gates will be required to deal with the evening flow. If four exits are provided, no reservoir space should be needed at the end of the internal aisle: apart, of course, from the space needed for the fan-out from the aisle or ramp to the four gates, assuming that all four gates are placed together. If fewer than four gates are provided, occasions could arise when a queue might extend back into the aisle, thus preventing vehicles from un-parking.

Let it be assumed that each exit lane is 3 m wide, and that a GIVE WAY sign will be provided at the junction with the main road.

Step 2 Determine whether the 'junction' between the access (main) street and each car park exit (minor) road is capable of handling the peak flows.

The query here is whether the maximum flow from the car park exit can enter the traffic stream on the main road without causing congestion. Assuming the junction to be a T-type, the problem resolves into determining the maximum flow (along the stem of the T) that can enter the main stream, assuming the heaviest flow conditions on the main road, i.e. when the commuters are on their way home. This determination can be made with the aid of the following formula:

$$q_{max} = \frac{Q\left(1 - \frac{Q}{S}\right)}{e^{(0{\cdot}0015Q - Q/S)}[1 - e^{-Q/S}]}$$

where q_{max} = maximum possible flow on the exit road, car/h
Q = flow ($\neq 0$) in the near-side lane of the main road, veh/h
S = saturation flow of near-side lane on main road, veh/h
and s = saturation flow ($\geqq 667$) on exit road, car/h.

The saturation flow, S, on the near-side part of the main road is obtained from

$$S = \frac{525 w_m}{N}$$

where w_m = carriageway width, metres
and N = number of traffic lanes.

Actual exit road saturation flow values measured along a 2·75 m wide exit into a 6·1 m road (kerb radius = 1·22 m; angle of entry = 90 deg; gradient = 0; estimated path radius = 6·1 m) are as follows:

Condition	*Saturation flow(s), car/h*	*Reducing factor, k*
1. Ideal, exiting cars turning left, no other traffic	1250	1·00
2. Limited visibility (required GIVE WAY sign), cars turning left, possibility of cars on main road	760	0·60
3. Blind exit, cars turning left, possibility of traffic on main road	710	0·55
4. Ideal (as 1 above) but with WAIT sign and stop-line	620	0·50

The saturation flow of a stream of vehicles following a curved path is less than for a straight section of similar width. If the exit lane from the car park is curved, an approximate representation of the saturation flow, *s*, can be obtained from:

$$s = \frac{1850k}{1 + \dfrac{100}{r_m^3}}$$

where r_m = estimated radius of curvature of path of cars, metres
and k = reducing factor (see the table above).

All of the above formulae are applicable to the situation where a single stream of left-turning cars enters a single lane of traffic on a two-way street, or right-turning cars enter a single lane of traffic on a one-way street. (Additional traffic in other lanes of the main road are ignored.)

For the problem considered here, the most critical situation occurs at the fourth exit, where the traffic flow for the main road will be 200+ 3(200) = 800 veh/h, and 200 veh/h on the exit lane. Thus

$$S = \frac{525(7)}{2}$$
$$\simeq 1840 \text{ veh/h.}$$

and

$$s = \frac{1850(0{\cdot}60)}{1 + \dfrac{100}{6{\cdot}1^3}} \simeq 770 \text{ car/h}$$

Therefore

$$q_{max} = \frac{\dfrac{800}{2}\left(1 - \dfrac{400}{1840}\right)}{e^{(0{\cdot}0015 \times 400 - 400/1840)}[1 - e^{-400/770}]}$$
$$= 545 \text{ car/h.}$$

This value of q_{max} considerably exceeds the gate capacity (200 car/h), so it appears (from a local congestion aspect) that there is no reason why the car park exits should not be located as intended.

Step 3 Check the size of the exit reservoir between the exit gate and the main road

Since there will be fairly heavy traffic on the main road, storage space will be needed on the exit road, after the exit barriers, where cars can wait until suitable gaps appear in the main road traffic stream. If sufficient space is not provided, the exit gate may not be able to achieve its maximum output of 200 car/h. However, the calculated value for q_m is so much greater than 200 car/h that experience would suggest that a reservoir of 2–3 car lengths will be adequate at each gate.

Step 4 Determine the number of entrance gates

In this instance let it be assumed that cars will enter the car park via entrance gates of the lifting-barrier take-ticket type which will be installed on slightly curved approaches (to suit the site). In a simulation carried out at the Transport and Road Research Laboratory, the maximum flows measured at different entrances were of the order:

Condition	*Maximum flow, car/h*
1. Tight left-hand turn and take ticket	350–450+
2. Straight-approach and take ticket	650–670
3. Tight left-hand turn only (no ticket taken)	575–970

In this instance, therefore, a capacity value of 450 car/h is appropriate, which suggests that at least two entrances are needed. (These entrances should be located such that they are upstream of the exits on a one-way street.)

It is appropriate to discuss here the problem of reservoir space *prior* to entrances. There are two distinct usages for these reservoirs: (*a*) they act as storage areas for cars arriving 'at random', and these then leave the reservoir to enter the car park via the in-gate at an approximately uniform rate of flow corresponding to the dynamic capacity of the parking system, and (*b*) they act as storage areas for any car waiting for a vacancy to arise in the full car park and which then enter at non-uniform rates as space is made available. The first instance indicates the need for increased *dynamic* capacity of the parking sytem, whereas the second indicates the need for increased *static* capacity.

There is, at this time, no way of calculating the reservoir space required in case (*b*) above. A statistical procedure has been devised[30] which can give useful answers with regard to case (*a*) when an accurate assessment of the rate of arrival of vehicles to the reservoir can be obtained. If certain assumptions are made regarding the distributions of arrival and departure times, then simple queueing theory can be utilized to estimate the length of queue which might build up, assuming that the dynamic capacity is known. For example, if it is assumed that the average rate of car arrival is equal to the rate at which parking takes place, then the following equation gives an indication of the 99 percentile queue length:

$$X = 2{\cdot}4(A)^{1/2}$$

where X = no. of reservoir spaces required to ensure that the reservoir area is overloaded less than 1 per cent of the time
and A = average arrival rate, car/h.

In the case of the problem described above, A may be taken as 450 car/h, so that the number of reservoir spaces required at a given entrance is 51. (The question then arises, of course, as to whether it is practicable to attempt to design to such a high (99 percentile) standard.)

Topography

The nature of the topography can also influence the location of a car park. For example, a site on a steep hill has the obvious disadvantage of requiring extra walking effort from its customers—and thus it may be shunned by, in particular, older and/or heavier people who will not like walking uphill, particularly from the shopping area when carrying packages, etc. On the other hand—and this particularly applies to towns whose residents are used to walking along grades—a sloping site may be very appropriate for a parking garage since it may allow direct access at different levels without the need for ramps.

Intended usage

A car park which is convenient to, for example, the shopping part of a central area may enjoy an excellent turnover during the daylight hours, but be practically empty in the evenings. It may be that, in certain circumstances, a little foresight could result in it being used on a day *and* evening basis. For example, if it is located so as to be convenient not only to the shopping area, but also to other diverse forms of traffic generators, e.g. cinemas, theatres, large hotels, restaurants, etc. then there is every likelihood that its usage—and its income—will be increased because of its appeal to evening visitors to the central area.

PERIPHERAL CAR PARKS

Peripheral parking is not at all a modern-day phenomenon. Its practical origins may be associated with the development of railway travel in the mid-late 19th century, when it was not uncommon for outlying-living travellers to leave their horse-drawn carriages at a suburban railway station and complete their journey into town by train; nowadays, of course, it is common practice for motorists to leave their cars at railway stations and complete their trips by train. What is often forgotten, however, is that there are hundreds of bus stops in the public transport network in any large town, and most of these are located in areas where there is ample opportunity for all day kerb-parking; hence, it is now regular practice for many drivers, of their own accord, to drive to convenient bus stops and take the regular bus service for the remainder of the trip into the central area.

The idea of *deliberately* attempting to persuade motorists to park their cars at prior-selected locations and then use another form of transport to travel to the central area (and back again) is, however, of much more recent origin. Its practical beginnings can be traced to the United States in the mid-1940's, and since then it has been continued in various cities with mixed success.

It is appropriate to comment at this time upon what is meant by a 'successful' peripheral parking operation, for this is something which historically has tended to distort its usefulness. From the point of view of the public transport operator a successful system is generally one which not only pays its way but, if possible, makes a financial profit. In contrast, the highway planner and engineer may well consider the operation successful if it takes significant numbers of vehicles off the roads leading into the town centre and helps reduce traffic congestion and parking problems—even though it may not be self-supporting in the process. *If there is one lesson to be learned from previous experiments in this area, it is that any authority which is considering a peripheral parking project should have a clear view of what it means by a 'successful' system before embarking on any experimentation.*

Nowadays it is possible to isolate a number of types of peripheral parking developments. These may be loosely differentiated as:
1. Park-and-walk. 2. Park-and-ride. 3. Kiss-and-ride.

Park-and-walk

This relates to the concept whereby the motorist completes the greater part of his journey to the city centre by car, parks it in a facility just outside the central area, and then walks (or takes perhaps a bus) to a destination which is a relatively short distance away.

Park-and-walk is primarily aimed at keeping the journey-to-work motor car out of the city centre. This concept, which may be applied to any size of town, is now being brought into use in very many British urban areas, and, properly implemented, is likely to prove very successful in all but the largest towns. Proper implementation in any given urban area requires (*a*) that the local authority be able to exercise strong control over parking both *within* the central area, and for a reasonable distance *outside* its periphery, and (*b*) the scheme be initiated as part of an overall transport plan for the urban area.

The key to successful operation is undoubtedly the first of the above two criteria. If the local authority does not have a firm grip on parking within the central area, then motorists will not use the designated outlying facilities. If, however, firm control is exercised, say through a pricing mechanism which makes it completely uneconomic for the commuter to leave his car within the town centre, then this type of parker will be forced outside its periphery. If parking control is exercised over the roads immediately outside the periphery, then the commuters will have no option but to use the peripheral parks—in effect, they become captive users of the car parks.

There is no technical reason why the park-and-walk motorist should not be charged for using peripheral car parks—provided that the fee imposed is significantly less than that charged within the town centre. It is generally found that motorists reasonably willingly accept such a charge. The income derived in this way may be used to pay for the operation of the car parks (which will be usually owned by the municipality).

Park-and-ride

This refers to the concept whereby motorists drive their cars to long-stay car parks which are located well away from the town centre, usually in suburbia, and then travel by public transport to their destinations.

Park-and-ride differs from park-and-walk not only in the fact that the car park is much farther away from the town centre, but also in that its success is much more dependent on the *voluntary* co-operation of the motoring public. While generally also aimed at the commuter, park-and-ride schemes can also (in very big cities) appeal to long-stay shoppers or business visitors.

In theory, park-and-ride schemes are excellent traffic planning measures. In practice, the degree of excellence achieved is dependent upon how they are used, and on the extent to which the commitments implicit in their usage are understood. Proper usage generally infers that a considerable advantage is to be gained by the user of the scheme, e.g. in the form of substantial savings in cost and/or travel time. Proper understanding of the commitments involved means, for example, realising that a 'successful' park-and-ride scheme may not necessarily be a financially viable one, and thus its operator(s) may well have to be subsidized so as to ensure its continued service.

Criteria for success. The successful operation of a park-and-ride scheme is dependent upon many, if not all, of the following features:

1. *The central area being serviced should normally be surrounded by an urbanized area of large population*

From a design aspect, the car park should be as far as possible from the town centre so as to remove the cars from the road; however, if the town is not a large one, then the interchange located well away from the central area may not be able to generate sufficient traffic to justify itself. For obvious reasons, therefore, the likelihood of a park-and-ride system being successful increases the greater the population of the catchment area of the town.

One extensive investigation of park-and-ride practices in the United States[31] found that of 19 cities that had established park-and-ride (bus) schemes, nine had abandoned them. Only two of the cities with abandoned schemes had populations above 500 000; of the cities retaining the park-and-ride schemes, only one had a population of under 500 000.

Another, more recent, compilation of results by a prominent firm of international consulting engineers[32] suggests that park-and-ride schemes are most beneficial in urban regions which exceed 2 m population. Its recommendations regarding the extent to which these schemes can usefully substitute for central area parking are given in Table 4.6.

TABLE 4.6. *Estimated outlying parking requirements along motorways or rapid transport routes into urban areas in the United States.*

Population of urbanized area × 10⁶	*Range of outlying parking spaces as % of central area spaces*
0·5	10–20
1	15–25
2	20–30
5	25–35

2. *The relationship between the peripheral car park and the town centre must be such that the total travel time by car and public transport is not significantly more than that by car alone*

It must always be remembered that the travellers' decision to park-and-ride is primarily determined by the 'trade-off' relative to the inconvenience and possible lost time associated with fringe parking, as compared with the high parking costs and congested traffic conditions experienced in and on the way to the central area. From a travel time aspect, the trip will always be slower if the motorist transfers to a bus, even if it be an express bus—unless (as rarely happens) the bus travels on a separate lane directly to the town centre. If, as is common practice, the buses become part of the general traffic stream then, *at best* (i.e. if the buses are non-stop) only the time required to park and transfer is lost. Furthermore, the car park should be easily accessible by car and located within, say, 100 metres of, and have direct access to, a direct route into the central area so that the total door-to-door travel time is minimized.

It is appropriate also to comment here upon the desirable frequency of service of the public transport undertaking, as this is also a major factor influencing the decision to park-and-ride, i.e. people are willing to 'waste' time on a bus but not at a bus stop. (One widely used rule-of-thumb is that the effect of 1 min waiting time at a bus stop is roughly equivalent—in terms of influence on public transport usage—to 3 min on a bus.) Good service frequency may mean, for example, that the park-and-ride scheme may require bus headways of not more than about 5 min during peak demand periods. If the service is discontinued during off-peak hours, early returning users will either have to utilize the local bus services or else a special bus service will have to be laid on, say, every hour. Rail park-and-ride services probably can be operated at lower frequencies because of their generally faster travel times and greater arrival time reliability, as well as convenience.

3. *The parking fee plus the two-way public transport fee must be appreciably less than the cost of parking in and about the town centre*

For obvious reasons, the cost of the round trip discernable to the park-and-rider must be significantly less than the cost of parking in or adjacent to the town centre. American experience(33) suggests that free parking is a necessity for the successful operation of a park-and-ride bus scheme, as otherwise motorists will (*a*) park in the surrounding streets and walk into the car park, or (*b*) bypass the special car park entirely and drive to a regular bus stop further in town where not only is free kerbside parking permitted, but a lower fare is charged on the bus. If, however, the park-and-ride operation involves a public transport vehicle on a reserved right-of-way (rail or bus), then there is a greater likelihood of a small parking fee being acceptable.

4. *The car park must not be close to any large local generator of parking demand*

The danger here is that the spaces designated for the park-and-ride patrons may be utilized by the local users (if there is a local shortage of parking) or vice versa (if the demand for park-and-ride is in excess of that provided). Nevertheless, experience has shown that outlying car parks used in bus schemes are best located on known paved land which is already used for public parking purposes. Ideal in this respect are parking areas at stadiums, auditoriums, shopping centres, churches, etc.; not only are the peak activities at these locations likely to occur during evenings or weekends, but they are also generally well serviced by good access roads.

5. *The car park must be properly operated. Continuing and ample publicity must also be given to the scheme.*

Proper operation implies that the parking facility be a safe place in which to leave the car all day—and in certain neighbourhoods this may well require the appointment of a car park supervisor. Good operation also requires an assurance to the motorist of a place to park, since excessive space hunting will discourage potential users. Substantial cover should also be provided at the car park so that patrons are protected from inclement weather while waiting for, and boarding, the public transport vehicles.

The formula for the success of a park-and-ride service requires one further major ingredient. It is that the system be given continuing and ample publicity so that the commuter, in particular, is made fully aware of the advantages associated with the service.

Kiss-and-ride

Kiss-and-ride (which refers to the practice whereby, for example, a wife drives her husband to/from a bus stop) is the most oft-forgotten part of peripheral parking programmes—even though it can be a major feature of a properly designed *park-and-ride* scheme. It poses its most severe problems where an express bus (or rail) service is the mode of transport, and buses (or trains) operate with very short headways. The dropping of passengers in the morning need not be a serious problem; this arises in the evening when waiting cars parked in the surrounding streets or on the internal roadways of the car park become a cause of serious congestion.

Experience has shown that it is practically impossible to regulate this form of congestion by legal means only. (What policeman or traffic warden with a family of his/her own is going to risk arresting or giving tickets to bevies of harassed housewives at the wheels of cars full of starry-eyed or rampaging children (as the case may be) while they wait for the bread-winner to return from work?) A much better solution is to provide adequate numbers of short-term parking spaces for these vehicles within the park-and-ride car park.

How many spaces?

Inevitably, the time comes when the decision has to be taken as to how many places are provided at any given peripheral car park.

The total number of *park-and-walk* places provided will be to a large extent determined by the policy adopted by the city council with regard to parking within the central area—and this, in turn, will be primarily affected by the size and composition of the urban area, the availability of existing parking, and the capacity of the road system servicing the central area. The number of spaces provided at any given car park will be then dependent on the total number of parking facilities, the demand created close to the facility, and the interaction between the car park and the traffic on the road(s) immediately servicing it.

There is no simple way of determining the number of spaces at any given outlying *park-and-ride* terminal. Obviously this depends on the success of the project, and on the nature and size of the area serviced by the facility. However, an indication of what might be required can be gained by looking at the needs created by a particular service interval.

Let it be assumed, for example, that a commuter park-and-ride service is to be serviced by double-decker express buses (each capable of carrying 76 passengers) operating at a service interval of 5 min during the peak hour. It is anticipated that the demand for this service will be such that every seat will be occupied when each bus leaves the car park. Then the maximum number of spaces needed is

$$76 \times \frac{60}{5} \times \frac{1}{1{\cdot}15} = 793$$

assuming that the average car occupancy is (typically) 1·15. If, as is likely, significant percentages of the patrons are kiss-and-ride passengers (say 5–30 per cent, based on American experience,[34] or walk-in passengers (say 5–15 per cent), then, of course, the number of long-term spaces would be considerably reduced.

To cater for the *kiss-and-ride* parkers at car parks, it will usually be necessary to provide special separate short-term spaces where cars can drop-off passengers or await their arrival. The number of pick-up spaces provided at the kerb will normally be much greater than the number of drop-off spaces, simply because passengers can be disembarked and the car driven away in approximately 1 min, whereas an awaiting vehicle may stay 5–15 min before the arrival of the public transport vehicle.

PREPARING THE TOWN CENTRE PARKING PLAN

It is now becoming increasingly clear that, for important social and economic reasons, most medium- and large-sized towns are simply not capable of providing for all the central area demands of the motor car within the foreseeable future. Thus, in order to fulfil their obligations in accordance with the transport and general environmental needs of the community as a whole, most British town councils—because of the immediacy of the traffic congestion problem—feel constrained to turn toward the implementation of stringent traffic control measures when devising their central area parking plans.

The term 'traffic control' as used here refers to the decision by the governing authority to restrict traffic demands to artificially low levels by deliberately imposing stringent parking restrictions in and about town centres—restrictions which result in a limiting of the total number of parking spaces provided, a limiting of the time during which a vehicle may stay in a regulated parking space, and the imposition of a relatively high charge for the use of a parking space. A primary reason why this method of traffic control is becoming the adopted policy in so many towns is that there is no practical alternative to it at this time: e.g. it can be put into operation within the existing legal and administrative framework; although it has a certain unpopularity, it is accepted by the public; and, most important, if properly implemented it has a good chance of succeeding in its objective of limiting the number of movements made by private car to levels which are compatible with desired community objectives.[35]

How many spaces in the central area?

Any 'true' determination of the potential number of parking spaces required in a town centre—or (perhaps more properly) the number which should be incorporated in its parking plan—presents a most difficult task, even for a city which has the results of a full-scale transportation study to call upon. It is practically an impossible one for a town which has few basic data at hand.

The problem is dominated by the difficulty of calculating the exact effects of the factors which can be expected to influence movement within the town. Some of these factors are as follows: 1. The future population of the catchment area (including people domiciled outside the town boundaries). 2. The car ownership level in the design year. 3. The number and proportions of person-trips generated by the central area (including work, business, shopping, educational, etc. trips). 4. The proportions of the daily travel which will take place during the normal daily travel hours and the peak travel times. 5. The capacity of the road system feeding the town centre. 6. The availability and quality of the public transport systems. 7. The relationship between the peak accumulation of parkers and the total number of parkers. 8. The parking duration times of the various categories of parkers. 9. Efficiency of parking space usage. 10. The time of year considered for planning purposes. 11. Anticipated cost of parking. 12. Increases in the floor area of central area uses under development. 13. Changes in attraction of the area after redevelopment and in the number of potential customers as a result of new industry and/or overspill schemes in the catchment area. 14. The anticipated parking policy, particularly in relation to the modal split and the desired environmental quality.

Examination of the items tabulated above emphasizes why it is that a comprehensive transport study is normally needed if a realistic appraisal of the future parking needs of any decent-sized central area is to be obtained (and why it is that the planning of parking facilities cannot be isolated from the planning of the overall transport system for a city). This list further emphasizes that the problem is not simply one of determining the number of parking spaces. What is also required is the correct proportion of on- and off-street parking, of space and time allocated to the short and long-term parker, and of charges levied for long and short-term parking.

Nevertheless, as a first step in the preparation of a town centre parking plan, it is useful to obtain a rough estimate of the number of spaces which could be required. Of the methods that have been used to provide estimates of central area parking demand, the following four are probably the more important. It should be appreciated that the methods, all of which are based on the concept of a high degree of motorization, can normally be expected to give quite different answers—so that a choice must then be made as to which is the most applicable to the particular circumstances under consideration.

Method 1. This procedure, an American one, postulates that the 'parking space coefficient' which should be applied to the proportion of the total town centre person-trips made by car, in order to obtain the number of spaces required, is related to the population of the catchment area. The method utilizes the curves shown in Figure 4.4 which are based on the following general formula[32]:

$$P = \frac{drsc}{oe} = \frac{(0{\cdot}70)\, rsc}{(1{\cdot}5)(0{\cdot}85)} = 0{\cdot}55\, rsc$$

where P = parking space coefficient
d = proportion of daily travel involving the central area which takes place between 7 a.m. and 7 p.m. = 0·70
o = car occupancy = 1·5 person/veh
e = efficiency of space usage = 0·85

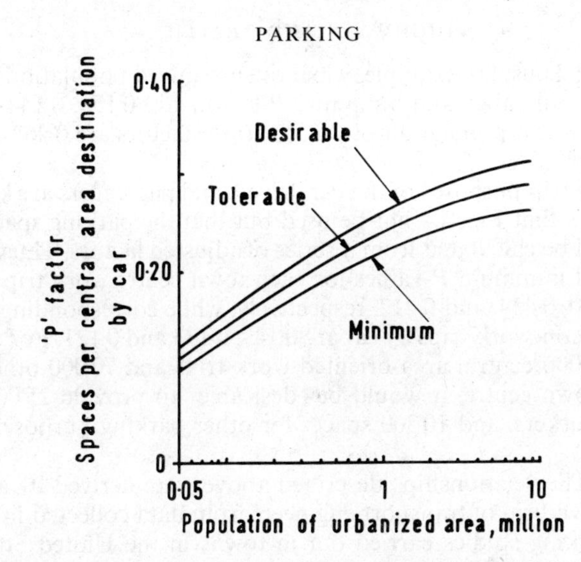

Fig. 4.4. Determining the parking space demand coefficient

r = ratio of peak to total daytime parkers (usually taken as 0·25 in small towns and over 0·40 in large ones)

s = seasonal peaking factor

and c = locational adjustment factor to reflect the concentration of demand in the core part of the central area.

The 'desirable' curve in Figure 4.4 assumes that s and c are each equal to 1.1; the 'tolerable' curve assumes $s = 1{\cdot}0$ and $c = 1{\cdot}1$; the 'minimum' curve assumes both s and $c = 1{\cdot}0$. The manner in which the figure is used is as follows:

Step 1. Estimate the number of central area person-trip destinations per day.

Step 2. Estimate the percentage of central area person-trips which will be made by car.

Step 3. Calculate the daily number of person-trip destinations by car. (Multiply 1 by 2.)

Step 4. For the urban population in question, read the appropriate P-value from the appropriate curve in Figure 4.4.

Step 5. Calculate the number of central area parking spaces required. (Multiply 3 by 4.)

If the 'desirable' curve is used in the above calculation, it means that both the particular needs of the core part of the central area and the seasonal fluctuations will be taken into account. Use of the 'tolerable' curve results in the provision of sufficient space for the typical weekday, while giving some consideration to the concentration of demand in the core area, i.e. it recognizes that certain spaces are beyond acceptable walking distance. Use of the minimum demand level results in the concentration factor not being taken into account. Thus, for example, when the urbanized population is 100 000, the desirable, tolerable, and minimum P-factors are 0·158, 0·144 and 0·131, respectively. For a population of 1 m, the three factors are 0·262, 0·238, and 0·217.

When the trip purposes of the car drivers and passengers are known, it is recommended that Fig. 4.4 not be used but that the parking space requirements should be calculated from a series of adjusted factors. Thus, desirable, tolerable and minimum P-values for each town centre work trip by car are given as 0·500, 0·454, and 0·412, respectively, while corresponding values for central area non-work trips by car are 0·147, 0·133 and 0·121. For example, if there are 50 000 central area-oriented work trips and 70 000 other trips by car to the town centre, it would be 'desirable' to provide 25 000 parking spaces for workers, and 10 300 spaces for other parking purposes.

Comment. The relationships described above were derived by an internationally known firm of transport engineers from data collected in the course of many parking studies carried out in towns in the United States. Their direct application to conditions in Great Britain cannot be recommended at this time, if only because they are based upon a much higher degree of central area motorization than, trends would suggest, will be tolerated in this country. Nevertheless, it illustrates how parking data collected from many studies can be used to establish patterns which can usefully be utilized by other towns which have not been able to carry out such studies.

Method 2. This method assumes that a relationship exists between the total number of vehicles registered in the city and the number who wish to park within the central area during the peak parking demand period. The number of parking spaces needed is then determined by estimating the number of vehicle registrations for the design year and multiplying it by the appropriate parking proportion value.

It is well known that as urbanized areas increase in size, less and less use per head is made of cars in order to travel into town centres. This is illustrated by the data in Table 4.7 which is taken from an analysis[(2)] of parking data collected in the United States.[(36)] This clearly suggests a strong tendency for the number of vehicles parking in the central areas of the smaller American towns to be approximately constant at 17 per cent; for towns of over 500 000 the parking proportion drops to about 10 per cent; when the population is over 1 m, the per cent parking drops to 6.

Comment. Because of its apparent simplicity, this is a most attractive way of estimating town centre parking needs. Nevertheless, usage of this technique at this time in this country can be criticized on many grounds of which perhaps the most important is that—as mentioned with respect to Method 1—the percentage figures quoted in Table 4.7 are based on a higher degree of motorization than would appear likely to be tolerated in Great Britain within the immediate future.

It is also appropriate to note that there would appear to be no reason

TABLE 4.7. *Percentage of vehicles parked in the central areas of American towns in relation to the number of vehicles registered in those towns*

Year	No. of veh per 1000 population	Population range, m	No. of veh	Max. no. of veh parked in the central area	
				Total	Per cent
1950	380	0·005–0·01	3000	490	16·3
1950	380	0·01 –0·025	6800	1180	17·1
1950	330	0·025–0·05	11 900	1 950	16·5
1950	320	0·05 –0·1	25 600	4 450	17·6
1950	320	0·1 –0·25	52 000	5 700	10·7
1948	260	0·25 –0·5	95 000	9 140	9·6
1947	240	0·5 –1	132 000	12 000	9·6
1954	300	>1	390 000	23 400	6·0

why there should be a firm relationship between the demand for parking spaces in city centres and other values such as the number of car registrations, the number of people working in the central area, the size of the area, etc. Town centres, particularly those in Europe, differ so much with regard to type and use of buildings, type of activity, employment, density, peculiarities and size of catchment area, quality of the public transport system, etc. that it is difficult to see why they should be unequivocally described by means of simple formulas. However, this may not necessarily be the case in America where more towns have grown up with the motor car and, hence, tend to be better adapted (by basic layout) to the needs of the motor age as compared with the older, more historical, cities of Europe.

Method 3. With this third procedure, the number of parking spaces is calculated on what might be termed a 'floor area' basis. In other words, the number of parking spaces is determined by summing estimates of the requirements of the individual parking generators within the central area.

Some parking standards specified at the present time by certain authorities in Great Britain[37] and in certain cities in Europe[38] in relation to the provision of planning permission for buildings in town centres are given in Table 4.8.

Comment. A major disadvantage associated with this procedure is that there is still widespread disagreement over what constitutes a reasonable relationship between parking space requirements and generator type. Moreover, the values generally used to indicate 'requirements' are, in fact, specified standards and these are not necessarily the same as the needs.

Another disadvantage of this method is that, in the end, some arbitrarily-chosen percentage factor (say 60 per cent) must be applied to the calculated total value in order to obtain a reasonable estimate of the required parking accommodation during the maximum demand period. The reason for this is that the maximum parking demands of different types of establishments normally will not coincide during the day.

TABLE 4.8. *Parking space standards set for selected town centre establishments*

Recommending Authority	Type of establishment: Residences	Offices	Shops	Stores
	1 parking space per:			
French Min. of Equipment	0·67–1 dwelling	40 m²	30 m²	—
Paris	1	50	50	—
Copenhagen	1–1·5	—	—	50–100 m²
London	1	186	232	—
Hamburg	1–2	100	50	—
Britain (range)	—	32·5–232	37–232	—

Method 4. This procedure relies upon the results of an assessment of the traffic capacities of the arterial roads leading into the central area during the peak inbound traffic period. The number of parking spaces is determined by subtracting the amount of through traffic from this capacity figure and (after allowing a suitable factor for space utilization) then adding an appropriate number of extra parking spaces to allow for vehicles entering the town centre after the peak period.

A variation[38] on this method simply suggests that the number of spaces (P) required in the central area is given by

$$P = 2CK/100$$

where C = capacity of the streets leading into the central area

and K = percentage of feeder road capacity which is not through traffic.

Comment. The basic principle described here is that which is now being implemented in very many European towns and particularly in Great Britain. A 'criticism' which could be levelled at this method is that, in this age of high motorisation, it generally results in parking 'discrimination' against journey-to-work motorists.

Developing the Plan—The Map Approach

When all the policy decisions and the estimates of needs have been made, the parking plan for a town centre must be devised. It is then that all the information previously gathered must be put together in a manner which will attempt to balance the mobility needs of the community with its social and economic capabilities.

There are many ways in which this can be tackled. Perhaps the most practical—certainly the most detailed—is that which has been devised by the Department of the Environment in Great Britain as a guide for local authorities. '*Parking in Town Centres*'[37] describes how a series of maps can be compiled as both a convenient and effective way of illustrating the interacting factors which, in most towns, have a bearing on parking problems. A major beneficial quality of this method of preparing the parking plan is that it enables the measure of the parking problem as a whole to be taken very quickly, and with the minimum use of resources. As such its usage is most applicable to towns which have not carried out a transportation survey, and, hence, do not have detailed data at their disposal.

The 'map' approach to the problem may be summarized in the following steps:

Step 1. Obtain an appraisal of the effects of parking on the town centre, on the convenience, pleasantness and efficiency with which the central and peripheral areas are used and function, together with the effect on accessibility and environment.

Step 2. Analyse the causes and pressures behind present parking usage and future parking demand.

Step 3. Decide how to make optimum use of the facilities available now and within the future period for which the plan is being devised.

In the following discussion on these steps, use is made of illustrative data from '*Parking in Town Centres*'. While the maps used in this discussion are shown here separately, in practice they would be prepared on transparencies so that they might be compared as overlays.

Map 1 is prepared in order to show the principal land use generators of parking demand, and the areas where competition for parking space is most intense. In Great Britain, the area used mainly by shopping parkers is best determined from a parking usage survey carried out on a Saturday afternoon, when offices are closed. The office parking areas are best determined in a similar way on an early closing day. Residential parking is best isolated by means of a usage survey carried out during the night. (These data will suggest where multiple use of public parking facilities might be appropriate, as well as the area where time-limit parking control is most needed.)

Maps 2 and 3 show the 'pinch-points' on the streets, i.e. where parking interferes with the movement of traffic and the operation of essential services. Data for Map 2 are generally available within the city's traffic section or, in the case of a small town, can be easily determined by inspection. Data for Map 3 are obtained in the course of the parking usage survey.

Map 4 is intended to isolate the areas where there is conflict between car parking and the environment.

Map 5 shows how the hinterland of the town centre can be subdivided into catchment areas, and the traffic from each related to the major approach road(s) carrying traffic to the central area. In this case, for example, the southern catchment area is seen to have the highest vehicle ownership potential; since the two approach roads within this zone converge at the outskirts of the central area, it suggests that the road capacity of the final leg should be examined to see if it is less than the potential parking demand from the catchment area.

Map 6 shows the numbers (if available) of commuters and shoppers

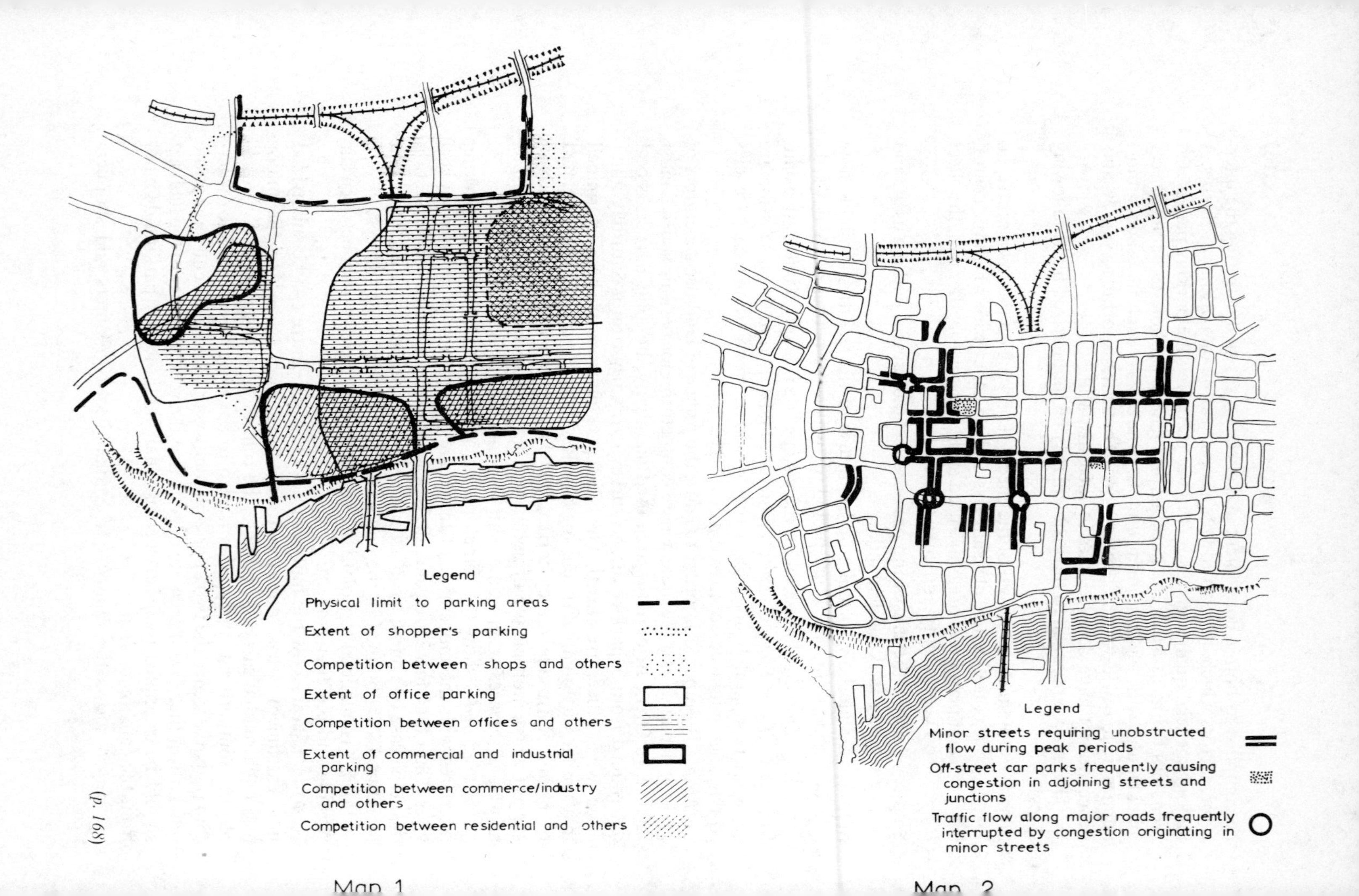

Map 1

Map 2

(p. 168)

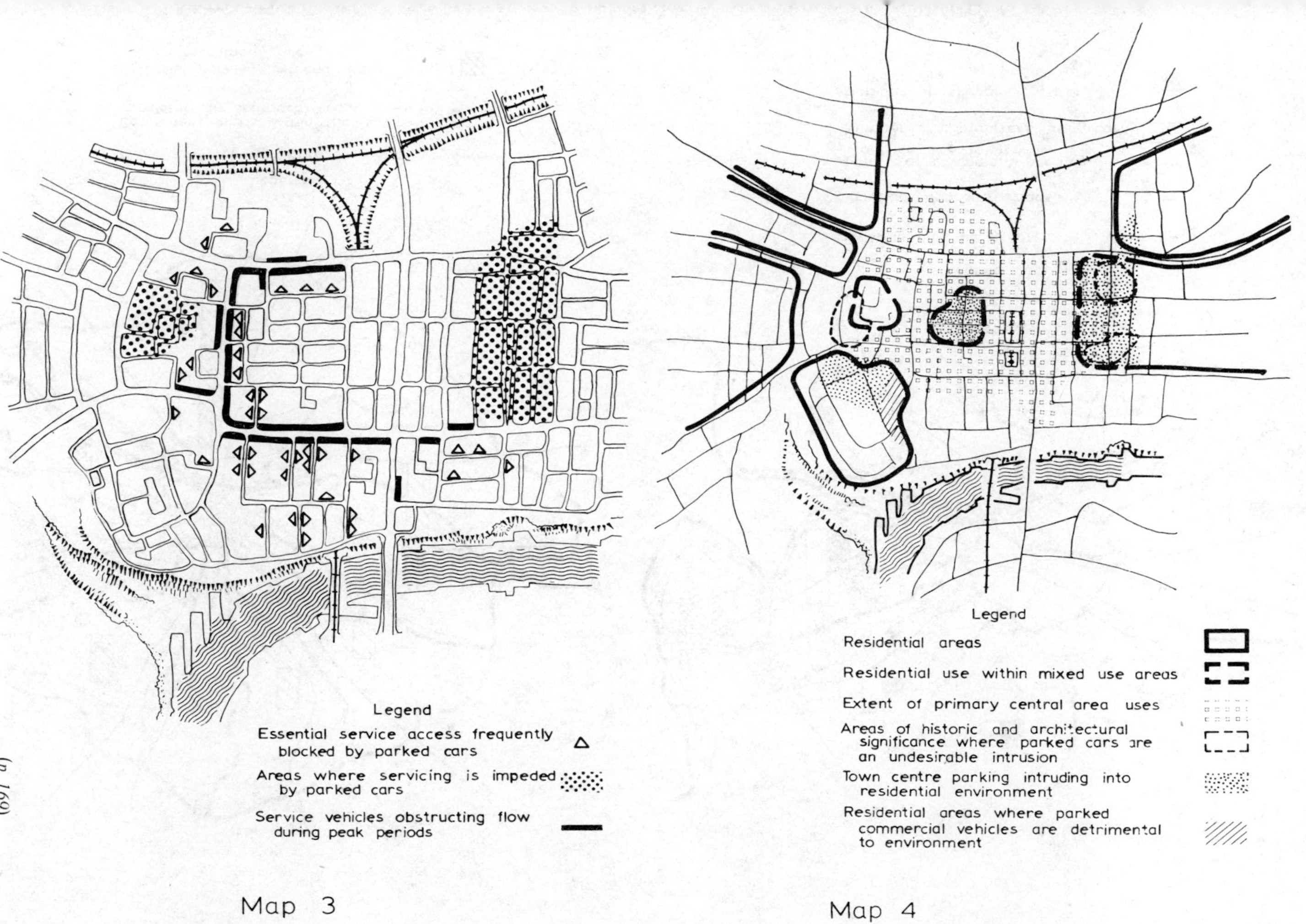

Map 3

Map 4

(p. 169)

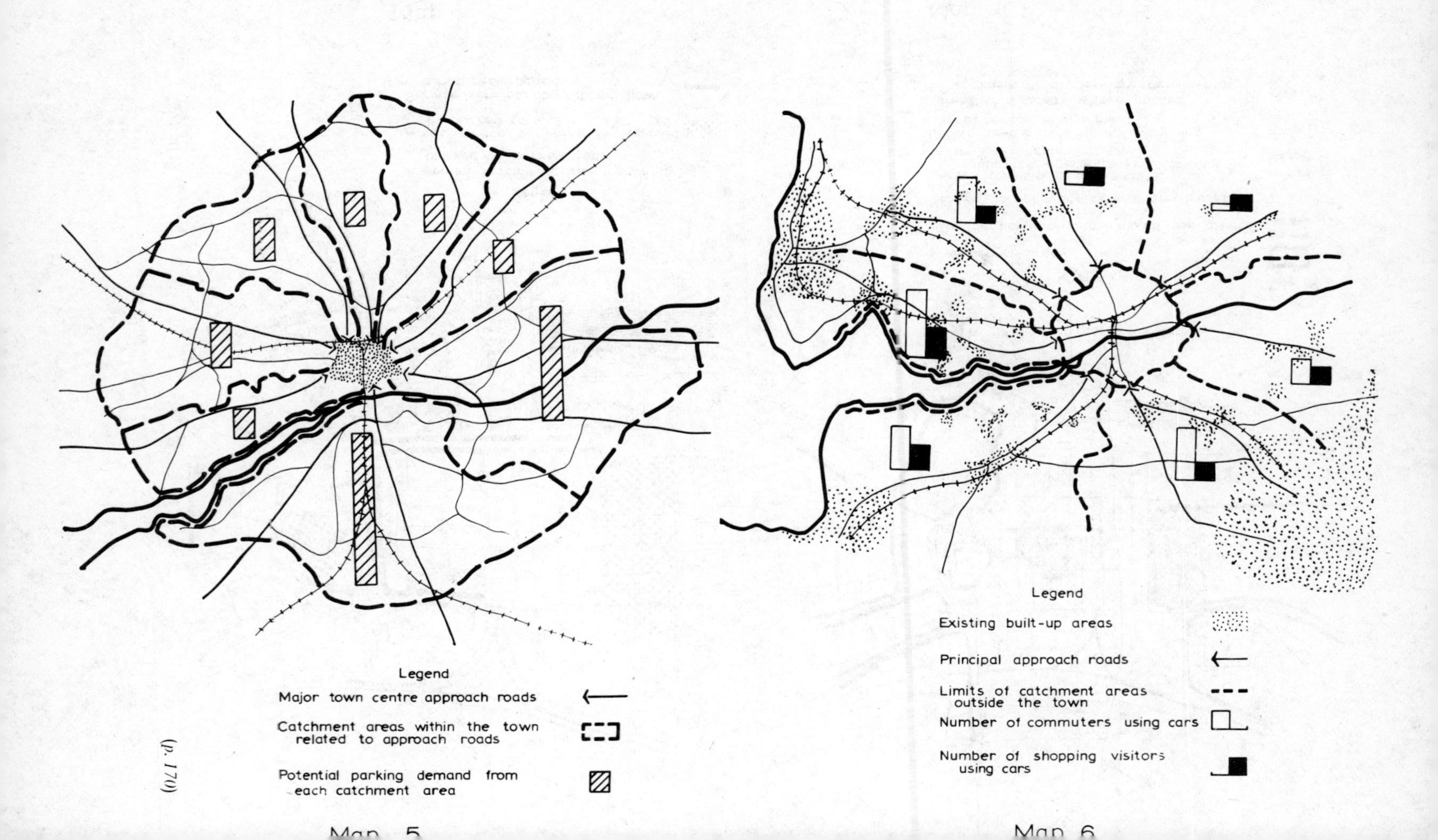

Map 5

Map 6

(p. 170)

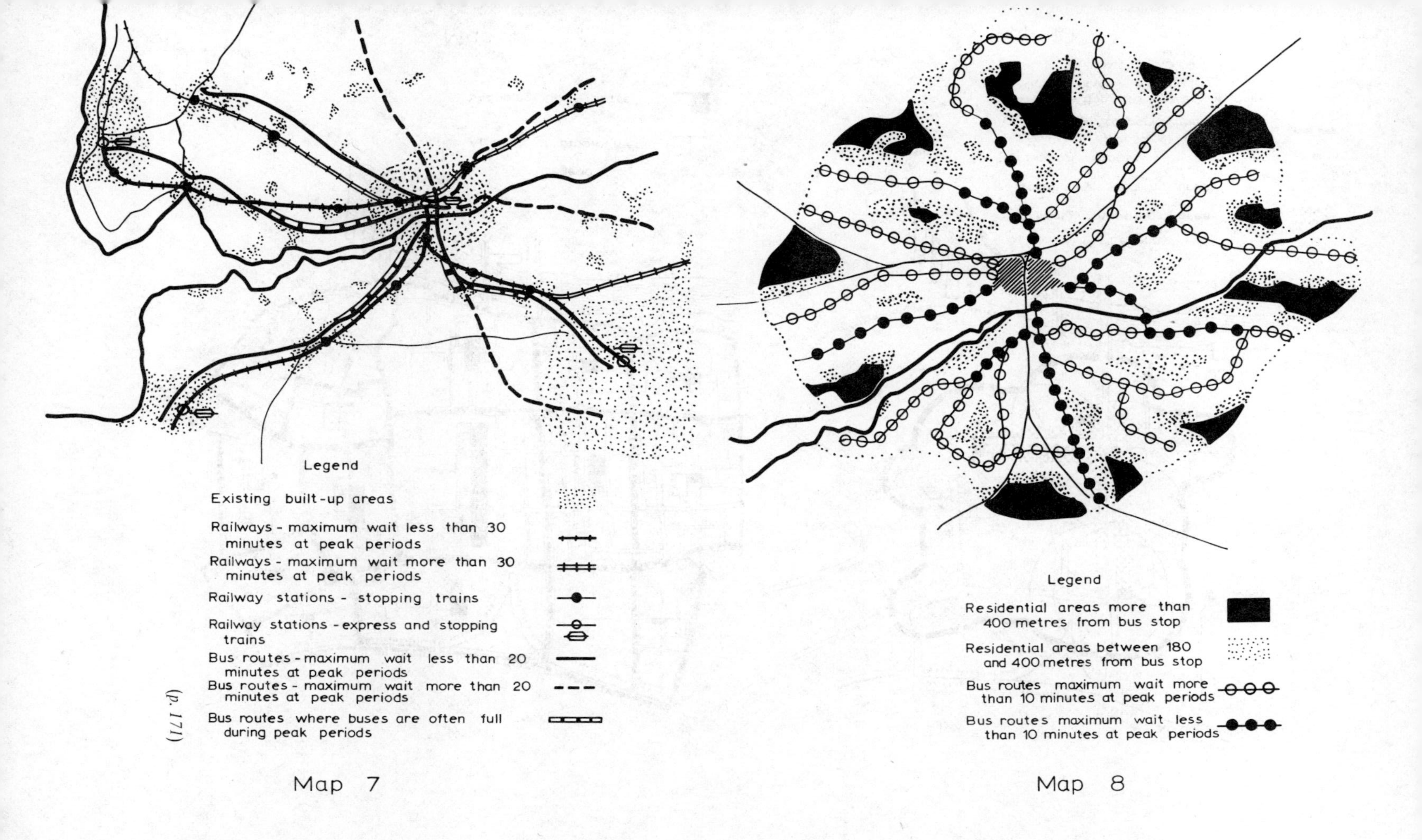

Map 7

Map 8

(p. 171)

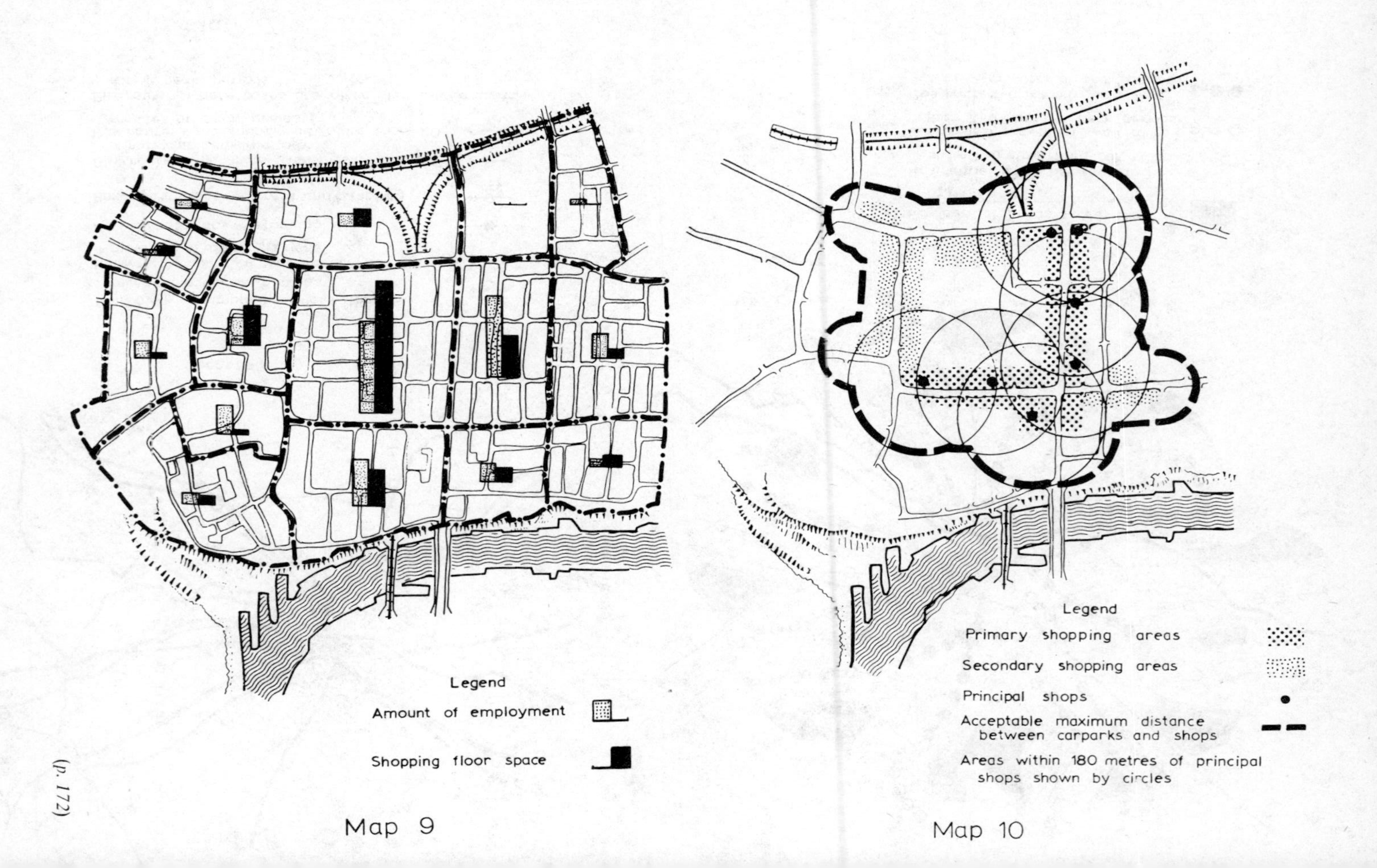

Map 9

Map 10

(p. 172)

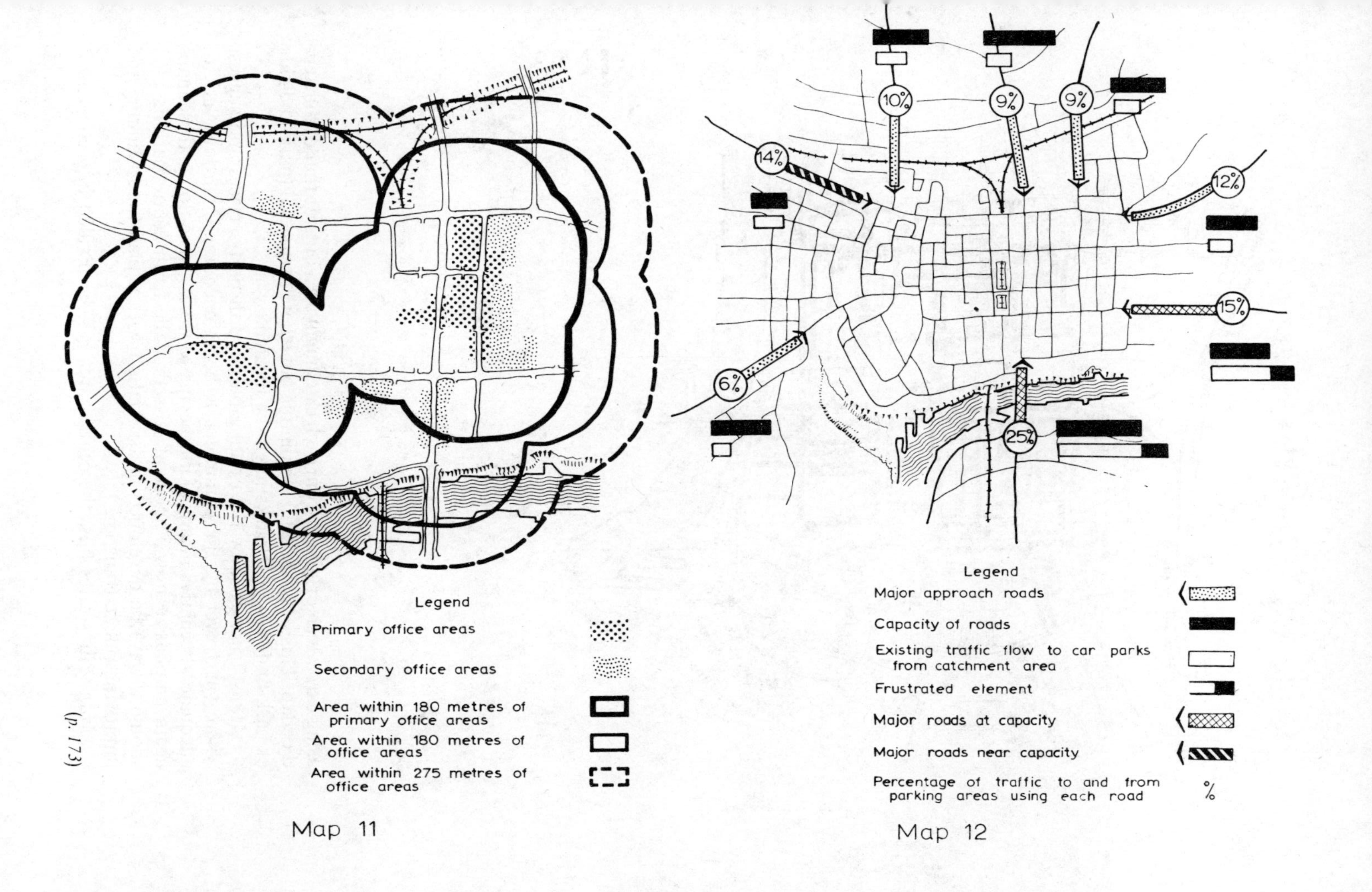

Map 11

Map 12

(p. 173)

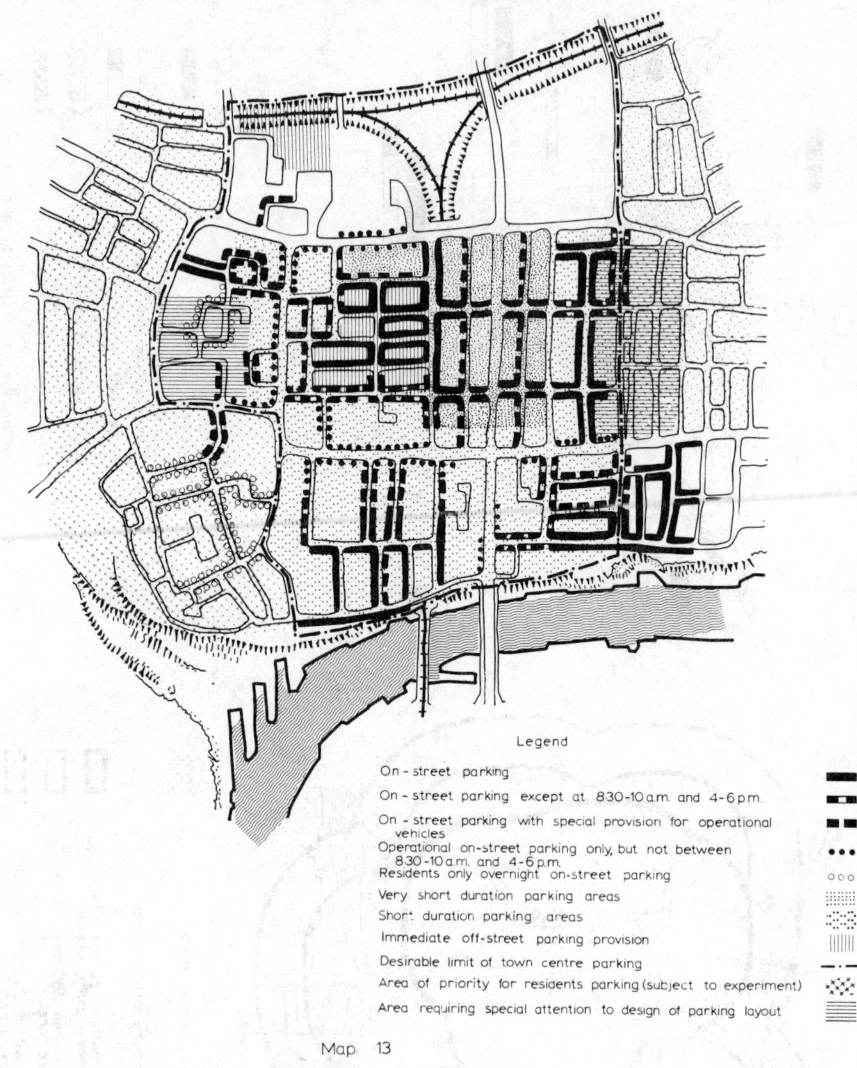

Map 13

using cars which originate from each catchment area outside that outlined in the map. This is most useful in pointing out where peripheral (park-and-ride or park-and-walk) car parks might be located.

Maps 7 and 8 show the existing public transport services and suggest where car travel by the commuter is being encouraged through lack of adequate alternative means of journeying to work. (In so doing, these maps again emphasize how important it is that the town centre parking plan should always be developed in association with a plan for sustaining and improving public transport.)

Map 9 illustrates how the central area can be divided into convenient destination zones (on the basis of figures for employment and floor space) in order to give guidance as to where parking demands are likely to be high.

Map 10 shows how the shopping parker can be given preference in the development of the parking plan. Note that in this example the acceptable walking distance for shoppers is given as 180 m; in fact, the acceptable distance from parking place to destination can vary considerably from town to town. An indication of present practice in any given town can be gained by observation in the central area on a Saturday afternoon.

If insufficient parking spaces can be made available to shoppers within what is judged to be an acceptable walking distance then it may be in the town's interests to consider the extent to which public transport facilities (e.g. special parker minibuses) might be used to offer a convenient alternative means of reaching the shops.

Map 11 is similar to Map 10 except that it deals with the problem of the journey-to-work parker. Again, ideally, provision should be made for these parkers within an acceptable walking distance. Where, however, this is not possible, consideration will need to be given to the development of park-and-ride services, e.g. by providing parking places at rail and/or bus public transport stations outside the central area.

Map 12 (which is an extension of the analysis initiated in Map 5) shows the existing traffic flows of the major approach roads, and indicates where there is spare capacity. (In this case, to simplify the illustration, the through traffic has been excluded.) Note that the south and south-east approach roads are filled to capacity during peak hours. Obviously, to provide additional parking space in the central area for long-stay cars approaching from the south could simply add to the congestion on the already crowded road (unless, of course, it is intended to carry out major road-works which will significantly increase its capacity), and so careful consideration will need to be given as to whether or not this added congestion is acceptable.

The town centre parking map. All of the foregoing analysis is designed to culminate in the development of the central area parking plan shown in *Map 13*. The sequence is to define as follows:

1. Existing off-street car parks, and land to be made available for off-street parking.
2. Where on-street parking is to be prohibited
   (*a*) at access points to buildings, loading bays and service areas
   (*b*) to obtain visibility at road intersections, pedestrian crossings, bus stops and taxi stands
   (*c*) on one side of the street only.
3. Those streets where flow is essential, at specific times such as at peak periods—*Map 2*.
4. Where only front service access to buildings is possible, and where rear service access is available—*Maps 2 and 3*.
5. Where priority should be given to service vehicles and other operational parking—*Map 3*.

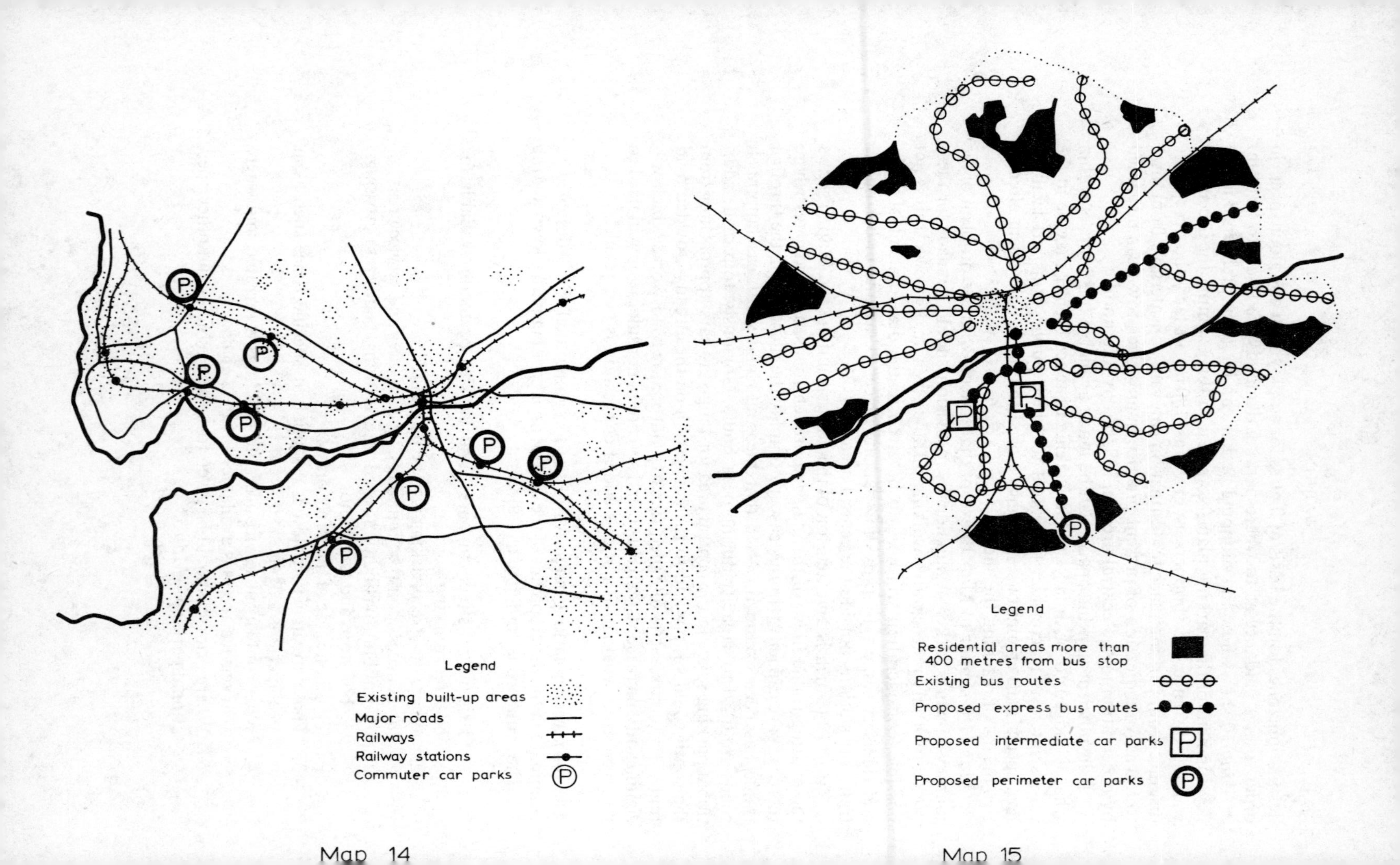

Legend
Existing built-up areas
Major roads
Railways
Railway stations
Commuter car parks
P
Legend
Residential areas more than 400 metres from bus stop
Existing bus routes
Proposed express bus routes
Proposed intermediate car parks
Proposed perimeter car parks
P

Map 14

Map 15

6. Where overnight parking for residents only is desirable—*Map 4.*
7. Those areas of historic and architectural significance where special attention needs to be given to the detailed design and layout of any parking provision—*Map 4.*
8. The areas where very short-term car parks should be given preference, e.g. near post offices, banks, etc.
9. The areas where shoppers and visitors to offices, etc. should be given preference—*Maps 1, 9, 10 and 11.*
10. The extent of the displacement of long duration parkers into the adjoining areas, related to acceptable walking distances—*Maps 10 and 11.*

Essential developments in the hinterland of the town cannot be isolated from central area proposals. Thus, *Map 14* is also developed in order to show where special parking facilities are needed to facilitate commuters at railway stations, while *Map 15* (which is derived from Maps 7 and 8) shows where express and improved bus services should be provided. Note that both outlying and intermediate car parks are proposed on the southern approach routes, and intermediate car parks only on the south-western approach route; these are intended both to alleviate congestion on the southern approach road (see Maps 5, 6 and 12) and to limit the overspill parking into the residential areas which adjoin the central area.

Note also (Map 15) that an express bus service which does not have a terminal car park is proposed for the eastern approach road; this would probably be best considered, initially at any rate, as an experimental service to see how many commuters will voluntarily use this facility in preference to their private cars. Similarly, the parking restrictions in the residential areas adjacent to the town centre (Map 13) should be considered as experimental, and subject to changes which will ensure the best solution in both traffic and environmental terms. If the experiments are successful, then similar provisions can be made at a later date in other parts of the town which would relate to the land use and traffic proposals for both the particular areas and for the town as a whole.

To conclude this discussion, it cannot be too strongly emphasized that *this town centre parking plan is in no sense static; modifications can be made from time to time to meet changes in traffic demand and land use, and to accord with other changes in the comprehensive town centre plan of which it forms a part.*

DESIGN OF OFF-STREET CAR PARKS

Off-street parking facilities can be either surface parks or garages. Whichever is used in a particular situation usually depends on the value of the land occupied by the facility and the intensity of usage anticipated. High-cost land frequently justifies a multi-storey garage i.e. the land cost per vehicle may be less than for a surface park. Conversely, expensive garage construction is rarely justified (except for environmental reasons) on low cost land where equal capacity can be obtained on a greater land area at a lower cost per space.

Surface parks

These are the most commonly used type of parking facility, primarily because of the ease with which they can be quickly and economically provided when suitable sites become available. The following is a brief

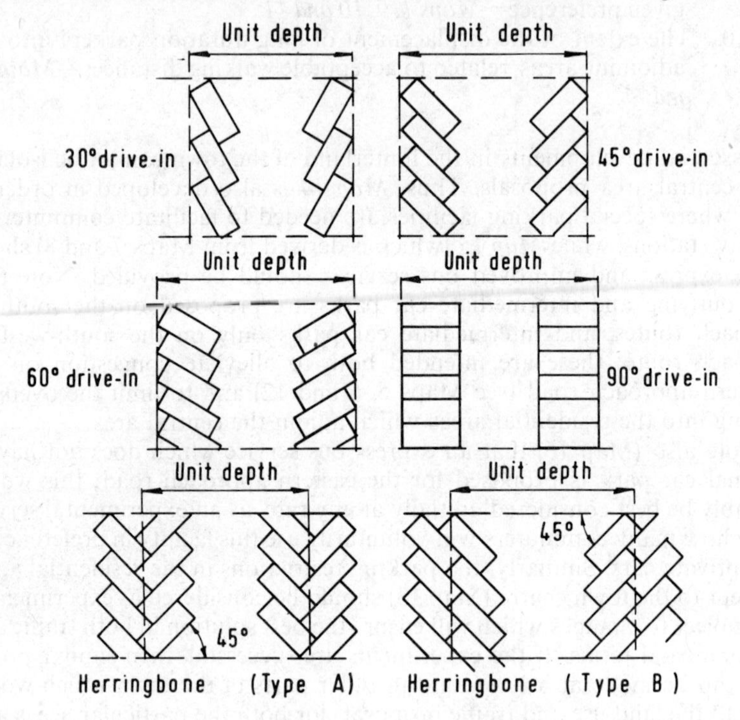

Fig. 4.5. Various parking angles commonly employed in off-street facilities

description of the basic physical features underlying the adequate design of such a car park. Since a surface park is essentially the same as one floor of a parking garage, many of these features will also apply to the design of a patron-parking garage facility.

Two basic and interrelated factors affecting the layout of a surface car park are the stall size and the aisle width. These are determined by the dimensions and turning radii of vehicles, the clearance between parked vehicles, the angle and direction of parking, and the minimum aisles' widths necessary for the free and safe flow of traffic.

Dimensions of cars vary substantially. Thus, it is necessary to design on the basis of a standard vehicle with dimensions greater than the largest common car models. One recent recommendation[39] suggests that such a European design vehicle should be 4·78 m long and 1·78 m wide.

TABLE 4.9. *Suggested dimensions for the geometrical layout of off-street car parks—90 deg. parking*

Type of parking	*Stall length, m*	*Stall width, m*	*Aisle width, m*	*Unit depth, m*
Long-stay	4·75	2·30	6·00	15·50
Short-stay	4·75	2·50	6·00	15·50

The width of stall is affected by the clearance needed for motorists to get in and out of their cars. This clearance is, in turn, dependent on such factors as the trip purpose, the age and sex of the motorist, whether or not shopping bundles will be carried, etc. Table 4.9 summarizes some suggested dimensions for the geometrical layout of 90 deg. off-street parking facilities. Figure 4.5 illustrates various parking angles that are commonly used in off-street parking facilities. Note that when the parking angle is 45 deg., two types of

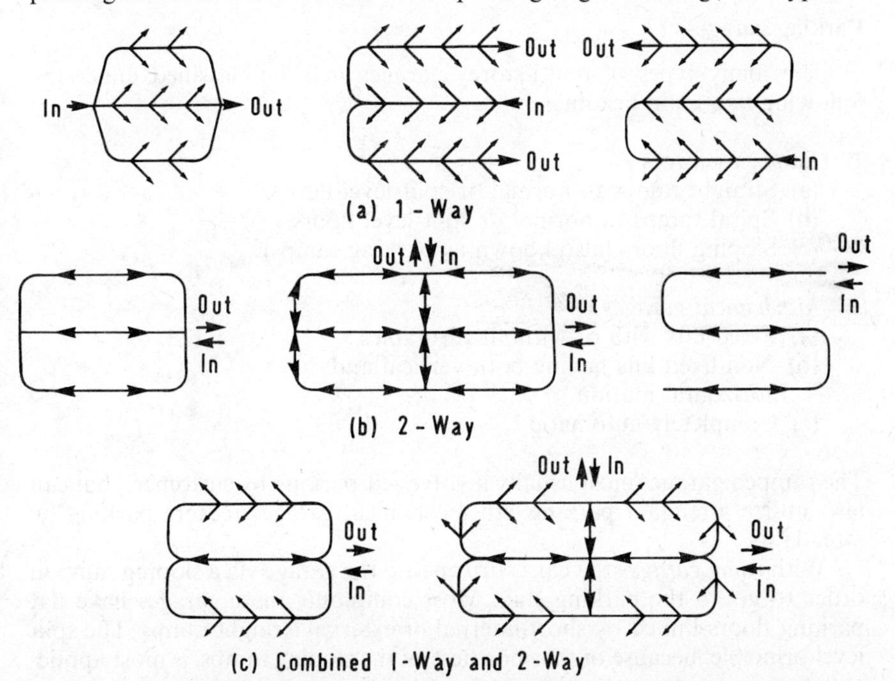

Fig. 4.6. Alternative traffic movements in a car park

an overlapping 'herringbone' arrangement can be used. Whichever angle and direction of parking is used in a given situation depends primarily on the shape and location of the parking area.

Fig. 4.6 illustrates some alternative traffic movements which can be used in car parks. As a general principle, however, it should be kept in mind that one-way circulation with vehicles driving directly into the stall will better ensure efficient traffic flow with the minimum of collisions.

Surface car parks should be well lit, well drained, kept clean and attractively landscaped. Good lighting at night reduces pilferage and increases confidence *re* the safety of patrons. Good drainage, cleanliness and landscaping makes the facility more attractive, as well as helping maintain the values of adjacent properties.

Raised pedestrian crossings may be necessary in large car parks. Widths vary, but walkways about 2·4 m wide will enable two people to walk abreast and allow for car fender overhang. In addition, some form of permanent bumper or wheelstop is useful at the end of each stall, both as a reference point and to prevent vehicles from over-running.

Parking stalls should be clearly marked on the surface of the parking area with paint; this will ensure orderly, efficient parking. Traffic signs and markings should clearly indicate the direction of movement in aisles, as well as entrances, exits fees, and other such regulations.

Parking garages

The many types of multi-storey garages may be classified under the following two main headings:

1. *Ramped garages*
   (a) Straight ramps to normal or split-level floors.
   (b) Spiral ramps to normal or split-level floors.
   (c) Sloping floors (also known as parking ramps).

2. *Mechanical garages*
   (a) Fixed lifts with or without turntables.
   (b) Non-fixed lifts having both vertical and horizontal motion.
   (c) Completely automatic.

The ramped garages most usually involve self-parking by customers, but can also utilize attendant parking. All mechanical garages require parking by attendants.

With *ramp garages* the car is driven into the garage via a sloping ramp in order to get to the parking stall. Most commonly these garages have flat parking floors linked by short internal or external straight ramps. The split level principle, because of the shortness of the sloping ramps, is most applicable to garages erected on congested sites. Spiral ramps, which can be externally or internally located, are most commonly used to provide a rapid exit for 'down' traffic from the various parking floors. If two-way spiral ramps are used, they should be designed so that the driver is adjacent to the wall; for right-hand driving (as in Britain) this means that the rule of the road must be to 'keep right'.

In a parking ramp garage, the parking floors are sloped continuously throughout the building. Since there are no inter-connecting ramps, this type of garage provides more parking stalls than a comparable split-level design[40]. Because, however, the cars are parked across the slope of the floors, the slope must be restricted and, hence, ramped floor garages tend to suit only large sites. Very often also they are linked with 'rapid exit' spiral ramps in order to minimize congestion at the lower levels. A disadvantage of a ramped floor garage is its relative unacceptability on aesthetic grounds, i.e. the long ramps give an undesirable effect in external elevation. To overcome this, a garage construction has been developed which uses warped (hyperbolic paraboloid) floors to give a horizontal external line to the building while containing the circulation features of a split level facility in combination with the efficiency of a parking ramp.

With *mechanical garages*, lifts take the place of ramps, the headroom required is for cars rather than people—which enables more cars to be parked per unit height of building—and automation is employed either wholly or partly. Hence, mechanical garages make the most effective use of available space and require the least floor area per car.

Mechanical garages can be either fully or partially automatic. One of the classic examples of the fully automatic type is the 'Park-O-Mat' in Washington, D.C. This has a frontage width of only 7·6 m, is 17 storeys high, and can contain 68 cars. The parking of cars is carried out by a single attendant who never moves from the office, i.e. this garage has two lifts, each serving parking stalls at the back and front of the lift shaft on each floor. When a customer arrives at the garage, the car is left locked in front of the lift, with the brakes released. The attendant, by push-button control, causes a 'dolly' to extend from the lift platform which moves the car onto the lift platform; on arrival at the selected floor, the lift dolly then moves the car backward or forward into an empty stall.

As might be expected, mechanical parking garages have high initial costs. Because of this, and some unfortunate 'teething' troubles experienced with them, very few of this type of facility have been built in Great Britain. Significant numbers, have however, been built abroad and are operated both efficiently and profitably.

Audit control

One of the first points to be considered when designing a new off-street parking facility is the control system which will ensure that the proper fees are collected.[41, 42, 43] The decision taken at this early stage can have a major effect on the entire design of a parking facility, e.g. on location, numbers of entrances/exits, etc.

The type of adult control utilized in any given situation is primarily a function of the trip purpose (i.e. type of parker) and, associated with this, the charging system utilized. Generally, the control procedures can be divided into the following five main types:

1. *Fixed charge, employing attendant at entrance.* This, the simplest system, is still widely employed at minor parking facilities. It simply utilizes an attendant who stands at the entrance and collects a fixed fee from every driver entering the car park. This method can result in hold-ups to traffic at

the entrance, and depends very much for its financial success on the honesty and integrity of the attendant.

2. *Fixed charge, fully automatic*. Typically, the entrance to (or exit from) a car park is controlled by a barrier arm which is automatically raised when an appropriate fee is paid into a coin machine. Machines can be made to accept coins of any currency, and can be adjusted by a simple mechanism so as to charge any fixed fee desired; a change-giving unit can also be incorporated. The capacity of the car park is controlled by add/subtract counters which illuminate a neon FULL sign and will not allow the entrance barrier arm to rise when all the parking stalls are occupied.

3. *Variable charge, employing attendant at exit*. This, the most common form of payment system, utilizes a 'ticket spitter' machine at the entrance, from which the driver takes a date- and time-stamped ticket when driving in. This ticket is collected by an attendant when the vehicle is exiting; the attendant calculates the charge due, takes the money, and then presses a push button to open the exit barrier.

4. *Variable charge, fully automatic*. With this system, a binary coded ticket is issued at the entrance, as in 3 above. When exiting, however, the driver inserts the ticket in a ticket-and-coin-acceptance unit; this causes the due fee to be calculated and automatically illuminated on a digital screen. On insertion of the correct money (a change unit is also provided), the exit barrier is raised and the vehicle can be driven away.

5. *Variable charge with token, fully automatic*. This is similar to 4 above, except that a single ticket-and-coin unit is made to cater for several exits by locating it within the car park, usually adjacent to a pedestrian entrance. Thus, when a driver walks into the car park to collect his vehicle, he first places the ticket received earlier from the ticket-issuer (upon entry) into the unit, pays the fee illuminated on the digital screen, and receives in return a 'short-life' coded token ticket. If the token ticket is placed in a small acceptance unit at the exit within a short-period of time, the barrier is opened and the car can be driven away with the minimum of delay at the exit.

One further system which might be noted here is that involving 'season' tickets. This system, which is used at parking facilities catering for regular parkers (e.g. at industrial establishments, universities, etc.), generally utilizes a plastic card—similar to a credit card—which is coded with a magnetic or an embossed message. When the motorist places the card in a reading unit at the entrance and/or exit, the barrier is raised.

Mechanical and electrical services

Despite the apparent simplicity of the engineering services in self-parking garages, they can account for between 10 and 20 per cent of the total cost of a structure, and for this reason alone, if no other, they should receive detailed consideration at the design stage. The service elements of importance in this respect are lighting, electrical installation, road and ramp heating, lightning protection, ventilation, fire prevention and fighting, drainage, and lifts. The following summary is primarily based on an excellent review of the problems involved that is available in the literature.[44]

Careful design of the *lighting layout* should ensure an installation that is economic to install and operate, and yet will provide a level of illumination that is adequate for safety and security purposes. Desirable standards of illumination in parking garages are as follows:

Parking stalls—20 lumen/m^2	Ramps—60 to 80 lumen/m^2
Access aisles—50 lumen/m^2	Entrances and exits—150 to 200 lumen/m^2

The value of good maintenance procedures relative to lighting must be emphasized. For example, it is not uncommon for an old flourescent tube on a cold winter's night, and in a dirty fitting, to emit only 25–30 per cent of its designed output—which explains why so many car parks appear gloomy and dangerous at night.

Because of the many variables involved in relation to the design of the *electrical installation*, the advice of a qualified electrical engineer should be sought at the planning stage, so as to ensure that an installation is provided that will be safe and efficient to use, as well as being adequate for its intended purpose throughout the life of the parking structure.

Because so many parking garages are open to the elements, *ice and snow* can create hazardous driving conditions on exposed circulation routes, particularly steep, sloping ramps. This can be overcome, where circumstances justify it, by warming the driveways with embedded electric cables. Heating can be automatically activated by control probes set in the carriageway surface which energize the cables when they sense an appropriate combination of surface temperature and moisture.

Whether or not a parking building requires major protection against the effects of *lightning* depends on (*a*) the numbers of people likely to congregate in or adjacent to it, (*b*) whether it is in an area where thunderstorms are prevalent, (*c*) if it is a very tall or isolated structure, and (*d*) whether it is either self-protecting or needs only minor expenditure to render it safe, viz. structures composed entirely of metal and adequately earthed are self-protected, while structures with frames of steel or reinforced concrete require only minor improvements. A procedure which gives guidance as to the need for protection is available in the literature;[44] however, it should be appreciated that there is no absolute method of assessment at this time. When, however, it is considered that the consequential effects of a lightning strike will be restricted to slight damage to the building fabric, then it may be more economical to accept the risk rather than incur the extra cost involved in protecting.

Ventilation is essential in a parking garage in order (*a*) to avoid the risk of fire and explosion arising from the presence of inflammable or explosive vapours, and (*b*) to minimize any damage to health which might arise from the presence of excessive concentrations of exhaust gases. As many multi-storey parking garages are open-sided, this is not as great a problem as it is often considered; generally it is safe to assume that a ventilation problem does not exist if an open space equivalent to 2·5 per cent of the total floor area is provided on each long elevation of the garage. The problem is perhaps most critical in underground car parks, and here the designer is well advised to get the advice of a qualified ventilation engineer. Generally, it will be found desirable to install appropriate alarm equipment, particularly at tunnel locations where cars have to queue, which will start up the ventilation plant when, for example, the level of carbon monoxide reaches a predetermined level.

The *fire problem* in parking garages is not primarily related to any danger of the building burning down (because it is normally very hard for the structure to catch fire), but rather to the hazard of vehicles in the garage going on fire as a result of, say, a lighted cigarette being left on a seat.[45] Here also the problem is most serious in underground car parks; these may, for example, require the installation of sophisticated detector devices which will 'sniff' smoke, or flames, or a significant rise in temperature and sound the alarm. Fire fighting equipment in use in parking buildings include wet rising mains and hose reels, dry rising mains, automatic sprinkler and drencher systems, and simple fire extinguishers. When all is said and done, it is probable that portable extinguishers are the most effective equipment for fighting fires likely to be encountered in parking buildings, as they are most easily used by the public; they can be complemented by dry rising mains for use by the fire department. The benefits of hose reels can be nullified by vandalism and they, as well as automatic sprinkler systems, can be subject to frost hazards. Most important is the fact that water is not suitable for fighting petrol fires, and improper usage (as could occur with hose reels and sprinklers) could result in the fire being spread indiscriminately and made worse.

Because they are so open to the elements, as well as subject to water carried in by cars, special attention must be paid to the *drainage* of parking garages. This generally involves, for example, the deliberate sloping of parking floors (a gradient of about 1·67 per cent is appropriate) so as to provide falls in all directions to gulley outlets.

The design of the *lift system* to serve a parking garage is another oft forgotten part of car park design, and the result is that lift service installed in a given facility can be inadequate and have querulous groups of people awaiting vertical transport during peak periods. An approach to satisfactory design may be gathered from the following calculations for a 6-storey parking garage with space for 100 cars on each floor, which assume that during the peak hour there is a complete turnover of vehicles:

Total no. of cars above ground floor	= 500
Ave. no. of persons per car	= 1·5
Total no. of persons	= 750
Assume 75% use lift	= 563
No. of people to move in 5 min = $\frac{563}{12}$	= 47
No. of stops	= 5
Approx. length of travel, m	= 12·2

Table 4.10 shows that in this instance two 8-person lifts should be able to provide a fairly reasonable service (provided there is not a sudden heavy surge of travellers, e.g. as might occur with journey-to-work parkers, or that the passengers are not laden with shopping, perambulators, etc.). Obviously, the provision of lifts with speeds lower than 45·75 m/min would also result in an unsatisfactory service during periods of peak demand.

TABLE 4.10. *Handling capacities of an 8-person, 544 kg lift*

No. of floors served	*Speed*			
	45·75 m/min		*61 m/min*	
	Capacity, persons/5 min	*Round trip travel time, sec*	*Capacity, persons/5 min*	*Round trip travel time, sec*
5	24	100	26	94
6	21	114	23	104
7	19	127	21	115

SELECTED BIBLIOGRAPHY

1. BUCHANAN, C. D., et al. *Traffic in Towns.* London, H.M.S.O., 1963.
2. BENDSTEN, P. H. *Town and Traffic.* Copenhagen, Danish Technical Press, 1961.
3. JONASSEN, C. T. *The Shopping Centre versus Downtown.* Columbus, Ohio, Ohio State University Bureau of Business Research, 1955.
4. REGIONAL PLANNING RESEARCH LTD. *Parking: Who Pays?* London, The Automobile Association, 1967.
5. MINISTRY OF TRANSPORT. *Urban Traffic Engineering Techniques.* London, H.M.S.O., 1965.
6. SMITH, W. S. Influence of parking on accidents, *Traffic Quarterly*, April 1947, pp. 162–178.
7. GARWOOD, F. Accident Statistics. *Road Research Laboratory Research Note LN/254.* Crowthorne, Berks., The Road Research Laboratory, 1962. Unpublished.
8. HEYMAN, J. H. Parking trends and recommendations, *Traffic Quarterly*, 1968, **22,** No. 2, 245–258.
9. SEBURN, T. J. Relationship between curb uses and traffic accidents. *Traffic Engineering*, 1967, **37,** No. 8, 42–47.
10. ALEXANDER, L. A. The downtown parking system, *Traffic Quarterly*, 1969, **23,** No. 2, 197–206.
11. THE CITY ENGINEER, et al. *City of Edinburgh Traffic and Transport Plan 1970–1974.* Edinburgh, The Corporation, Sept. 1970.
12. CLEVELAND, D. E., (Editor). *Manual of Traffic Engineering Studies.* Washington, D.C., Institute of Traffic Engineers, 1964. 3rd ed.
13. STOUT, R. W. Trends in CBD parking characteristics, 1958 to 1968, *Highway Research Record* No. 317, 40–47, 1970.
14. COMMITTEE 3F(64) OF THE INSTITUTION OF TRAFFIC ENGINEERS. Local Street Parking Criteria, *Traffic Engineering*, 1967, **37,** No. 6, 32–33.
15. MINISTRY OF TRANSPORT. *Parking—The Next Stage.* London. H.M.S.O. 1963.
16. MOHR, E. A. A performance index for parking facilities, *Traffic Engineering*, 1954, **28,** No. 7, 11–14, 38.

17. O'Flaherty, C. A. and Atkinson, C. H. Characteristics of parkers at meters in a central area, *Journal of the Institution of Municipal Engineers*, 1969, **96,** No. 2, 19–23.
18. Kennedy, N., Kell, J. H., and Homburger, W. S. *Fundamentals of Traffic Engineering*. Berkeley, Calif., The Institute of Transportation and Traffic Engineering, University of California, 1966. (6th ed.)
19. Box, P. C. Curb parking effect, *Traffic Digest and Review*, 1970, **18,** No. 2, 6–10.
20. British Standards Institute. Parking Discs, *BS 4631*. London, The Institute, 1970.
21. Morgan, E. J. and Lewis, C. G. A before-and-after study of the effect of the introduction of disc parking in Harrogate, *Journal of the Institution of Municipal Engineers*, 1971, **98,** No. 2, 35–41.
22. Carty, T. F. Vault-type meters stop vandals, *The American City*, 1966, **81,** No. 3, 123, 126.
23. Skinner, H. B. Vault-type meters increased revenues, reduced collection costs, *Public Works*, 1967, **98,** No. 9, 91.
24. Harrison, M. Vandalism to parking meters, *Traffic Engineering and Control*, 1969, **11,** No. 3, 142–146.
25. Cave, F. J. Resident parking problems and schemes, In *Making Parking Pay—The Economics of Car Parking*, London, The British Parking Association, 1972.
26. Bennett, S. J. The problem of heavy goods vehicles in urban areas, In *Making Parking Pay—The Economics of Car Parking*, pp. 29–30, London, The British Parking Association, 1972.
27. Department of the Environment. *Lorry Parking*. London, H.M.S.O., 1972.
28. O'Flaherty, C. A. This parking business, *Journal of the Institute of Municipal Engineers*, 1968, **95,** No. 12, 368–379.
29. Ellson, P. B. Parking: Dynamic Capacities of Car Parks. *RRL Report LR221*. Crowthorne, Berks., The Road Research Laboratory, 1969.
30. Ricker, E. R. *The Traffic Design of Parking Garages*. Saugatuck, Conn., The Eno Foundation, 1957.
31. Lovejoy, F. W. Resumé of fringe parking practice, *Highway Research Board Bulletin* 19, 57–60, 1949.
32. Wilbur Smith and Associates. *Parking in the City Centre*. Washington, D.C., The Automobile Manufacturers Association, 1965.
33. Deen, T. B. A study of transit fringe parking usage, *Highway Research Record No.* 130, 1–19, 1966.
34. *Park 'n' Ride Rail Service, Jersey Ave. Station, New Brunswick, N.J.* Final Report on the Mass Transportation Demonstration Grant Project, New York, N.Y., The Tri-State Transportation Commission, May 1967.
35. O'Flaherty, C. A. *Town, transport and parking*. Proceedings of the British Parking Association Seminar on 'Off-Street Parking' held in London. April 15, 1970. 2:4–6. London. The British Parking Association. Aug. 1970.
36. Bureau of Public Roads. *Parking Guide for Cities*, Washington, D.C., Government Printing Office, 1956.
37. Ministry of Housing and Local Government. *Parking in Town Centres*, Planning Bulletin No. 7, London, H.M.S.O., 1965.
38. Bentfeld, G. Effect of parking policies on traffic volumes. General Report on the papers present during Theme VI at the *10th International Study Week in Traffic and Safety Engineering held at Rotterdam, 7–11 Sept., 1970*, Proceedings **10,** 51–58, 1971.

39. BPA Technical Committee. Metric dimensions for car parks—90° parking, *Traffic Engineering and Control*, 1970, **11,** No. 12, 615.
40. Glanville, J. Developments in multi-storey car parking in Great Britain. Paper presented at *5th World Meeting of the International Road Federation* held at London, Sept 1966.
41. Glanville, J. Car park control, *Traffic Engineering and Control*, 1969, **10,** No. 9, 473–478.
42. Dowling, R. J. and Seymer, N. V. Planning and design of parking facilities. In *Off-Street Parking*, pp. 15–20, London, The British Parking Association, Aug. 1970.
43. Cameron Cooper, T. C. Automatic Car Park Control Systems. Paper presented at the *5th World Meeting of the International Road Federation*, London, Sept. 1966.
44. Glanville, J. and Richards, D. L. The design of mechanical and electrical services for car parking buildings. In *Off-Street Parking*, pp. 25–35, London, The British Parking Association, Aug. 1970.
45. Butcher, E. G. Fire and car park buildings, *Traffic Engineering and Control*, 1970, **12,** No. 7, 377–388.

5 The economics of road improvements

Economic assessments and the application of economic concepts to justify road improvements are not at all new to highway engineering. It is reported[1] that in a well-known treatise of 125 years ago there was written, 'A minimum of expense is, of course, highly desirable, but the road which is truly cheapest is not the one which has cost the least money, but the one which makes the most profitable returns in proportion to the amount which has been expended upon it'. It is true, nevertheless, that it was not until relatively recently that highway engineers were able to really put into practice these economic concepts with which they had been familiar for many years. In recent years, with a substantial road system established to a relatively high degree of quality, the highway administrator-cum-engineer has been confronted with the necessity for indicating scheme priority in expenditure of still limited funds to further improve or replace portions of the system. It is the need to assign these priorities that has primarily led to the modern resurgence in highway economic assessment studies and applications.

It must not be thought, however, that the only important criterion in any highway planning decision is the economic one. Important it is, of that there is no doubt. However, the reader should be aware that there is still considerable controversy *re* the extent to which the methods of economic analysis used by engineers can take into account such factors as traffic service, effect on local planning objectives, effect on local traffic patterns and street networks, impact on the life of the community, effect on land uses, environmental considerations, and national objectives. It cannot be too strongly emphasized that the highway is simply a functional element in the area plan and thus its justification, location or quality cannot simply be considered in the light of strictly economic considerations, but must always be qualified by its overall service to the community.

In this chapter the approach taken is to present the basic features which engineers attempt to take into account when carrying out economic analyses for road improvements.

BASIS OF ROAD IMPROVEMENT STUDIES

Analysis of the economic benefits to be obtained from a proposed road improvement scheme as related to the economic costs of the scheme can be made with the following specific objectives:[2]

1. To determine whether a scheme is economically justified.
2. To aid in the choice of engineering features of design.

3. To provide one way of determining priority of one scheme as related to others. (This is their primary function in Great Britain.)
4. To assist in tax or cost allocation studies or decisions.
5. To develop information which will aid in evaluating a specific improvement as against other proposals in public works or community projects.

The analysis for these economic consequences involves such factors as motor-vehicle operating costs, time value of individuals, safety of individuals and property, changes in the economic value of land, buildings, businesses, and resources, as well as the economic cost of the capital investments and the maintenance and operating costs of physical property. Educational, social, and general community values are not included in economic analyses since they cannot be reduced to supportable and realistic monetary values. The following discussion therefore is primarily concerned with those factors and their consequences which can be readily converted to monetary values—in particular, those which are of special interest in British highway economic studies.

Cost of highway improvements

The cost of a highway improvement really consists of both the total initial cost and the subsequent maintenance costs.

Initial-cost items. The items making up the total initial cost of the road improvements can be summarized as follows:

1. *Land.* This includes not only the amount actually paid for the land, but the cost of all legal fees connected with the transfer of land and, where applicable, the cost of rehousing displaced occupiers. Also included in this cost is any amount paid in recompense for environmental damage caused to an owner/occupier as a result of, for example, a new motorway being constructed within sight and sound of his residence. It is sometimes suggested that if the highway authority has had the land in its possession for some time it should either be excluded from the cost calculations or else put in at a reduced rate. This concept is not to be supported; the fact that the land for the improvement was acquired a long time ago should not be allowed to affect the calculations of the cost of the improvement, but instead an attempt should be made to value the land at current prices, i.e. its current value in its most probable alternative use or uses.

2. *Accommodation works.* These are carried out on land and property adjoining the road and which remain the property of somebody other than the highway authority. The principal cost items are fencing and walling.

3. *Statutory undertakings and services.* These are the costs of removing and relaying gas and water mains, electric and post office cables, etc. They can be considered as similar to accommodation works in that they are costs incurred in carrying out works on somebody else's property.

4. *Bridges and subways.* These items are very expensive and exceptional when compared to the cost per square metre of highway.

5. *Site preparation.* This can be divided into two parts. The first, site clearance, is usually only an important item when it involves demolition in heavily built-up areas or the filling in of cellars. The second, earthworks, is a major item; it includes the cost of retaining walls since they are closely allied to earthworks.

6. *Carriageway.* This covers the cost of the surfacing, roadbase and sub-base courses.

7. *Kerbs and channels.*

8. *Roadway other than the carriageway and kerb.* Main items are footpaths and verges, but often also include central reservations, cycle tracks, shoulders and any fencing not considered part of the accommodation works.

9. *Drainage.*

10. *Miscellaneous and overheads.* Miscellaneous includes many small items, but is mainly lighting and traffic signs. Overheads include supervision and administrative expenses—these are costed as a fixed percentage on all trunk road schemes—salary of the resident engineer, and 'watching and lighting'.

An indication of the relative importance of the various components can be gained from the total-cost data in Fig. 5.1, which summarize the results of an analysis of road improvement schemes carried out in the late 1950's. Some important conclusions derived from this study are as follows:

(*a*) While land costs were involved in 93 per cent of the road improvements, in two-thirds of them they accounted for 5 per cent or less of the total cost. In urban schemes, however, land accounted for more than half the cost in nearly 12 per cent of the cases, and in nearly 50 per cent it accounted for 10 per cent of the total. In rural areas land accounted for more than 10 per cent of the total costs in only 7 per cent of the cases studied. The type of improvement where land costs were of least importance was that of widening from single to dual carriageways where, in 62 per cent of the cases, land costs were less than 2·5 per cent of the total cost.

(*b*) Accommodation works were relatively more important in rural areas than in urban ones; they occurred in 95 per cent of the former schemes and in only 75 per cent of the latter. On average they accounted for 3·5 per cent of the total costs, but in more than 16 per cent of the schemes they accounted for over 10 per cent of the total cost; the majority of these latter schemes were, however, small.

(*c*) Costs connected with statutory undertakings constituted 4·5 per cent of all costs. They were relatively unimportant in rural areas, but in nearly one-tenth of all urban schemes they accounted for more than 20 per cent of the total cost. They were of less importance in building new roads than in alterations to existing ones, accounting on the average for 10 per cent of the total cost for the latter while amounting to only about 3 per cent of the cost of new roads.

(*d*) The incidence of bridge costs, in schemes covered by the analysis, varied considerably between different types of improvement. They entered into less than one-fifth of all the schemes, but when they did they amounted to nearly 15 per cent of the total costs.

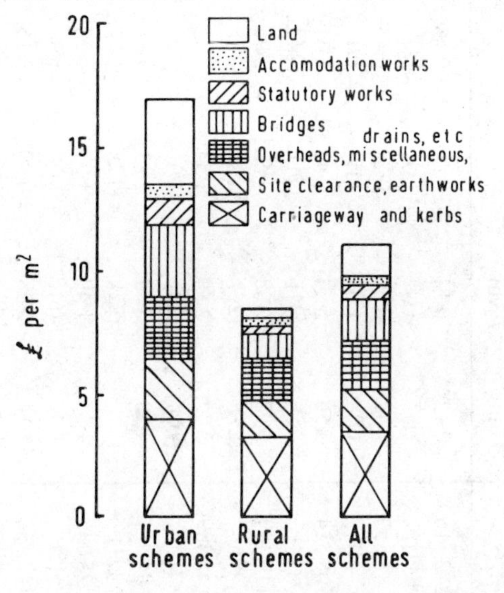

Fig. 5.1. Division of total initial costs into main components[3]

(*e*) When the results were analysed on a regional basis it was found that, due to the nature of the terrain, the cost of earth works in Wales was about double the average cost and in Eastern England they were half the average. All costs were high in the London area.

(*f*) There was little connection between cost per unit area and the size of the scheme. The only significant variation was that schemes of less than 1850 m^2 were more expensive per unit area than larger ones.

A measure of road costs that is commonly used in economic assessments is the cost per unit length. Table 5.1 shows actual construction costs (1969 prices) for a unit length of different types of highway. Data such as these are very useful when evaluating new road proposals; cost values per unit area are more valuable, however, when considering improvements that involve varying amounts of widening.

Maintenance. In road improvement economic studies, maintenance costs are often regarded as a yearly item even though the expenditure may actually be incurred at infrequent intervals. Costs taken into account should be at a level to completely offset wear and tear and maintain the highway intact, and should include such 'additional' items as cleaning, lighting and verge maintenance. In the case of small improvements these additional costs may be small in comparison with other costs, and possibly may be neglected; in a large improvement, however, they may well be a significant factor in the economic calculations and should be included. With new roads it usually happens that the transfer of traffic from the old to the new road will reduce the maintenance costs on the original road and this saving should be offset

TABLE 5.1. *Component costs of primary road construction (excluding land)*[15]

Component	*Average price per kilometre of primary road, £ (with percentages in brackets)*				*Average price per square metre of primary road carriageway, £ (with percentages in brackets)*			
	Twin 11 m	Twin 7·3 m	Single	All	Twin 11 m	Twin 7·3 m	Single	All
Site Clearance	2 688 (1)	2 110 (1)	1 744 (1)	2 516 (1)	0·12 (1)	0·14 (1)	0·24 (1)	0·13 (1)
Accommodation works	9 081 (2)	8 791 (2)	5 207 (4)	8 853 (2)	0·42 (2)	0·60 (2)	0·72 (4)	0·45 (2)
Earthworks	122 004 (25)	81 030 (21)	25 034 (21)	108 514 (24)	5·56 (25)	5·54 (21)	3·42 (21)	5·51 (24)
Carriageway construction	131 078 (27)	88 891 (23)	34 178 (28)	117 309 (26)	5·97 (27)	6·08 (23)	4·66 (28)	5·96 (26)
Drainage	48 874 (10)	40 984 (10)	11 314 (9)	45 489 (10)	2·22 (10)	2·80 (10)	1·54 (9)	2·32 (10)
Structures	100 627 (5)	86 564 (6)	20 696 (17)	94 051 (21)	4·58 (20)	5·92 (22)	2·83 (17)	4·78 (21)
Miscellaneous	27 052 (5)	21 786 (6)	9 023 (7)	25 085 (5)	1·23 (5)	1·49 (6)	1·23 (7)	1·28 (5)
Preliminary and general	46 932 (9)	34 880 (9)	6 523 (5)	42 468 (9)	2·14 (9)	2·38 (9)	0·90 (5)	2·16 (9)
Contingencies	6 537 (1)	25 351 (6)	10 242 (8)	11 029 (2)	0·30 (1)	1·73 (6)	1·40 (8)	0·56 (2)
Totals	494 873 (100)	390 387 (100)	123 958 (100)	455 314 (100)	22·54 (100)	26·68 (100)	16·95 (100)	23·17 (100)

against the maintenance costs of the improvement to give the net additional maintenance costs.

Table 5.2 summarizes estimated maintenance costs at 1969 prices for a unit length of different types of pavement for periods of 25 and 50 years. The maintenance operations taken into account in the calculations for this table include (*a*) strengthening of the road structure to restore loss of strength due to cracking or to restore loss of riding quality, etc., due to deformation, (*b*) surface treatment to restore loss of resistance to skidding due to 'polishing' and other effects, (*c*) renewal of road markings and 'cats-eyes', (*d*) resealing the joints in concrete surfacings, (*e*) resurfacing the hard shoulder, raising kerbs, and gulleys, etc., and (*f*) local patching. A further important feature included in the values in Table 5.2 is the cost of delays to traffic resulting from road closures associated with operations (*a*) to (*f*).

Evaluation of road user benefits

Highway improvements can bring benefits to the community in many ways, the more obvious ones being related to the road user. These are reflected by reduction in the cost of vehicle operation, savings in time and accidents, and increased comfort and convenience, with accompanying increases in the volumes of traffic that can be handled. In association with the benefits to road users, however, there can also be general benefits, e.g. those that accrue in the form of increases in land values, and the benefits to amenity caused by reductions in noise and fumes on relieved roads.

From traffic engineering studies it is possible to estimate the changes in traffic volumes, speeds, delays and accident rates that may be expected from any particular highway scheme. As soon as these have been estimated in terms of vehicle-hours, vehicle-kilometres, and accidents saved, they can be translated into monetary terms for use in the economic studies. Much more difficult to evaluate, however, are the intangible factors of comfort and convenience, increasing amenity and, indeed, land prices.

The following discussion is therefore primarily concerned with those factors which can be evaluated in monetary terms and which are used in highway economic studies in this country. Some of the intangible factors are also presented and briefly discussed.

Labour, vehicle-time, and operating costs. Studies carried out in this country and abroad have established that the main factors affecting road user costs are the type of vehicle, type of area in which the vehicle is operated (e.g. urban vs rural), speed of vehicle operation, type of highway and the quality of road surface, and the horizontal and vertical alignment of the roadway and, of course, the value placed on the road user's time. The British investigations have resulted in the relationships shown in Table 5.3 regarding the operating costs of different types of vehicles, expressed as a function of vehicle running speed.

The different vehicles referred to in Table 5.3 are ones selected as being representative of their class. Vehicle fuel costs, as used in the calculations,

TABLE 5.2. *Maintenance (including delay) costs for 1·6 km of pavement for periods of 25 and 50 years*[4]

Type of road	*Form of construction*	*Type of construction*	*Costs up to 25 years, £*			*Costs up to 50 years, £*		
			Main-tenance	*Delay*	*Total*	*Main-tenance*	*Delay*	*Total*
1. Rural motorway	Flexible	Composite base (lean concrete)	66 580	3 015	69 600	129 000	8 000	137 000
		Fully flexible base	59 700	2 000	61 700	123 000	6 800	129 800
	Concrete	Hot-poured joint sealants	33 120	1 145	34 265	66 710	4 450	71 160
		Cold-poured joint sealants	35 940	575	36 515	75 350	2 660	78 010
2. Peri-urban (dual-carriage-way) road	Flexible	Composite base (wet-mix) Composite base (lean concrete) Fully flexible base	34 600	14 300	48 900	72 800	82 600	155 400
	Concrete	Hot-poured joint sealants	12 300	18 340	36 640	28 600	117 190	145 790
		Cold-poured joint sealants	14 140	5 300	19 440	34 270	53 130	87 400
3. Rural secondary road	Flexible	Wet-mix base	16 400	900	17 300	24 200	3 750	27 950
		Lean concrete base	13 200	1 000	14 200	21 100	4 000	25 100
		Fully flexible base	14 000	3 500	17 500	21 100	6 200	27 300
	Concrete	Hot-poured joint sealants	6 920	4 830	11 750	9 730	16 700	26 430
		Cold-poured joint sealants	7 970	2 330	10 300	13 060	8 470	21 530
4. Housing estate road	Flexible	Wet-mix base	8 500	—	8 500	10 200	—	10 200
		Lean concrete base	9 000	—	9 000	10 700	—	10 700
		Fully flexible base	1 700	—	1 700	4 000	—	4 000
	Concrete	Reinforced slabs	1 240	—	1 240	2 700	—	2 700
		Unreinforced slabs	1 900	—	1 900	4 350	—	4 350

TABLE 5.3. *Composite operating costs per vehicle-kilometre on all-purpose roads*[5]

Vehicle type	*Cost per veh-km*, p
Car	$0{\cdot}71+66/v_c+0{\cdot}000022v_c^2$
Light van	$1{\cdot}01+91/v_v+0{\cdot}000028v_v^2$
Other goods vehicle	$2{\cdot}37+96/v_g+0{\cdot}000063v_g^2$
Public service vehicle	$2{\cdot}38+624/v_B+0{\cdot}000058v_B^2$
Ave. traffic composition (80% cars, 9% vans, 9% other comm., 2% p.s.v.)	$1{\cdot}13+76/v_{\text{ave c}}+0{\cdot}000021v_{\text{ave c}}^2$

were primarily derived in the course of a study carried out in 1967[6] which showed that fuel consumption could be expressed by a formula of the general form:

$$\text{Fuel consumption per kilometre} = a+b/v+cv^2$$

Also included in the formulas were allowances for oil (assumed as a fixed cost per kilometre), tyre costs (assumed to vary with speed in the same way as fuel costs), and vehicle maintenance (two-thirds being assumed to be a fixed cost per kilometre, and one-third to vary with speed in the same way as fuel costs). Only that part of vehicle depreciation (40%) which is due to use was taken into account in the derivation of the equations. The hourly values of occupants' time per vehicle used were as follows:

Car—56p/h	Other goods vehicles—70p/h
Light van—75p/h	Public service vehicles—575p/h

The above time values assume that 17 per cent of all car travel is in working time. For both cars and other vehicles, working time is valued at relevant average wage rates; non-working time for cars and (passengers in) passenger service vehicles is valued at three-quarters of the appropriate wage rate.

The 'average' traffic composition formula given in Table 5.3, which is expressed in terms of the average speed of cars, is derived from the previous three formulas and makes allowances for the variations in speeds of different classes of vehicle as car speed increases.

To allow for the fact that vehicle operating costs on motorways are less than on all-purpose roads, it is necessary when determining motorway cost values to reduce the constant term in the formulas shown in Table 5.3 by 0·05p/veh-km for cars and light vans, and 0·1p/veh-km for other goods vehicles and public service vehicles.

It is appropriate to comment here upon one of the most controversial aspects of the formulas quoted in Table 5.3—the value placed on a person's time. This is necessary as there is no doubt that the saving in travel time is one of the greatest, if not the greatest benefit to be gained from a highway improvement.

There is wide acceptance of the premise that a saving in commercial vehicle time has value corresponding to a reduction in those operation costs, such as the hire of drivers and the hourly rental of equipment, which are directly related to time. A saving in travel time may also result in greater vehicle usage so that benefits accrue from increased usage and not from savings in costs. For costing purposes, this, in turn, means that fixed costs, such as depreciation, etc., may be spread over a greater travel distances.

Not all passenger vehicle trips are for the purpose of business; they may be shopping trips or pleasure trips as well as business journeys or journeys to work. When they are for business purposes, as in the case of commercial travellers or other such service workers, they are usually regarded in the same light as commercial vehicle trips. (This, of course, may be queried as it assumes that a business traveller who saves, say, 10 min on a journey will automatically do 10 min extra work—which is doubtful at the least.) The most controversy arises over the values to be attached to journey-to-work trips and to leisure trips. While values of 75 per cent working time were applied to both of these types of trips when determining the relationships shown in Table 5.3, it is important to point out that at least one recent important study[(7)] has applied a value of only 25 per cent to leisure trips, while another investigation found that Department of the Environment employees[(8)] placed a value on time saved on their journey-to-work trips of between $\frac{1}{3}$ and $\frac{1}{2}$ their equivalent wage rate. Nevertheless, despite the obvious differences as to what are the 'correct' values to use, there is general agreement that some price should be attached to time-saving.

These last points also, of course, focus attention on the relative importance of *comfort and convenience*. That most motorists place a value on comfort and convenience is well demonstrated by the significant diversions to new and modern highway facilities, even though greater distances and times can be required. Certainly there is comfort value, over and above the saving in operating cost, in being able to drive without frequent brake applications, stops or starts or unexpected interference to travel. There is further value in the conservation of health through driving in a relaxed manner without the tensions necessary where roadside interference is imminent. These facts must be drawn upon to explain *in part* instances on modern highway facilities, such as motorways into urban areas, where the traffic volumes turned out to be substantially greater than those predicted from the origin and destination surveys.

Positive identification of values for assignment to various degrees of comfort and convenience are not yet possible. Present practice in highway economic studies in Britain is to omit this factor altogether, except of course inasmuch as it is already taken into account in the operating costs.

One further feature of the equations in Table 5.3 is that they do not include any values for fuel tax, purchase tax and licence fees. These are considered to be indirect taxes which, although they are a cost to the individual road user, do not represent a real cost to the community but only a transfer within the community, i.e. the revenue represented by these taxes would probably have been raised anyway. Insurance premiums were also excluded from the calculations since to include these and to value savings in

TABLE 5.4. *Breakdown of average costs per accident*[9]

Class of accident	Location	Costs, £ (1968 prices)							Costs, £ (1970 prices)
		Gross loss of output	Medical treatment	Funerals	Damage to property	Administration	Non-resource costs	Total	Total
Fatal	Urban	10 190	60	80	140	50	5270	15 790	18 000
	Rural	11 560	120	100	450	50	6160	18 440	21 000
	Motorways	13 460	150	110	400	50	7180	21 350	24 000
Serious	Urban	210	140	—	150	40	560	1 100	1 200
	Rural	260	170	—	400	40	690	1 560	1 800
	Motorways	280	190	—	390	40	760	1 660	1 900
Slight	Urban	5	15	—	130	30	10	190	210
	Rural	5	20	—	270	30	15	340	380
	Motorways	5	20	—	300	30	15	370	420
Ave. personal injury	Urban	240	50	2	130	30	250	700	1 400
	Rural	590	80	4	330	30	520	1 550	2 300
	Motorways	1 130	90	10	340	30	810	2 410	3 500
Damage only	Urban	—	—	—	80	10	—	90	100
	Rural	—	—	—	100	10	—	110	130
	Motorways	—	—	—	120	10	—	130	150

accidents—this is discussed later—would be double-counting. Interest charges are included in the vehicle-time costs.

Accident costs. Any increase in safety of vehicle operation is of benefit to the road user. Part of this benefit is measurable in monetary terms; the greater part of it, the suffering and sense of bereavement, is not. In Britain, the approach taken used to be to estimate the net costs due to the loss of output caused by death and injury *with due allowance in the case of persons killed for what they might otherwise have consumed*, medical expenses, damage to property and the administration expenses of motor insurance. The reasoning now employed, however, is that the costs determined should be those which relate to the benefits which arise when accidents are prevented, i.e. the accidents which should be costed are those that do *not* occur but which, without the introduction of the road improvement, would have occurred; thus, since the individual concerned is still alive, *his consumption should not be deducted when assessing the benefits* of preventing accidents. In either case, of course, the costs are those which may be attributed to known accidents and do not take into account the costs of accidents which are not reported. They do, however, take notional account of the pain and suffering 'costs' imposed on other members of the community as a result of road accidents (see 'non-resource' costs column in Table 5.4). Costs of accidents in 1968 and 1970 are shown in Table 5.4. Also shown in this table are the constituent features of the 1968 determination.

TABLE 5.5. *Average number of casualties per accident*[9]

Type of casualty	*Class of accident*		
	Fatal	*Serious*	*Slight*
Urban			
Fatal	1·03	—	—
Serious	0·23	1·11	—
Slight	0·25	0·25	1·18
Rural			
Fatal	1·16	—	—
Serious	0·71	1·36	—
Slight	0·50	0·51	1·42
Motorways			
Fatal	1·35	—	—
Serious	0·85	1·50	—
Slight	0·69	0·66	1·52

In practice the figure usually used to assess the cost of accidents is the average value of all accidents per personal-injury accident, i.e. the total cost of personal-injury and damage-only accidents divided by the number of personal-injury accidents. In 1970 the average values for the different types of road were £1400, £2300 and £3500 on urban roads, rural roads and motorways, respectively. The reasons for using an average in this form are that records of damage-only accidents are often very inadequate and, except over a wide area, there will be in any given year only a few accidents of any degree of severity (certainly of fatal and serious ones) and any changes in these numbers will probably not be significant.[10]

If, however, the number of casualties per accident is much different from the average (see Table 5.5), then the cost which should be used in the accident calculations is the cost per casualty. For this purpose the average cost per casualty has been calculated (1968 prices) as £14960, £810 and £25 for fatalities, serious injuries and slight injuries, respectively.

Land value effects. The impact of highway improvements on land uses and values causes very much concern to highway economists. One of the claims often advanced by highway protagonists is that highway improvements enhance land values. This can be illustrated in the following manner.

Consider a highway improvement which is scheduled for a particular place at a specific time. Immediately this is announced it disturbs a previously existing theoretical state of equilibrium. The improvement raises the comparative accessibility values of sites within its zone of influence. It exerts a gravitational pull which channels demand for land in that direction by drawing it away from other areas. A type of chain reaction is set in motion; the highway improvement attracts a factory, the factory attracts employees who seek home sites, the resulting population growth attracts retail stores, service industries, etc.

Generally speaking, however, it is not yet possible to take such long-term land-value effects into account in highway economic investigations. While studies have been carried out in this area, the results obtained are as yet too inconclusive to allow them to be used in engineering practice. (Indeed, it can be argued that the land value effects should *not* be taken into account at all, since they themselves are derived from the traffic benefits which have already been included in the economic assessment.)

Traffic and accident data required for economic studies

Before economic studies can be carried out it is necessary to obtain information about the traffic flows, speeds, delays, accidents, etc., on the roads concerned. Present conditions should be obtained by direct observation and these, when taken in combination with knowledge based on previous research, can be used to predict changes in the future. The following discussion assumes that analysis is to be carried out on a single link of a route.

Traffic flows. The first and most basic step is to measure the existing traffic flow, and its composition, in each direction.

The annual traffic flows on existing facilities may be conveniently estimated by means of sample counts. These counts can be of two types: 1. The counts that are made so as to be sufficiently representative of the whole year, i.e. so that all times of day, all days of the week and all the seasons occur in appropriate proportions. 2. The counts that are not made during representative periods, but in which this is allowed for in the analysis by applying correction factors.

Both methods provide suitable results but for practical reasons the second is the one most usually carried out. For example, one procedure widely used on existing roads in Great Britain is to carry out the volume survey on an appropriate 16-hour day in August, and then to expand the results using the following formulae:[11]

Rural areas	A(150 + 873·60H), when $H < 0{\cdot}30$
Commuter routes	A(289·77 + 479·70H), when $H < 0{\cdot}25$

where A = flow on an average 16-hour August day

and H = proportion of the August traffic composed of heavy and medium goods vehicles.

If $H > 0{\cdot}30$ (or 0·25), then 0·30 (or 0·25) should be used in the grossing-up equations.

Methods of estimating average daily traffic flows have been tabulated by the transport and Road Research Laboratory and are shown in Table 5.6. Each of these methods can be, under particular circumstances, the 'best' one to use, and the decision in any particularly instance is made on the basis of local knowledge and availability of facilities.

The accuracy classification used in Table 5.6 is explained as follows:

Category	*Error exceeded with probability of* 1 *in* 10	*Interpretation*
A	Up to 5 per cent	Very satisfactory
B	5 to 10 per cent	Satisfactory for all normal purposes
C	10 to 25 per cent	Good enough for a rough guide
D	25 to 50 per cent	Unsatisfactory
E	Over 50 per cent	Useless

It should be emphasized that these measures of accuracy are not universally applicable; there may be, for instance, circumstances in which a method classified as C may be relied upon to give perfectly satisfactory results. Again each case should be considered individually in the light of local conditions.

In the case of a new road, origin-destination information will need to be gathered in order to re-assign 'current' traffic to the new road (see Chapter 3—Highway Planning).

Once the current annual traffic volume determinations have been made, it will be necessary to project them ahead to (*a*) the date when the road is opened to traffic, and (*b*) the design year into the future. One way of doing this—and this method is most applicable to rural roads—is simply to expand the measured data using the expansion factors given in Table 5.7.

Alternatively, it may be desired to express the future flow in terms of a single representative hourly flow. An approximation to a representative flow

TABLE 5.6. *Methods of estimating annual average daily flow*[10]

No.	Method: Description	Accuracy classification	Possible variations	Comments on analysis	Remarks
(i)	Count for 1 hour, on a weekday, between 9 a.m. and 6 p.m.	D	Count could be lengthened by any convenient amount		These methods and II(i) are the only ones available if an answer is required at short notice
(ii)	Count on one weekday from 6 a.m. to 10 p.m.	C or D			
(iii)	Count from 6 a.m. to 10 p.m. on a successive Friday, Saturday and Sunday	C	Could be extended to 4 days by including Monday	Estimate week's total as 5 × Friday + Saturday + Sunday	
(iv)	Count from 6 a.m. to 10 p.m. on 7 consecutive days	C			
(v) (vi) (vii) (viii)	As (i) to (iv) but carried out on 4 occasions at 3-monthly intervals (For (i) and (ii) use different hours and days)	C C B B	The number of occasions could be 2, 3 or 6 instead of 4, with appropriate alteration in spacing	As (i) to (iv) to estimate weekly totals; then average the four weekly totals	

TABLE 5.6 (*cont.*)

(ix)	Count from 6 a.m. to 10 p.m. every 52nd day for a year (7 counts in all)	B			These methods are especially useful when counts have to be made at a number of points in the same area. Numerous variations on these methods could be prepared to meet special conditions but it is impossible to list them all here
(x)	Count from 6 a.m. to 10 p.m. every 26th day for a year (14 counts in all)	A or B			
(xi)	Count from 6 a.m. to 10 p.m. every 13th day for a year (28 counts in all)	A			
(xii)	As (ix) but divide the part of the day of interest into 7 equal parts (e.g. of 2 hours each) On each of the 7 days count successively parts 1, 4, 7, 3, 6, 2, 5	C	Other similar arrangements of parts are equally suitable, e.g. 4, 7, 3, 6, 2, 5, 1		
(xiii)	As (x) but divide the day into 14 equal parts and count successively parts 1, 4, 7, 10, 13, 2, 5, 8, 11, 14, 3, 6, 9, 12	C			
(xiv)	As (xi) but again divide the day into 14 equal parts and count successively parts 1, 6, 11, 2, 7, 12, 3, 8, 13, 4, 9, 14, 5, 10 and then repeat the cycle	B			

I Using manual counters only

TABLE 5.6 (*cont.*)

(i)	Continuous count for one week	C			
(ii)	4 continuous counts of 1 week at 3-monthly intervals	B	Replace 4 counts by 2, 3 or 6 at appropriate intervals		
(iii)	Continuous count for 1 year	A			
	II Using automatic counters only				
(i) to (xiv)	As I, but with continuous automatic count for whole year	A	The continuous count could be reduced to 1, 2, 3, 4 or 6 equally spaced 1-week counts. Accuracy would then be C, C, B, B, B	The manual counts should be analysed in the same way as in I(i) to I(xiv), but the results of this should be solely to give the average percentage composition of the traffic. The actual flows of each type of vehicle should be obtained by applying these percentages to the total flow obtained from the automatic count	For automatic counts of less than a full year, the associated manual counts need not be taken, though a useful check is obtained if they are
	III Manual and automatic counts				

Notes: The total number of vehicles counted divided by the number of hours of counting gives the average hourly flow for the times of day covered. The hours 6 a.m. to 10 p.m. may be extended if desired, but usually it will be sufficient to add 10% for the night hours 10 p.m. to 6 a.m.

which gives sufficiently accurate results if the speed in the peak period is >32 km/h is given by:

$$r = \frac{\Sigma q_i^2}{\Sigma q_i} = \bar{q}(1+\mu^2)$$

where r = representative hourly flow
q_i = hourly flow
$\bar{q}$ = annual mean hourly flow
and μ = coefficient of variation of hourly flows.

When data relating to August flows are available, then the representative flow can be determined from

$$r = A\left(0{\cdot}1076H + 0{\cdot}0260 + \frac{1{\cdot}60}{150+874H}\right)$$

where A and H are as defined before.

TABLE 5.7. *Estimated future growth in flow of vehicles*[11] (*Note: 1964 date=100*)

Year of measurement	*Vehicle-kilometres*	*Year of measurement*	*Vehicle-kilometres*
1964	100	1982	245
1966	116	1984	257
1968	134	1986	267
1970	152	1988	277
1972	170	1990	285
1974	188	1992	294
1976	205	1994	302
1978	219	1996	310
1980	233	1998	318

Speed and delay. The next step is the estimation of present speeds or journey times over the roads and/or intersections to be improved or affected by the improvement. It is very important that the speed and time determinations should refer to all stretches of road likely to be affected. In the case of a new highway or a bypass, all existing roads should be studied from which substantial amounts of traffic are likely to be diverted; it is likely that speeds on these roads will rise if traffic is diverted from them.

Speed. Methods of obtaining speeds of traffic on existing facilities are described in detail in the chapter on Highway Planning and so will not be gone into here. Ideally, the most up-to-date speed data at the site being evaluated should be obtained whenever possible. If only the average speed of cars is known, the mean speeds of different classes of commercial vehicles can be estimated from the relationships shown in Fig. 5.2.

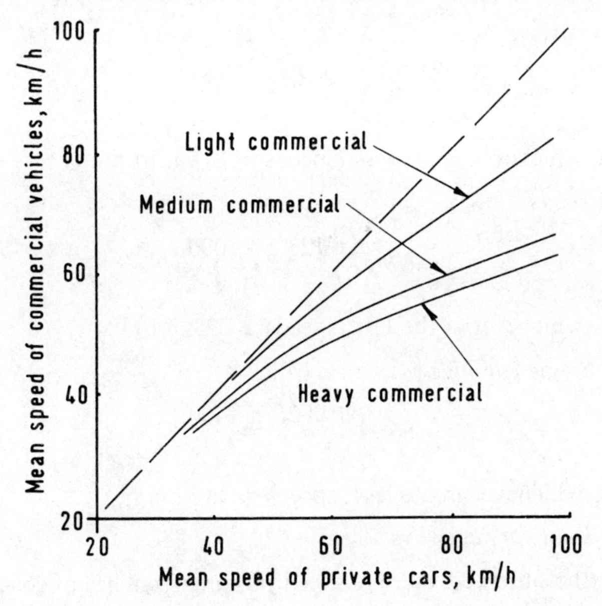

Fig. 5.2. Relationship between mean speed of private cars and commercial vehicles on roads with gradients <4 per cent

Between when the speeds are measured and the time when the road improvement is opened, it can be expected that flows will increase and that measured speeds will have to be amended to allow for this. The following approximate formulas can be used to adjust the running speed of an 'average' vehicle in this situation:

Central urban streets $$\Delta v = \frac{-0{\cdot}163\Delta q}{w-a}$$

Faster urban roads $$\Delta v = \frac{-0{\cdot}123\Delta q}{w-a}$$

Rural roads $$\Delta v = \frac{-0{\cdot}082\Delta q}{w}$$

where Δq = predicted change in traffic flow, veh/h
Δv = corresponding change in running speed per vehicle, km/h
w = present carriageway width, metres
and a = effective reduction in carriageway width due to parked vehicles (see Table 4.1).

Measurements taken at various locations have shown that the following speed-flow formulae can be used to estimate the average speed of all traffic on *new* roads:

Central urban streets

$$v = 49{\cdot}9 - \frac{0{\cdot}163(q+430)}{w-a}$$

or 38·6 km/h, whichever is the less (speeds > 16 km/h)

Faster urban roads

$$v = 67{\cdot}6 - \frac{0{\cdot}123(q+1000)}{w-a}$$

or 56·3 km/h, whichever is the less (speeds > 32 km/h)

Rural roads: Single carriageways

$$v = 86{\cdot}9 - \frac{0{\cdot}082(q+1400)}{w}$$

or 72·4 km/h, whichever is the less (speeds > 38·5 km/h)

Rural roads: Dual carriageways

$v = 74$ km/h (the observed average speed on long stretches of road)

Motorways

$v = 82$ km/h (the observed average speed).

The likely increase in speed resulting from *widening* a road by Δw metres can be estimated from the following formulae:

Central urban streets

$$\Delta v = \frac{0{\cdot}163\Delta w(q+430)}{(w-a)(w+\Delta w-a)},$$

provided that speeds with and without the improvement lie within the range 16 to 38·5 km/h

Faster urban roads

$$\Delta v = \frac{0{\cdot}123\Delta w(q+1400)}{(w-a)(w+\Delta w-a)},$$

provided that speeds lie between 32 and 56·25 km/h

Rural roads

$$\Delta v = \frac{0{\cdot}082\Delta w(q+1400)}{w(w+\Delta w)},$$

provided that q (the total traffic flow) < 2000 veh/h, and the resultant speeds lies between 38·5 and 72·5 km/h.

If the speed obtained is greater than the upper limit quoted, this upper limit should be used instead. On roads subject to a speed limit, speeds may be less than predicted by the formulae.

The increase in speed which may be anticipated as a result of the *horizontal realignment* of a 2-lane rural road can be deduced from:

Class of vehicle	*Reduction in mean speed for each 62·15 deg. of average curvature per km*
Car	4·94 km/h
Goods vehicles	2·69 km/h
All vehicles	3·72 km/h

By average curvature is meant the sum of the deflections of the horizontal curves divided by the length of the section in kilometres. (Any increases in speed determined by these relationships are additional to those due to decrease in distance subsequent to the alignment.)

Increases in speed resulting from *change in gradients* instead of travelling on the level can be estimated from Table 5.8; they are based on observations made on average rural roads under normal traffic conditions.

TABLE 5.8. *Approximate percentage reduction in speed averaged over up-and-down grades on different gradients*

Gradient, %	*Car,* %	*Med. comm. veh,* %	*Heavy comm. veh,* %	*Ave. veh,** %
2	0	1	2	0
3	0	6	10	2
4	0	10	17	3
5	10	19	27	14
6	25	32	48	37

*For a traffic composition of 71% cars, 12% light commercials, 7% medium commercials, and 10% heavy vehicles.

On narrow roads with many heavy vehicles the reduction in speed of smaller vehicles may be greater than those indicated in Table 5.8.

Delay. The above equations relate only to speeds of vehicles when actually moving between major intersections; obviously, however, any economic analysis must also take into account changes in the amount of time during which vehicles are forced to stop or slow down considerably. The effect on delay of improving *signal-controlled intersections* by widening the approaches can be estimated by the Transport and Road Research Laboratory traffic signal formula;[12] this is described in detail in the chapter on Traffic Management. A simplified graphical procedure for estimating the delay at a traffic signal-controlled junction is also available in the technical literature.[13]

The delay which an individual vehicle experiences at a *roundabout* can be regarded as consisting of two components. The first of these might be called the moving delay, since it is caused by the vehicle having to slow down, travel the extra distance about the roundabout and then accelerate to the normal speed of the road. The second component of vehicle delay is the

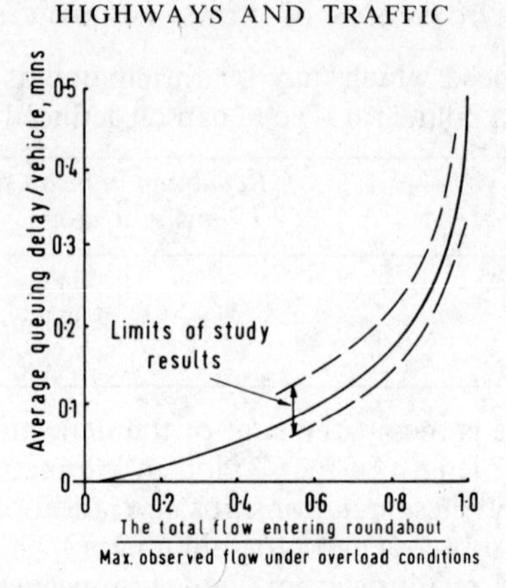

Fig. 5.3. Queueing-delay/flow curves for six roundabouts[10]
Note: These curves refer only to conditions when locking did not occur.

obvious one caused by the vehicle having to queue in order to get into the roundabout and by obstruction from other vehicles while in the roundabout.

Table 5.9 shows the results of some calculations for the first delay component. Values for the second delay component can be seen in Fig. 5.3. which gives the average delay per vehicle as a function of the ratio of the flow to the capacity of a weaving section of a roundabout. It is the sum of these two delay components that is then used in the economic assessment.

TABLE 5.9. *Calculated average delay per vehicle (seconds) at conventional roundabouts, due to 'layout' effects*[11]

Normal speed for road, km/h	*Diameter of roundabout, m*		
	<61	61–122	>122
48	5	6	8
64	6	8	10
80·5	—	12	15

Delays to traffic on minor roads at *priority intersections* are tabulated in Table 5.10. The calculations on which the data in this table are based assume that traffic along the minor road divides at the junction in the proportion 0·5 straight ahead, 0·25 turning left and 0·25 turning right.

TABLE 5.10. *Delays at priority junctions to vehicles on the minor road, sec/veh*

Flow on minor road, veh/day	*Flow on major road, veh/day*					
	5000	10 000	15 000	20 000	25 000	30 000
100	10	15	20	30	49	295
500	12	18	24	35	75	395
2 000	12	18	28	67	370	990
5 000	12	23	138	395	985	>1800
10 000	17	190	985	>1800	>1800	

Accidents. Information on accidents is most necessary in any economic evaluation of the effects of highway improvements. The accidents which should be considered can be divided into two categories, i.e. personal-injury and non-injury (damage only) accidents. The former, however, are generally more serious and important while the reporting of the latter is sometimes unreliable; hence the usual practice is to concentrate attention on the personal-injury accidents and to make some allowance for associated non-injury accidents.

The first step in forecasting possible changes in personal-injury accidents is to establish from accident records the average annual occurrence of such

TABLE 5.11. *Changes which have influenced personal-injury accident rates*[10]

Description of change	*Effect, %*	*Type of accident*
Reconstruction of short lengths of road on new line	− 95	Injury accidents
Reconstruction of length of road on same line	+120	Injury and damage only
Providing dual carriageways in place of 2-way roads	− 30	Injury accidents
Provision of by-passes to small towns	− 25	Injury accidents
Improved alignment at bends	− 80	Injury accidents
Provision of roundabouts	− 50	Injury accidents
Squaring rural right-hand splay intersections	+ 90	Non-pedestrian accidents
Staggering of cross-roads	− 60	Injury accidents
Provision of stop signs	− 80	Injury accidents involving emerging vehicles
Provision of slow signs	− 75	Injury accidents
Automatic traffic signals	− 40	Injury accidents
Addition of 'all-red' period at 4-way intersections with traffic signals	− 40	Injury accidents
Improving slippery surfaces	− 80	All accidents on wet roads (injury and damage only)
	− 45	All accidents
Removal of tram tracks	− 10	Injury accidents
Improvements of street lighting	− 30	Injury accidents in dark
Reconstruction of bridges and culverts	− 70	Injury accidents
Improving visibility at bends	variable	Injury accidents
Improving visibility at intersections	− 30	Injury accidents
Major resurfacing improvements on rural roads (other than slippery roads)	+ 65	Injury accidents
Provision of pedestrian guard rails	− 10	Pedestrian accidents
'No-waiting' regulations	up to 30	Injury accidents
One-way systems	− 30	Injury accidents on all roads concerned
Replacement of trams by buses in London	− 9	Injury accidents on roads concerned

accidents over several years on the stretches of road to be improved and which will be affected by the improvement. This number is then corrected for past and future traffic increases by assuming that personal-injury accidents will increase at the same rate as the total traffic. The result is an estimate of the number of accidents before the improvement is opened at the future traffic flow.

The change in accidents resulting from the improvement may then be estimated by either or both of two different ways. The first method uses the data from before and after studies, which give an indication of the percentage change in accidents which can be expected from various changes. Table 5.11 gives a list of some such data; these have been judged significant at the 5 per cent level in bringing about changes in the existing accident rate at particular locations. The second way of estimating changes in accidents due to road improvements is by using known data regarding accident rates on similar types of road. The use of known accident rates is preferable when dealing with by-passes, entirely new facilities, or radical road improvements, since the improvement can then be regarded as the transfer of traffic from conditions bearing one accident rate to conditions bearing an entirely different rate. Some details regarding accident rates on different types of facilities are given in Table 5.12; summaries of many other investigations are readily available in the literature[14].

TABLE 5.12. *Personal-injury accidents per million vehicle-kilometres on various types of roadway in Britain*[11]

Type of road and development	*Accidents/10⁶veh-km*
Shopping centres	5–8·1
Residential	2·5–4·4
Country roads	0·9–1·6
3-lane roads	1·3
Dual carriageways—rural	1
Dual carriageways—urban	3
Motorways	0·4

Generated traffic. The benefits to new traffic generated by the road improvement should be valued at half the change in user costs, including time savings, per generated vehicle[11].

COST-BENEFIT METHODS

The field of economics literally abounds with concepts and procedures by which road improvement economic studies might be theoretically, if not practically, carried out. It is not possible within the limited scope of this text to describe these in detail and the reader is referred to the literature[1,2,16,17,18,19] for many excellent treatises and references. The following discussion is primarily concerned with those procedures which are of particular interest to highway engineers in this country.

Net Present Value method

This procedure is also known as the *Present Worth* method. To assess whether a scheme is worthwhile, the method requires the net present value of the improvement to be calculated. This represents the present market value of the project, i.e. the surplus over and above the capital costs of the scheme. To use the net present value method the following information is required: 1. An assessment of the capital and maintenance costs, and user benefits attributable to the project. 2. The economic life of the project. This may/may not coincide with its physical life. (A 30-year economic life is a common assumption in Great Britain.) 3. The value of the rate of interest, or discount, to be applied to the scheme. This determines the relative weights which are attributed to benefits and costs at different time points in the life of the project. (The discount rate currently applied to government investments in Great Britain is 10 per cent.)

With this information available, the following calculation is carried out:

$$NPV = \sum_{t=0}^{t=n} \frac{A_t}{(1+r)^t} - K$$

where A_t = user benefits, net of maintenance costs, accruing in each time period,
K = capital expenditure,
r = rate of discount to be applied,
and t = number of years ahead at which each benefit will accrue.

If the capital expenditure is to take place over more than 1 year, then it has to be discounted back to its present value before being used in the above equation.

A positive *NPV* indicates that the project is acceptable in that it is able to repay the interest charges implied by the chosen discount rate and yet yield a surplus. If the net present value is negative, then the project is unacceptable and should be postponed or rejected.

The *present value* method has been put forward by the Transport and Road Research Laboratory[11] as being an appropriate way 'of bringing scientific technique to bear as an aid to administrative judgement'. The steps involved in the *TRRL* procedure are as follows:

Step 1. (*a*) For each year for which the benefits are being determined, calculate (i) the difference in operating costs in conditions with and without the improvement, on all roads or sectors of road affected by the improvement, (ii) the reduction in accident costs as a result of the improvement, and (iii) the gains to generated traffic. (*b*) Sum (i), (ii), and (iii) to obtain the total benefits in each year. (*c*) Discount the benefits to their present value. (*d*) Sum the benefits over the life of the scheme (very often taken as 30 years).

Step 2. (*a*) Calculate the following costs for each of the years in which they occur: (i) capital cost, (ii) maintenance costs, and costs of delay to traffic during construction. (*b*) Sum these to give total costs in each year. (*c*) Discount total costs to present values using the same base year as for the benefits.

Step 3. Subtract the discounted costs from the discounted benefits in order to obtain the net present value of the investment.

Step 4. Express the net present value as a proportion of the discounted capital cost. This gives the present value/cost ratio. (If there is a budgetary restraint on the amount of capital available, then this step can be used to order different schemes and accept them until the budget capital supply is exhausted.)

Step 5. For the first year of full operation, express the net benefit i.e. total benefits less maintenance costs, for that year as a percentage of all the capital costs incurred by that date. This is the annual rate of return for the first year. (If the annual rate of return is less than the discount rate used in the calculation, the project should not be initiated as planned, as the net present value will be increased by postponement; rather the project should be started during the year when the annual rate of return equals the discount rate, as this is when the maximum net present value will be achieved, i.e. postponement beyond this date will mean that the costs saved—the discount rate applied to the capital cost—will be less than the benefits foregone.)

Examples of the use of this technique are readily available in the literature.[11]

Short-term rate-of-return method

This procedure is a very simplified one. In general the rate-of-return is obtained by calculating for *'current' traffic* (i.e. when the improvement will be open to traffic), the first year savings in labour, vehicle-time and operating costs over the road to be improved or affected by the improvement, adding the expected annual savings in accident costs, deducting average annual maintenance costs and expressing the result as a percentage of the total capital cost of the improvement at the time it is opened. In equation form the first year rate of return may be expressed as

$$R = \frac{O + A - M}{C} \times 100$$

where R = rate-of-return, %,
O = savings in labour, vehicle-time, and operating costs,
A = savings in costs of accidents,
M = additional maintenance costs,
and C = capital cost of improvement.

Internal rate-of-return method

The procedure which will now be discussed produces a long-term rate-of-return which can be defined as that rate which discounts all future yield to equal the initial investment. It attempts to allow for some of the limitations in the short-term rate-of-return method described above by recognizing that the future must be taken into account in the analysis. In its simplest form the method can be described as follows.

The capital cost of the improvement is first determined in the manner described previously. Next the benefits (i.e. labour, vehicle-time, operating and accident savings) occurring in each year in the future are estimated separately and from these are taken the highway costs (i.e. maintenance costs) likely to be experienced in the same years. The values obtained are then substituted in the following formula and the rate-of-return calculated:

$$C = \frac{(B_1 - M_1)}{1+r} + \frac{(B_2 - M_2)}{(1+r)^2} + \ldots + \frac{(B_n - M_n)}{(1+r)^n}$$

where C = value (i.e. capital cost) of the investment in the year zero,
B = value of the benefits which occur in a particular year,
M = maintenance costs incurred in that year,
n = number of years into the future for which the rate-of-return is to be calculated,
and r = rate-of-return per year.

This economic formula is very old and is often used by engineers. From the formula it can be seen that if the calculated rate-of-return is greater than the rate of interest obtainable by investing the capital cost on the open market, then the road investment is worthwhile.

Benefit-cost ratio method

The benefit-cost method of analysis, which is the procedure favoured by highway authorities in the United States, attempts to assess the merit of a particular scheme by comparing the annual benefits from the scheme with the increase in annual costs. The increase in annual highway cost due to the proposed improvement is equal to the annual maintenance on the new highway plus its estimated annual amortized cost (irrespective of whether the facility is financed from public loan or not), less the annual cost on the old facility. The comparison is made by a simple formula in which the estimated annual road user benefits are divided by the increased annual highway costs in order to obtain the required benefit-cost ratio. Thus:

$$B/C \text{ ratio} = \frac{\text{annual benefits from scheme}}{\text{annual costs of scheme}}$$

$$= \frac{R - R_1}{H_1 - H}$$

where R = total annual road user cost for the basic condition; this is usually the existing road,
R_1 = total annual road user cost for the proposed scheme; this includes travel on existing facilities affected by the scheme,

H = total annual highway cost for the basic condition,

and H_1 = total annual highway cost for the proposed scheme.

In using the ratio to appraise a proposed project, a B/C value of less than 1 is taken to indicate a poor investment; this means that the benefits derived by the highway users are less than the funds to be invested in the highway. However, because of the general deficiency of highway construction funds, any project is usually required to show a ratio of considerably more than unity. High-priority schemes may well have ratios of 10 to 1 or more.[20]

In using the procedure for the selection of an alternate location or design, each alternative is normally compared with the basic, or existing, condition with the same total volume of traffic being used in all cases. The preferred alternative indicated by the analysis is the one for which the B/C ratio is the highest. Where several major alternatives are under study, however, a second analysis is made, using the 'preferred' alternative as the base, to determine if an added increment of investment might yield a proportionately larger increase in road user savings on another alternative.

SELECTED BIBLIOGRAPHY

1. A.A.S.H.O. *Road User Benefit Analysis for Highway Improvements.* Washington, D.C. American Association of State Highway Officials, 1959.
2. WINFREY, R. Concepts and applications of engineering economy in the highway field, *Highway Research Board Special Report* 56, 1959, 19–33.
3. DAWSON, R. F. F. Analysis of the cost of road improvement schemes, *Road Research Technical Paper* No. 50. London, H.M.S.O., 1961.
4. RESEARCH COMMITTEE ON DESIGN AND CONSTRUCTION. The Cost of Constructing and Maintaining Flexible and Concrete Pavements Over 50 Years. *RRL Report* LR 256, Crowthorne, Berks., The Road Research Laboratory, 1969.
5. DAWSON, R. F. F. Vehicle Operating Costs in 1970. *TRRL Report* LR 439, Crowthorne, Berks., The Transport and Road Research Laboratory, 1972.
6. EVERALL, P. F. The Effect of Road and Traffic Conditions on Fuel Consumption. *RRL Report* LR 226, Crowthorne, Berks., The Road Research Laboratory, 1968.
7. COMMISSION ON THE THIRD LONDON AIRPORT. Papers and Proceedings, Vol. 3, Stage III. London, H.M.S.O., 1970.
8. BEESLEY, M. E. The value of time spent in travelling: some new evidence, *Economica*, 1965, **32,** 174–185.
9. DAWSON, R. F. F. Current costs of road accidents in Great Britain. *RRL Report* LR 396, Crowthorne, Berks., The Road Research Laboratory, 1971.
10. ROAD RESEARCH LABORATORY. *Research on Road Traffic.* London, H.M.S.O., 1965.
11. DAWSON, R. F. F. The Economic Assessment of Road Improvement Schemes. *Road Research Technical Paper* No. 75, London, H.M.S.O., 1968.
12. WEBSTER, F. V. Traffic signal settings, *Road Research Technical Paper* No. 39. London, H.M.S.O., 1958.
13. DICK, A. C. A monogram for solution of the R.R.L. traffic signal delay formula, *Traffic Engineering and Control*, 1964, **6,** No. 2, 106–108.
14. ROAD RESEARCH LABORATORY. *Research on Road Safety.* London, H.M.S.O., 1963.

15. James, J. G. Quantities and Prices in New Road Construction, 1969: A Brief Analysis of 60 Successfull Tenders. *TRRL Report* LR 513, Crowthorne, Berks., The Transport and Road Research Laboratory, 1972.
16. Grant, E. L. and W. G. Ireson. *Principles of Engineering Economy.* New York, Ronald Press, 1960.
17. Institution of Civil Engineers. *An Introduction to Engineering Economics.* London, The Institution, 1962.
18. Winfrey, R. Highway Economics. In Woods, K. B. *Highway Engineering Handbook.* New York and Maidenhead, McGraw-Hill, 1960.
19. Gwilliam, K. M. Economic evaluation of urban transport projects. The state of the art, *Transportation Planning and Technology*, 1972, **1,** No. 2, 123–142.
20. Loutzenheiser, D. W., W. P. Walker and F. H. Green. Resumé of A.A.S.H.O. report on road user benefit analyses, *Highway Research Board Special Report* No. 56, 1959, 36–42.

6 Geometric design

Geometric design is primarily concerned with relating the physical elements of the highway to the requirements of the driver and the vehicle. It is mainly concerned with those elements which make up the visible features of the roadway, and it does not include the structural design of the facility.

Features which have to be considered in geometric design are, primarily, horizontal and vertical curvature, the cross-section elements, highway grades and the layout of intersections. The design of these features is considerably influenced by driver behaviour and psychology, vehicle characteristics and trends, and traffic speeds and volumes.

Proper geometric design will inevitably reduce the number and severity of highway accidents while ensuring high traffic capacity with the minimum of delay to vehicles. While these are the main factors to be considered, care must also be taken that the highway presents an aesthetically satisfying picture to both the driver and the onlooker. The aim should be to design a facility that blends harmoniously with the topography, and not one that leaves an ugly scar on an otherwise pleasant landscape. When geometric design is improperly carried out, it may result in early obsolescence of the new highway, with consequent economic loss to the community.

DESIGN SPEED

A most important consideration in geometric design is the design speed of the highway. This can be defined as the highest continuous speed at which individual vehicles can travel with safety on the highway when weather conditions are favourable, traffic volumes are low, and the design features of the highway are the governing conditions for safety. As such the choice of a particular design speed will have a considerable effect on the design of such geometric features as horizontal and vertical curvature, safe stopping and passing sight distances, and acceptable highway grades.

The primary factors affecting the choice of design speed are the type of highway and the terrain through which it will run. Other qualifying factors are traffic characteristics, cost of land, speed capabilities of vehicles and aesthetic features. Proper consideration of these factors should result in a design speed being selected that is acceptable to all but a very few drivers on the new facility.

Rural roads

In practice the design speeds for rural roads are often selected by administrative decision, although in certain instances they may be selected from an analysis of actual speeds on similar stretches of existing highways. Table 6.1 shows the design speeds recommended by the Department of the Environment for rural roads in Britain. The design speeds for dual carriageways in this table are also those used for rural motorways.

TABLE 6.1. *Design criteria for new rural roads in Britain*[1]

Design speed, km/h	*Carriageway width, m*	*Design capacity, pcu/16h-day*	*Min radius, m*		*Min radius for curves without transitions, m*
			Desirable	*Absolute (with max 0.07 super-elevation)*	
120	Dual 14·60	66 000	1 000	510	1 500
120	Dual 11·00	50 000	1 000	510	1 500
120	Dual 7·30	33 000	1 000	510	1 500
100	10·00	15 000	650	350	1 350
100	7·30	9 000	650	350	1 350
80	7·30	9 000	400	230	1 200
60	—	—	250	130	600

There is still debate as to whether speeds greater than 120 km/h should be used for design purposes on motorways. It is often pointed out that a higher design speed not only safeguards against early obsolescence of the highway, but also provides an increased margin of safety for those driving at high speeds. That there is some validity in this statement is reflected by the fact that the design speed of high-type roads is now 120 km/h as compared with 56 km/h in 1927, this change being primarily due to the continued increase in vehicle performance.

The choice of a design speed for a dual carriageway highway is much less subject to cost considerations than that for other rural roads. In practice, lower design speeds are often accepted on single carriageway roads in order to keep construction costs within certain limits. There is danger in this philosophy since, although drivers will obviously accept lower values in what are clearly difficult locations, repeated studies have shown that they do not adjust their speeds to the importance of the facility. Instead they endeavour to operate at speeds consistent with the traffic on the facility and its physical limitations.

Observations in rural areas have shown that the speeds exceeded by the faster car drivers are closely related to the mean speed of cars. Studies have shown that the 85 and 95 percentile speeds are approximately 1·19 and 1·32 times the average speeds of cars.[2] Hence, if the desires and travel habits of drivers are taken into account and each individual highway is treated on its own merits, these relationships can be used to select the proper design speed.

When designing a substantial length of highway, it is most desirable that a constant design speed be used. However, this may not always be feasible because of topographical or other physical limitations. Where these occur the change in design speed should not be introduced abruptly, but instead a transition stretch of adequate length should be inserted. Within this stretch drivers should be encouraged to reduce speed gradually by means of adequate signing notifications.

Urban roads

The design speed of an urban motorway should normally be as high as possible. However, the basic consideration in choosing the design speed is the running speed desired for capacity flow during the peak hours. For instance, American experience with urban motorways indicates that, if the design speed is 80 km/h, the running speeds will be lowered to about 48 km/h during the peak traffic hours. Little is gained by using a design speed greater than is necessary for this purpose, since only a comparatively few drivers are benefited during periods of light traffic. From an economic point of view this can rarely be justified, especially since higher speeds are more easily regulated in urban areas.

Recommended design speeds for urban roads in Britain are as follows:

Primary distributors:	Urban motorway	80 km/h
	All purpose	60 km/h
District distributors, local distributors, important access roads:		50 km/h

Primary distributors as noted in the above, form the primary network for the town as a whole; all longer-distance traffic movements to, from and within the town are normally canalized onto these routes. *District distributors* distribute traffic within the residential, industrial and principal business districts of the town; they form the link between the primary network and the roads within environmental areas (i.e. areas free from extraneous traffic in which considerations of environment predominate over the use of vehicles). *Local distributors* distribute traffic within environmental areas; they form the link between district distributors and access roads. *Access roads* give direct access to buildings and land within environmental areas.

DESIGN FOR CAPACITY

Basic considerations

The term capacity is used here to define the ability of a road to accommodate traffic under given circumstances. Factors which must be taken into account in determining the governing circumstances are the physical features of the road itself and the prevailing traffic conditions. In addition, it must be considered whether capacities for uninterrupted or interrupted traffic flow are being evaluated. This is important since uninterrupted flow conditions apply to road sections along which intersectional flows do not interfere with continuous movement; on the other hand, interrupted flow capacities refer to highways with junctions at-grade where intersecting flows interfere with each other.

Prevailing road conditions. Capacity figures for uninterrupted flow on highways have to be modified if certain minimum physical design features are not adhered to. Poor physical features which tend to cause a reduction in capacity are as follows:

1. *Narrow traffic lanes.* Lane widths of 3·65 m are now accepted as being the minimum necessary for *heavy* volumes of mixed traffic.

2. *Inadequate shoulders.* Too narrow, or lack of, shoulders alongside a road cause vehicles to travel closer to the centre of the carriageway, thereby increasing the medial traffic friction. In addition, vehicles making emergency stops must of necessity park on the carriageway. This causes a substantial reduction in the effective width of the road and thereby reduces capacity.

3. *Side obstructions.* Vertical obstructions such as poles, bridge abutments, retaining walls or parked cars that are located within about 1·75 m of the edge of the carriageway contribute towards a reduction in the effective width of the outside traffic lane.

4. *Imperfect horizontal or vertical curvature.* Long and/or steep hills and sharp bends result in inadequate sight distance. Drivers are then restricted in opportunities to pass and hence the capacity of the facility will be reduced.

In addition to the above physical features, the capacities of certain rural roads and the great majority of urban roads are controlled by the layouts of intersections on these roads. Some physical features having considerable influence are the type of intersection, i.e. whether plain, channelized, roundabout or signalized, the number of intersecting traffic lanes and the adequacy of speed-change lanes.

Prevailing traffic conditions. Unlike the physical features of the highway, which are literally fixed in position and have definite measurable effects on uninterrupted and interrupted traffic flows, the prevailing traffic conditions are not fixed but vary from hour to hour throughout the day. Hence the flows at any particular time are a function of the speeds of vehicles, the composition of the traffic streams and the manner in which the vehicles interact with each other, as well as the physical features of the roadway itself. An understanding of how capacity is affected by traffic conditions can be obtained by examining Fig. 6.1.

The term concentration used in this figure is the same as the term 'density' used in the American traffic literature, and is defined as the number of vehicles occupying a unit length of a traffic lane at a given instant. Concentration is usually expressed in vehicles per kilometre.

A not too inaccurate picture of the relationship between speed and concentration is given by the curves in this figure. One of the curves is based on an early study conducted in the United States on speed-flow relationships,[3] the other on speed-flow and headway-speed data collected by the Transport and Road Research Laboratory.[4] Both of these curves indicate quite clearly that as the concentration per lane increases the average speed of traffic decreases. The kink introduced at the top of each curve reflects the manner in which vehicle speeds flatten out for each road at an independently determined free-speed. A flat portion such as this must be expected on any speed-concentration curve, since the mean speed of traffic will be unaffected by concentrations below a certain limiting value; on wide roads this may be as much as 30 veh/km. In addition, the average rate of speed decrease can be expected to vary with the physical features of the road; for instance the slope will be steeper for narrow roads, but more gradual for wider roads.

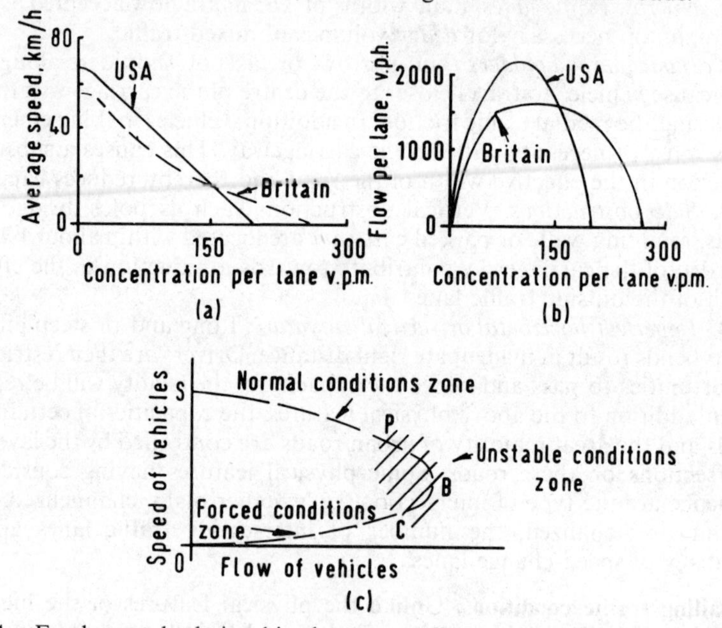

Fig. 6.1. Fundamental relationships between traffic concentration, speed and flow on a highway

The relationships between flow and concentration for the same data are shown in Figure 6.1(b). The term flow as used in this context refers to the number of vehicles that can pass a given point in unit time. It is usually expressed in vehicles per hour. The basic features of these curves are most important. As the concentration of vehicles on the road tends towards zero, obviously the flow must also tend towards zero. On the other hand, when the concentration is very high, vehicles travelling in a given direction are in a saturated condition of road occupancy, each vehicle being nearly bumper to bumper with the vehicle in front. This represents an almost impossible operating condition so that the flow of traffic again tends towards zero. Practically, some headway must be allowed between vehicles for flow to take place. The headway distribution at any particular time on any given road is dependent on the traffic composition, driver perception and reaction times, brake application time, braking distance and a suitable factor of safety. Finally, at some concentration value between the two extremes, the flow is a maximum value which can be described as the *ultimate* capacity of the road.

From these two figures, it is possible to obtain a relationship between speed and flow for the given road situations. Figure 6.1(c) indicates the form that such a relationship might take, and illustrates the following very basic features relating to traffic movement:

1. The portion *SP* on the curve, termed the zone of normal conditions, represents the situation where free driving occurs on the highway. Within this zone, mean speeds are higher and more variable, and traffic flow increases as speed decreases. This arm marks a relationship between speed and flow that reflects primarily on the prevailing highway geometric standards. The higher the standards of highway design, the freer are the driving conditions and the flatter the curve *SP*. On the other hand, the lower the standards, the steeper the curve *SP* and, eventually, the lower the capacity of the highway.

2. The portion *OC* is known as the zone of forced conditions. Within this zone mean speeds are much lower and vehicles move under conditions of forced driving. A decrease in speed under these saturated conditions is associated with a decrease in traffic flow. The shape of the arm *OC* depends primarily on the interaction between vehicles. The concentration is very high and hence the control exercised on each vehicle by the one in front has a most important influence on the flow. To all intents, therefore, the curve *OC* is relatively independent of the standard of geometrical design of the road.

3. The portion *PBC* can be described as the zone of unstable conditions. In this zone, flows are very high but driving conditions are very unstable, i.e. it is possible for vehicles to be moving along under free driving conditions and then, under the influence of some restricting factor which may or may not be determinable, forced conditions of driving may supervene even though the flow may be the same as before. In fact, within this zone, traffic conditions seldom remain sufficiently stable to obtain reliable survey figures during operational measurement of traffic flows—which is why the ultimate capacity (at point *B*) is never chosen for design purposes, but rather some point about *P*.

The influence of the geometrical layout on uninterrupted flow is greatest within the zone of normal conditions. Within this range, the individual driver is most content with driving conditions towards point *S* in Fig. 6.1(c). His speed can be high and little influenced by other vehicles. The fact that traffic flow is very small is not a relevant consideration in the driver's mind. The highway engineer, on the other hand, is most interested in flows towards the point *P*, in order to get the most economical return from the highway investment.

The dilemma posed by these two differing points of view is only resolved when the curve *SP* is as flat as possible, so that there is only a relatively small drop in speed as traffic flow increases. This shape of curve is only obtained when the geometric features of the highway are built to a very high standard.

British capacity studies

The procedure adopted in Britain has been to study capacity in terms of the relationship between the average running speed of vehicles and the traffic flow. This decision was based on the acceptance of loss of time as the factor with which drivers are most concerned, and this is easily determined from measurements of running speed values.

The results of one of the more important British studies on various types of road are shown in Fig. 6.2. The speed referred to in this graph is the average running speed, which is the average speed of the vehicles while in motion. There are certain basic relationships to be ascertained from this figure. For instance, the speed decreases on each road as the flow increases, indicating that these measurements lie within the zone of normal conditions. Secondly, traffic flow at any particular speed increases as the roads get wider. For all roads, the rate of speed decrease gets smaller as the road width increases, indicating the lessening effects of lateral obstructions and the greater opportunities for overtaking. Note that the capacity of one wide two-way road is greater than the combined capacities of two narrow two-way roads of the same total width.

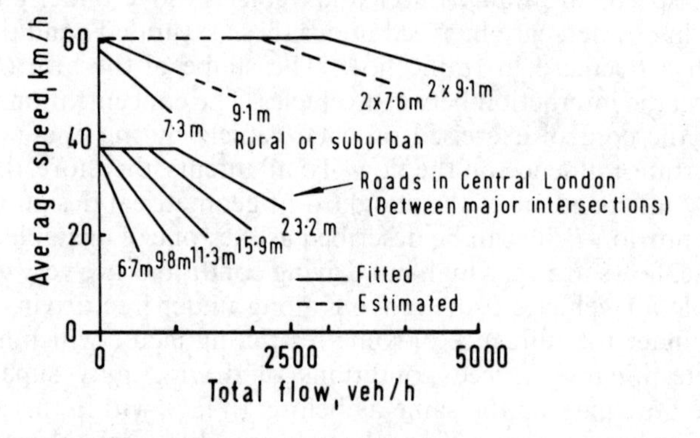

Fig. 6.2. Speed-flow relationships for traffic in Britain[5]
Traffic conditions: Single carriageways = 60 per cent commercial vehicles
Dual carriageways = 30 per cent commercial vehicles

A further point to be emphasized is the very significant difference between speeds on rural and urban roads. This can be attributed to the effects of such factors as parked cars, pedestrian conflicts, heavier concentrations of commercial vehicles and, most important, speed limits and the greater caution of drivers in built-up areas.

From this study, the practical capacity of a road was defined as that flow which produced a minimum journey speed acceptable to drivers. This means that for each individual road there is a practical capacity that can be used for design purposes, and this design value is entirely dependent on the average speed that is selected as being acceptable for the road concerned. Values suggested as appropriate average running speeds were 24 and 26 km/h for urban and rural roads. Some capacity figures obtained at these speeds are given in Table 6.2. In this table, the 4- and 6-lane rural roads are dual carriageways on which the traffic is composed of 70 per cent passenger cars and 30 per cent commercial vehicles. The traffic on the single carriageways is composed of 40 per cent passenger cars and 60 per cent commercial vehicles.

TABLE 6.2. *Suggested practical capacities of urban and rural roads*[5]

Total number of lanes	*Capacity, v/h per lane*	
	Urban, 24 km/h	*Rural, 56 km/h*
2	150	—
3	230	230
4	350	750
6	380	750

Department of Environment recommendations for the design of new rural roads are given in Table 6.1. In this table, the capacities are expressed in terms of total 16-hour (i.e. 6 a.m. to 10 p.m.) daily passenger car volumes and are defined as the maximum traffic volumes that will ensure comfortable free-flow conditions for the movement of traffic. To allow for mixed traffic conditions, the equivalent traffic values of various types of vehicle in relation to passenger cars on rural roads are given in Table 6.3. The capacity values were arrived at by taking the following maximum hourly capacities and assuming them equal to 10 per cent of the 16-hour day:

7·3 m *single carriageway*	900 *p.c.u./h total for both directions*
10·0 m *single carriageway*	1500 *p.c.u./h total for both directions*
Dual carriageways	1000 *p.c.u./h per lane in the direction of heavier flow, assuming one carriageway carries 60% of the total flow in both directions*

Governmental recommendations for the design of non-motorway urban roads are available in two forms. The first of these refers to the case where it is intended to widen an already existing roadway. Figure 6.3 shows recommendations regarding the variation of speed with flow for typical urban streets with no parked vehicles present, and this diagram can be used to estimate the effects of widening a carriageway. For example, if a 9·1 m wide carriageway carrying 1300 p.c.u./h is widened by 1·5 m, an increase in average running speed of 3·7 km/h may be expected. Conversely, if the speed is maintained constant there will then be an increase in capacity of 350 p.c.u./h. Care should be taken in using this diagram, however, since conditions can vary considerably and so the relationship expressed here may not give the same result as those observed on a particular street in a practical situation.

Recommendations referring to practical (design) capacities for both new and existing urban roadways are given in Table 6.4. These capacity recommendations should make it possible to attain average speeds under peak-

TABLE 6.3. *Equivalent passenger car units (pcu's) to be used in capacity determinations at different locations.*

Type of vehicle	*Urban standards*	*Rural standards*	*Roundabouts*	*Traffic signals*
Private car, taxi, motor cycle combinations, light goods veh. (1·5 tonnes unladen)	1·00	1·00	1·00	1·00
Motor cycle (solo), motor scooter, moped	0·75	1·00	0·75	0·33
Medium or heavy goods vehicles (>1·5 tonnes), horse-drawn vehicle	2·00	3·00	2·80	1·75
Bus, coach, trolley bus, tram	3·00	3·00	2·80	2·25
Pedal cycle	0·33	0·50	0·50	0·20

hour conditions of about 56–64 km on the high quality roads of types A and B. No speeds are specified at which the capacities on types C and D are

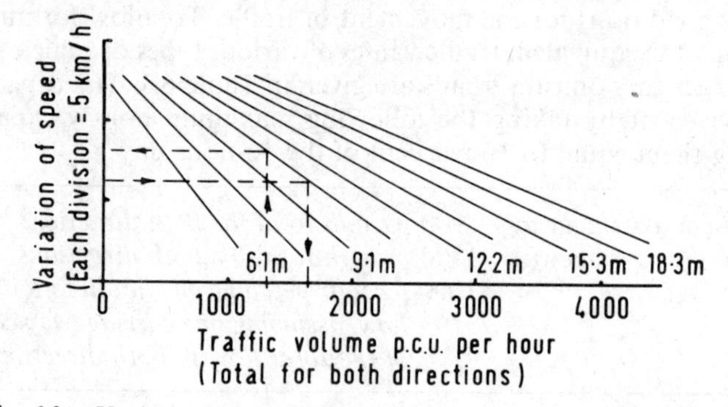

Fig. 6.3. Variation of speed with flow for two-way urban streets

Note: These curves refer to flows between major intersections and no waiting vehicles on the carriageway[6]

achievable. The capacities given in Table 6.4 should be applied so that for divided roads the design is based on the peak-hour demand in the heavier direction of travel; on undivided roads the design capacities are based on the total peak-hour flow for both directions of travel. The capacities of dual 3-lane carriageway roads of type C1 will be higher than for roads divided by refuges, and so the recommended values in Table 6.4 may be increased by 200, 450 and 700 p.c.u./h for carriageway widths of 18·3, 20·1, and 21·9 m, respectively.

The values given in the table do not apply to one-way streets. Recommended values for these facilities are tabulated in the chapter on Traffic Management.

TABLE 6.4. *Recommended design capacities for urban roads*[6]

Effective width of carriageway (excluding refuges or central reservation) m		*2-lane*			*3-lane*		*4-lane*			*6-lane*		
		6·1	6·7	7·3	9·1	10·0	12·2	13·4	14·6	18·3	20·1	21·9
		Capacity in p.c.u./h										
Type of road	*Description*	*Total for both directions of flow*					*For one direction of flow*					
A	Urban motorway, no frontage access, grade separations at intersections								3000			4500
B1	All purpose road, no frontage access, no standing vehicles, negligible cross traffic	1200	1350	1500	2000	2200	2000	2200	2400	3000	3300	3600
C1	All purpose street, 'no waiting' restrictions on parking of vehicles, high capacity junctions	800	1000	1200	1600	1800	1200	1350	1500	2000	2250	2500
D1	All purpose streets, capacity severely restricted by waiting vehicles and intersections	300 to 500	450 to 600	600 to 750	900 to 1100	1100 to 1300	800 to 900	900 to 1000	1000 to 1200	1300 to 1700	1500 to 2000	1600 to 2200

To allow for mixed traffic conditions between major junctions on city streets, equivalent traffic values of various types of vehicle in relation to passenger cars are given in Table 6.3. It should be noted that these ratings are not intended to reflect the influence of these vehicles at roundabouts and traffic signals; separate vehicle weightings should be used for traffic at these locations.

At-grade intersections. The capacities of certain rural and suburban roads and the great majority of urban streets are influenced tremendously by the traffic restrictions imposed by closely spaced at-grade intersections, parking and unparking manoeuvres, public transport vehicles, pedestrian movements, and generally low geometric design standards. Of these, intersections generally form the major deterrents to free traffic flow. For instance, one study by the Transport and Road Research Laboratory showed that six major intersections per 1·6 km caused a loss in capacity of about 40 per cent along a major city street.

For capacity purposes, at-grade intersections can be divided into the following main types:

1. Uncontrolled and priority intersections, where one road takes precedence over another. At either type of intersection little or no delay is caused to the traffic on the major road.

2. Space-sharing intersections where weaving of traffic can take place. Roundabout intersections are the most typical examples.

3. Time-sharing intersections where the right-of-way is transferred from one traffic stream to another in sequence. While police-controlled intersections can be considered as belonging to this category, the most important from a capacity point of view are those controlled by automatic traffic signals.

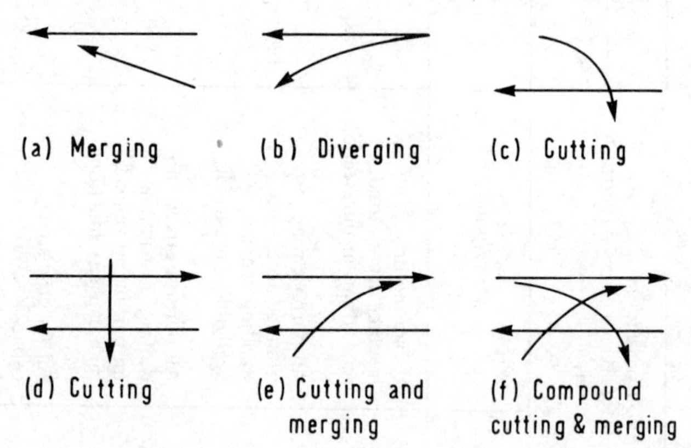

Fig. 6.4. Traffic movements at uncontrolled and priority intersections

Uncontrolled, and priority intersections. At these intersections the traffic flow from the minor road enters the main road only during 'spare time' gaps on the major road. The manoeuvres involved in so doing are illustrated in Fig. 6.4. The capacity of either type of intersection is dependent on the ratio of the flows on the major and minor roads, the minimum gap in the main road traffic stream acceptable to the merging or crossing traffic, and the maximum delay that is acceptable to the minor road traffic.

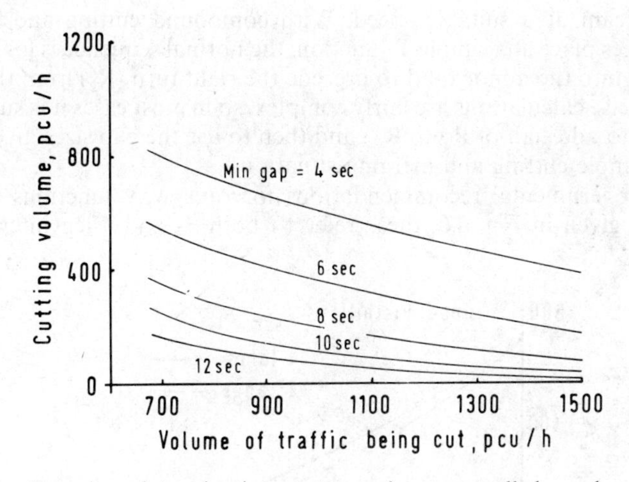

Fig. 6.5. Capacity determinations at rural uncontrolled and GIVE WAY intersections[1]

As is to be expected when traffic builds up on the major road, fewer gaps become available and delays on the minor road increase accordingly, theoretically to infinity. The maximum flow that can enter the intersection from the minor road has been theoretically deduced as follows:[5]

$$q_m = \frac{q_M[1 - \beta_M q_M]}{e^{q_M(\alpha - \beta_M)}[1 - e^{-\beta_m q_M}]}$$

where q_m = maximum flow from the minor road,

q_M = flow on the major road,

β_M = minimum headway in the major road traffic stream,

β_m = minimum headway in the minor road traffic stream,

α = minimum acceptable gap in the major road traffic stream,

and e = base of the natural logarithm.

The sum of q_m and q_M gives what is defined as the ultimate capacity of the intersection for the particular flow ratio.

The outcome of this theoretical work in relation to the capacities of *rural* road junctions is given in Fig. 6.5. Note that these curves relate to *cutting* and *cutting and merging* vehicles emerging from a single lane on the minor road; these movements take place in suitable-duration gaps in the traffic stream, with the larger gaps being utilized by several vehicles cutting at one time. For one-way flows (e.g. across one carriageway of a dual-carriageway road), average minimum gaps of 4–6 sec should be chosen, while gaps of 6–8 sec are applicable to two-way carriageways; in either case the larger gap should be used if the design speed is high. Unlike the simple cutting situation, the cutting and merging situation requires traffic gaps of 8–12 sec to allow a turning vehicle to accelerate across a two-way flow to join the far traffic stream at a suitable speed. With compound cutting and merging, which takes place at a simple T-junction, the normal sequence is for the right turn (R_1) into the minor road to precede the right turn (R_2) from the minor road; precise calculations are fairly complex, so in most cases it is sufficiently accurate to add half of R_1 to R_2, and then to use the gap sizes given above for the simple cutting and merging situation.

The governmental recommendations for GIVE WAY junctions in *urban* areas are given in Fig. 6.6; these refer to both 3- and 4-leg intersections.

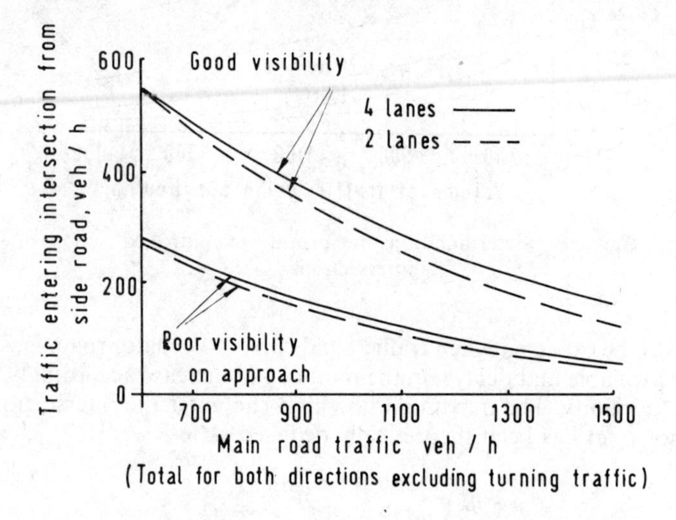

Fig. 6.6. Capacity determinations at urban uncontrolled and GIVE WAY intersections[6]

The lower pair of curves in this figure assumes poor visibility from the side road, a single lane approach and a minimum acceptable gap of 8 sec. These conditions apply when visibility is restricted on the approach but is adequate when the vehicle is at the stopped position at the junction. Under conditions of good visibility which permit vehicles from the side road to enter suitably long gaps in groups, without each vehicle necessarily having to stop at the major road, the intersection capacity is increased considerably; this is reflected by the upper set of curves in Fig. 6.6. These curves are based on a minimum acceptable gap of 6 sec.

Good visibility from the minor road is defined as being to the right and left between points 1·05 m above road level *over* the triangular area defined by (i) a line x metres long measured back along the centreline of the side

TABLE 6.5. *Desirable visibility distances at priority junctions in urban areas*[7]

Type of major road	*Speed limit, km/h*	*Min visibility distance, y metres*
All purpose primary distributor	80	152·5
	64	122
District or local distributor	48	91·5
Access road	48	61

Note: The x-distance should normally be 9·1 m but can be reduced to 4·6 m if the side road is lightly trafficked, e.g. for cul-de-sacs and other such access roads.

road from the continuation of the nearer edge of the major road carriageway, (ii) a line y metres long measured along the nearer edge of the major road carriageway from its intersection with the centreline of the side road, and (iii) a straight line joining the ends of the above lines. Values of x and y, given in Table 6.5, refer to double splays in the case of 2-way roads and to a single splay in the direction of the approaching traffic if the major road is 1-way (or a dual carriageway with no gap in the central reservation). The curves in Fig. 6.6 are intended only as a general guide as:

1. Where traffic is unequally distributed between the two carriageways, the capacity will be somewhat reduced.
2. Where there is a high proportion of heavy vehicles, or an up-gradient on the side road, longer gaps will be required and capacity will be reduced.
3. The visibility conditions for the lower pair of curves assume adequate visibility at the stopped position, and poorer visibility at that point may reduce capacity substantially.
4. The effect of gaps in main road flow caused by nearby signals or pedestrian crossings may result in an increase in the capacity of the junction.

The recommended (*diverging*) capacity of a slip road exit on a rural road is given as 1200 p.c.u./h provided the decelerating lane is well-designed and signing gives adequate warning well in advance of the junction. No definitive data are available at this time regarding merging capacities, and U.S. design values are currently used until local criteria are determined.

Roundabouts. The capacity of a *conventional roundabout* is directly affected by the capacity of each weaving section incorporated within the intersection. If any of the weaving sections is overloaded, then locking of the roundabout may occur and it can be said that the capacity of the roundabout is exceeded.

Within a particular weaving section, true non-stop weaving will only occur when the headways between the vehicles are of sufficient lengths and frequencies that safe merging and diverging movements can take place. Discontinuous flow, due to stop-go movements of the weaving vehicles, occurs when these headways are not available, or when the weaving section length is so short that the paths of the weaving vehicles cross at large intersecting angles.

The main factors controlling the capacity of a conventional weaving section are the geometric layout, including entrances and exits, and the percentages and composition of the weaving traffic. According to studies conducted by the Transport and Road Research Laboratory the maximum flow through a weaving section at speeds between 14·5 and 24 km/h is given by:

$$Q_M = \frac{354w\left(1+\dfrac{e}{w}\right)\left(1-\dfrac{p}{3}\right)}{\left(1+\dfrac{w}{l}\right)}$$

where Q_M = maximum flow in dry weather p.c.u./h,
w = width of section, varying between 6·1–18·3 m,
e = average entry width, varying between $0{\cdot}12w$–$0{\cdot}30w$ m,
l = length of section, varying between $0{\cdot}76w$–$2{\cdot}54w$ m,
and p = ratio of weaving traffic to total traffic, varying between 0·4 and 1·0.

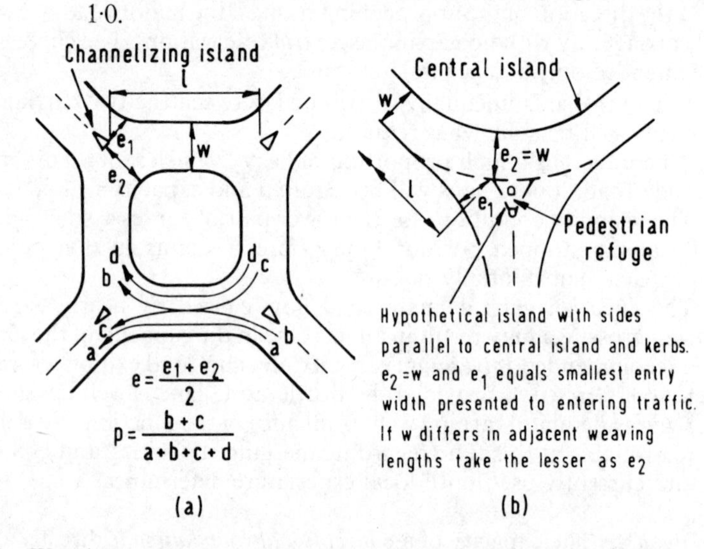

Fig. 6.7. Weaving section dimensions at a roundabout with and without channelization islands

Figure 6.7(a) illustrates these dimensions at an idealized roundabout. Since, in practice, the weaving sections often tend to differ from this layour, it can be necessary to adopt 'effective' dimensions for use in the formula. The manner in which these dimensions are obtained is shown in Fig. 6.7(b).

The equivalent passenger car unit values used in conjunction with the above formula are given in Table 6.3.

To avoid overloading of a weaving section during periods of high unstable traffic flow, and since the possible capacity during wet weather is about 10 per cent lower, it is generally accepted that the design capacity should be taken as 80 per cent of the value computed by the equation. This factor should be applied after the following corrections have been made for site conditions:

1. Reduce the computed capacity by 16 per cent when more than 300 pedestrians per hour cross the exit road from the section being considered.

2. Reduce the calculated capacity by 5 per cent if the existing vehicles have to turn through an angle greater than 75 degrees when leaving the weaving section, and by $2\frac{1}{2}$ per cent if the exiting angle is between 60 and 75 degrees.

3. Reduce the calculated capacity of the *preceding* section by 5 per cent if the angle of entry to the section being studied is less than 15 degrees. If the angle of entry is between 15 and 30 degrees reduce by $2\frac{1}{2}$ per cent.

TABLE 6.6. *Average values of length-to-width ratios at weaving sections of existing roundabouts*[12]

Site location	*Road classification*	
	Trunk	*Non-Trunk*
Rural	4·8	4·4
Suburban:		
Semi-rural	3·8	3·9
Built-up	3·8	3·9
Urban	3·4	

Average l/w-values for existing conventional roundabouts, given in Table 6.6, suggest that land availability is the controlling design factor in urban areas, and hence speeds are relatively low. As the main disadvantage of a conventional roundabout in a rural area is the reduction in speed, l should be preferably longer (by 33 to 50 per cent more) than necessary for capacity, while w should be 3 to 3·6 m wider than (but never more than double) the mean entry width; this will, of course, result in a design capacity well in excess of that actually required.

A detailed example of the calculations involved in conventional roundabout design is available in the literature[6].

The introduction, in 1963, of the 'Give way to traffic from the right' rule at roundabouts in Great Britain[8] made a significant contribution to the operating efficiency of conventional roundabouts, particularly in urban areas. At the same time, however, a rule which gave priority to vehicles leaving the junction also caused the actual amount of weaving to be reduced and so long weaving sections became of less benefit in attaining capacity. Furthermore, empirical experiments[9, 10] showed that at existing roundabout intersections the capacity of a junction *as a whole* (i.e. the total discharge of vehicles through the junction from queues on all roads) could be considerably improved by widening the entry approach, providing a smaller central island, and installing such traffic management devices as bollards and carriageway delineators to deflect traffic well to the left on entry. For new designs, a layout of this nature has the advantages of requiring less land than a conventional roundabout in order to achieve a given capacity, and a lower construction cost. This design is also quicker for drivers to negotiate, and is less liable to be blocked when negotiated by abnormal loads.

Fig. 6.8 is a typical example of what is known as a *mini-roundabout* layout. The main features of this design are (a) an increase in the number of lanes at the GIVE WAY line, (b) a central island with a diameter of about one-third that of a hypothetical circle inscribed within the outer carriageway boundaries, but normally not less than 8 m, (c) a minimum stopping distance of 25 m between the GIVE WAY line and the point of conflict with a vehicle from the left, i.e. dimension X, (d) a width between traffic islands and the roundabout, i.e. dimension Y, which preferably is not less than the total lane width at the entry preceding it, i.e. dimension Z, (e) an entry taper that is about twice as sharp as the exit taper, and (f) a deflection island, B, that is intended to ensure that straight through movements do not occur.

The possible capacity of a mini-roundabout can be estimated by[9]

$$Q = k(\Sigma W + A^{1/2})$$

where Q = total entry volume (from all fully loaded approach roads), p.c.u./h

ΣW = sum of the basic road widths (not half-widths) used by traffic in both directions to and from the intersection, m

A = area of junction widening, i.e. the area within the intersection outline (including islands) which lies outside the area of the basic cross-roads, m^2

and k = efficiency coefficient which depends on site conditions, viz:

3-way junctions	$k = 80$ p.c.u./h.m
4-way	$k = 70$
$\geqslant$5-way	$k = 65$

For practical purposes the capacity in normal peak conditions should be taken as 80 per cent of that given by the above equation.

At the time of writing there are over 100 mini-roundabouts in use in Great Britain, 28 of which have been constructed in Newcastle-Upon-Tyne. On the basis of data gathered in Newcastle, the following simpler formula has been developed[11] as a means of getting a rule-of-thumb guide to saturation capacity:

$$Q = kD$$

where Q = total entry volume, p.c.u./h

k = 150 for 3-way, and 140 for 4-way, junctions, p.c.u./h.m

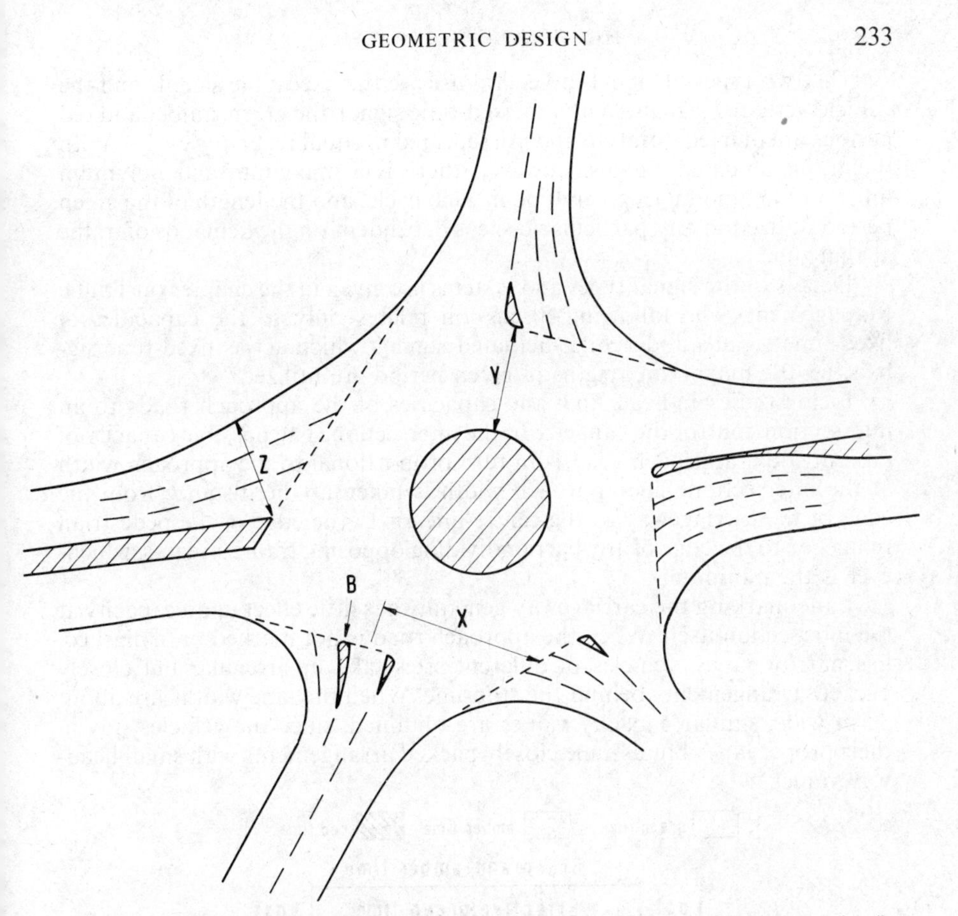

Fig. 6.8. Mini-roundabout layout at a 5-way junction of roads subject to a 48 km/h speed limit

and D = diameter of hypothetical circle inscribed within the outer carriageway boundaries, m.

The p.c.u. values used in either of the above equations are those used for traffic signal design (see the following section). It is not unlikely, however, that a new rating system will be devised for this new type of junction at some future time.

Present experience with mini-roundabouts has been mostly confined to roads subject to 48 km/h speed limits, and to some others with 64 km/h speed limits. The limited evidence available at this time re their safety records suggests that they have lower numbers of fatal and serious accidents at the expense of, possibly, slightly increased numbers of total accidents.

Signalized intersections. The main factors affecting the capacity of an intersection controlled by traffic signals are the characteristics of intersectional layout, the traffic composition and needs, and the setting of the signal control.

The two types of signal in general use are the fixed-time signals and the vehicle-actuated signals. With the fixed-time signal, the green, amber and red periods are of fixed duration and are repeated in equal recurring cycles. With the traffic-actuated signals, however, there is a maximum and minimum limit on the amount of green time in each cycle, and the length of the green period utilized in any particular cycle is dependent on the demands of traffic at that time.

Details of the signal types and systems are given in the chapter on Traffic Management. The following discussion relates only to the capacities of fixed-time signals and vehicle-actuated signals which act as fixed-time signals, i.e. the maximum lengths of green period are utilized.

Field studies indicate that the capacities of the approach roads to an intersection control the capacity of the intersection. The possible capacity of an individual approach road is in turn proportional to the approach width at the intersection. The approach width is taken as the distance from the edge of the carriageway to the centre-line, or to the edge of the pedestrian refuge, or to the edge of any barrier dividing opposing traffic streams, whichever is the minimum.

Lane-marking the carriageway generally has little effect upon capacity at the intersection itself. When the approach road is not marked or, if marked, has narrow lanes, vehicles of different sizes take up irregular but closely packed arrangements behind the stop-line. When the lane widths are about 3·5 m wide, similar capacity values are obtained, since the vehicles stay in their proper lanes but assume closely packed arrangements with small headway values.

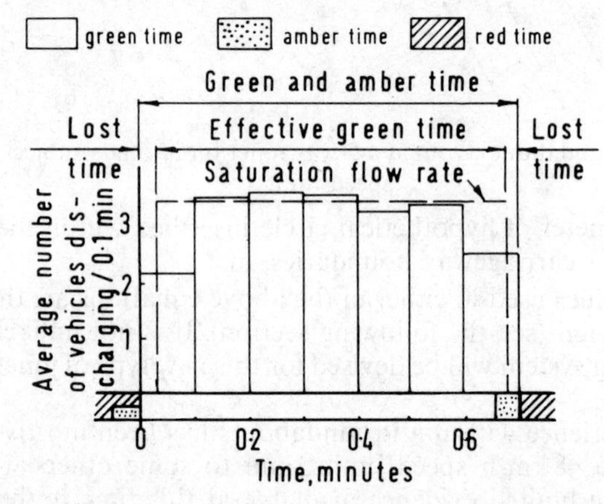

Fig. 6.9. Typical discharge rate from an approach road during a fully saturated green period at a signalized intersection

At an intersection, the maximum number of vehicles that can enter from an approach road is affected by the amount of green time available to that approach. When the green period commences, vehicles in the traffic queue take some time to accelerate to a constant running speed. When this speed has been reached, the queue discharges into the intersection at a more or less constant rate as illustrated in Fig. 6.9. At the end of the green period, some of the vehicles in the queue make use of the amber period to enter the intersection, and the discharge rate falls away to zero during this period.

The term *Saturation Flow* is used to describe the constant maximum rate of discharge from the approach road, and it is usually expressed in veh/h of green time for existing intersections and p.c.u./h for new intersections. That part of the actual green plus amber time which, when multiplied by the saturation flow rate, gives the maximum number of vehicles that can enter the intersection is known as the effective green time. The difference between the actual green plus amber time and the effective green time is termed lost time.

Observations at intersections throughout Britain have resulted in the saturation flow values given in Table 6.7. For approach widths between 5·18 m and 18·29 m, the saturation flow is estimated by the equation

$$S = 525w$$

where S = saturation flow, p.c.u./h,
and w = approach width, m.

TABLE 6.7. *Saturation flow values for signalized intersections at average sites*

Approach width, m	3·05	3·35	3·66	3·96	4·27	4·57	4·88	5·18
Saturation flow, p.c.u./h	1850	1875	1900	1950	2075	2250	2475	2700

These p.c.u.-values (see Table 6.3) assume that no parked vehicles, no right-turning vehicles, and no two-wheeled vehicles are present at the intersection (a right-turning p.c.u. = 1·75 p.c.u.). Furthermore, they must be adjusted for site location according to the values recommended in Table 6.8.

TABLE 6.8. *Effect of approach road location on saturation flow*

Site designation	*Description*	*Per cent of average saturation flow*
Good	Dual carriageway, no noticeable interference from pedestrians or right-turning vehicles. Good visibility and adequate turning radii. Exit of adequate width and alignment.	120
Average	Average site. Some conditions good and some poor.	100
Poor	Average low speed. Some interference from standing vehicles and right-turning traffic. Poor visibility and/or alignment. Busy shopping street.	85

If the site location is on a slope, then the saturation flow values should be reduced by 3 per cent for each 1 per cent uphill grade averaged over the 61 m section of approach road just prior to the stop line. Similarly the saturation flows should be increased by 3 per cent for each 1 per cent of downhill gradient. These corrections apply to gradients ranging from −5 to +10 per cent.

If passenger cars are regularly parked near the intersection, the approach width must be reduced in the following manner:

$$\text{Loss of width, m} = 1{\cdot}68 - \frac{0{\cdot}9(Z - 7{\cdot}62)}{k}$$

where Z = clear distance of the parked car from the stop line, m. If $Z < 7{\cdot}62$ m, it is taken as equal to 7·62 m

and k = green time, sec.

If the parked vehicle is a commercial one, then the loss of width should be 1·5 times that obtained by the equation.

EXAMPLE Determine the possible capacity, in vehicles per hour, of an approach road to an intersection that is being reconstructed just outside the central area of a town. The total carriageway width is 14·6 m and it is level just prior to the intersection. Commercial vehicles park regularly outside an adjacent builder's yard at a distance of 38 m back from the stop line. During the peak traffic period the traffic is composed of 75 per cent passenger cars, 22 per cent commercial vehicles and 3 per cent buses. One third of the passenger cars and one half of the commercial vehicles turn right at the intersection at this time. The number of buses turning right is negligible. The green time on the approach road will be 40 sec of a 90-sec cycle, while the amber period will be 3 sec and the lost time per phase is estimated at 2 sec.

Solution. This intersection can be considered to be an average one, so that no adjustment for site location is necessary. Since commercial vehicles stop regularly on the carriageway, the loss of width is estimated as follows:

$$\text{Loss of width, m} = 1{\cdot}5\left[1{\cdot}68 - \frac{0{\cdot}9(38 - 7{\cdot}62)}{40}\right]$$

$$= 1{\cdot}5\ \text{m}$$

$$\text{Effective approach width} = 7{\cdot}3 - 1{\cdot}5$$

$$= 5{\cdot}81\ \text{m}$$

$$\text{Saturation flow} = 525 \times 5{\cdot}8$$

$$= 3045\ \text{p.c.u./h of green time.}$$

However, saturation flow exists for only 41/90 of the cycle time. Therefore,

$$\text{Possible capacity} = 3045 \times \frac{41}{90}$$

$$= 1387\ \text{p.c.u./h}$$

To obtain the possible capacity in vehicles per hour, assume that the possible flow = Q v/h.

Adjustments:

Straight-through and left-turning passenger cars $= \frac{50}{100} Q$ at 1 p.c.u. $= 0{\cdot}500Q$

Right-turning passenger cars $= \frac{25}{100} Q$ at 1·75 p.c.u. $= 0{\cdot}4375Q$

Straight-through and left-turning commercial vehicles $= \frac{11}{100} Q$ at 1·75 p.c.u. $= 0{\cdot}1925Q$

Right-turning commercial vehicles $= \frac{11}{100} Q$ at $(1{\cdot}75)^2$ p.c.u. $= 0{\cdot}3369Q$

Straight-through buses $= \frac{3}{100} Q$ at 2·25 p.c.u. $= 0{\cdot}0675Q$

Therefore, Possible capacity $= 1{\cdot}5344Q$ p.c.u./h

$= 1387$ p.c.u./h

Therefore, $$Q = \frac{1387}{1{\cdot}5344} = 904 \text{ v/h.}$$

Traffic signals versus roundabouts. The problem often arises whether an intersection should be signalized or whether a roundabout should be constructed. While each situation has its own characteristics, the choice of form at most junctions may be based on the following[13]: 1. The land requirement of a conventional roundabout is usually greater than for traffic signals, especially if the flows on one pair of arms (e.g. the minor road) are relatively low. (Even so, it may be easier to acquire corner plots for a roundabout than to take up the same area in the long narrow multi-plot strips required for the parallel widening of a signalled junction.) 2. If traffic entering from one arm greatly exceeds traffic leaving by it, there will be often a shortage of gaps in the circulating stream in both conventional and mini-roundabouts, which can lead to excessive delays to side-road vehicles. 3. When the number of off-side turning vehicles (right-turning in Britain) is large, say greater than 30 to 40 per cent, either a special turning phase is required at a signalled junction or the vehicles must queue in the approach—in which case a roundabout is to be preferred. 4. 3-way or 5-way junctions are not really satisfactorily treated by traffic lights, especially when the flows are balanced. Mini-roundabouts are most successful at 3-way junctions. 5. The accident rate at conventional roundabouts is only about two-thirds the rate under signal control.

SIGHT DISTANCES

The ability to see ahead is probably the most important feature in the safe and efficient operation of a highway. Ideally the geometric design should ensure that at no time is one vehicle invisible to another within normal eyesight distance. Because of topographical and other considerations this is not always economically possible, so it is necessary to design roads on the basis of sight distances that are the minimum necessary for the safety of drivers.

Sight distance can be defined as the length of carriageway that is visible to the driver. To the highway engineer engaged in designing a road there are two forms of sight distance that are of particular interest.

If safety is to be built into the highway, then sufficient sight distance must always be available to drivers to enable them to stop their vehicles prior to striking an unexpected object on the carriageway. This sight distance is known as the safe stopping sight distance.

If efficiency is to be built into the highway, then sufficient sight distance must be provided for drivers to overtake and pass slower vehicles in complete safety. This sight distance is called the safe passing sight distance.

Safe stopping sight distance

Whether on a 2-lane, 3-lane or multi-lane roadway, the most important consideration is that at all times the driver travelling at the design speed of the highway must have sufficient carriageway distance within his line of vision to allow him to stop his vehicle before colliding with a slowly moving or stationary object appearing suddenly in his own traffic lane. Calculation of the minimum distance required to stop the vehicle is based on an evaluation of the driver's perception and reaction times, and the distance necessary to stop the vehicle after the brakes have been applied.

Perception-reaction time. Perception time is the time which elapses between the instant that the driver sees the object on the carriageway and the instant that he realizes that brake action is required. The length of perception time varies considerably since it depends upon the distance to the object, the natural rapidity with which the driver reacts, the optical ability of the operator, atmospheric visibility, the type, condition and location of the roadway, and the type, colour and condition of the hazardous object.

Reliable data on perception time are very meagre due to the difficulty of relating laboratory test conditions and results to actual situations. Most studies indicate, however, that the perception time at high speeds is usually less than at low speeds, since fast drivers are usually more alert. Also perception times in urban areas tend to be lower than in rural areas, due to the varied conditions present in built-up locations which cause drivers to be continually alert.

Brake reaction time is the time taken by drivers to actuate the brake pedal after perceiving the object. Since it begins after the driver has been alerted, more reliable data are available concerning this time. Tests indicate that the average driver's brake reaction time is about 0·5 sec. In order to provide a safety factor for drivers whose reaction times are above average, a time of 0·75 sec is usually assumed for design purposes.

In practice, driver perception time is usually combined with brake reaction time in order to arrive at a total perception-reaction time that is suitable for highway design purposes. Measurements of this combined time showed that many drivers required as much as 1·7 sec under normal roadway conditions. Since there is little control over the calibre of most motorists on the road today, it is clear that a value in excess of this must be used for design purposes. In line with this concept the American Association of State Highway Officials recommend a perception-reaction time of 2·5 sec as being desirable for rural design purposes.[14] A value often used in urban areas is 1·5 sec.

Expressed in formula form, the distance travelled by a vehicle during the perception-reaction time is as follows:

Perception-reaction distance, m $= tv = 0{\cdot}278tV$

where $v =$ initial speed, m/sec
$V =$ initial speed, km/h
and $t =$ perception-reaction time, sec.

Braking distance. The distance needed by a vehicle to stop on a level road, after the brakes have been applied, can be visualized as depending primarily on the initial speed of the vehicle and the friction developed between the wheels and the carriageway surface. This distance can be estimated by utilizing the principle that the change in kinetic energy is equal to force multiplied by distance. Thus,

$$\frac{1}{2}\frac{Wv^2}{g} = Wf \times d$$

Therefore,

$$d = \frac{v^2}{2fg} = \frac{V^2}{254f}$$

where $g =$ acceleration due to gravity $= 9{\cdot}81$ m/sec^2
$v =$ initial speed, m/sec
$V =$ initial speed, km/h
$d =$ braking distance, m
$f =$ coefficient of friction developed between the tyre and the surface of the carriageway.

There is considerable variation in the value of f used for design purposes. If the comfort of the motorist is considered to be the sole criterion, then deceleration rates greater than 4·9 m/sec^2 should not be used, as it is at about this rate that passengers are caused to slide from their seats. This deceleration rate is equivalent to a developed coefficient of friction of nearly 0·5 which, as will now be discussed, tends to be too high a value from a safety point of view.

Friction factor. The choice of the value of f is a most complicated one because of the many variables involved. Some physical elements which cause considerable variation in test results are the resilience and hardness of the tread rubber, tread design, amount of tread, efficiency of the brakes, and the condition and type of carriageway surface. Other important considerations are whether the friction value should be selected on the basis of dry, wet, icy or muddy conditions, and whether or not the wheel is rotating after the driver slams on the brakes.

The quality of the tyre tread is something over which the highway engineer has little control, so most investigators assume that the tyre is relatively smooth.

The choice of whether the friction factor should be measured under conditions of incipient sliding or after the wheel is locked and skidding is taking place is usually resolved in favour of the locked wheel concept. When an unexpected object appears on the road, most motorists automatically step on the brakes, thus causing the brakes to lock on very many vehicles. Until such time therefore that brake systems are widely used that will prevent the wheels of cars from locking after the sudden application of brakes, design calculations must be based on the lower and locked wheel values.

The choice of road surface conditions is also a straightforward one. Ideally, design should be based on icy conditions, since repeated studies have shown that braking distances on icy surfaces are from three to twelve times as great as those for dry surfaces. In Britain, however, this is not a practical design concept because of the relatively mild climate. More important in this country are the friction values developable on wet carriageway surfaces. Values measured on wet surfaces are substantially lower than those measured on dry surfaces since water tends to act as a lubricant between the road and the tyre. In this respect also it should be noted that the developable friction factor decreases as the vehicle speed increases. This can be related to the variation in contact time between the tyre tread and the road surfacing. At low speeds the contact time is relatively long and there is ample time for water film to be expelled from between the rubber and the surfacing. This is facilitated by a well-patterned tyre with effective drainage characteristics. On the other hand, when the vehicle speed is increased, the relative contact time is smaller, the expulsion of the moisture becomes less complete, and the developable friction coefficient becomes smaller.

Friction values measured at given speeds on particular carriageway surfaces also show considerable variations. For instance, friction measurements on a concrete road at one location may result in very different values from those taken on a concrete road at a different site under similar conditions. Attempting to take into account the many variables involved, as well as the prime consideration of safety, the American Association of State Highway Officials recommend friction values of 0·36, 0·33, 0·31, 0·30 and 0·29 at 48, 64, 80, 96·5 and 112 km/h, respectively for use in designing safe stopping sight distances on most highways.

The question also arises as to whether it is realistic to assume that vehicles travel at the full design speed of the highway when the carriageway is wet, i.e. the condition chosen for measuring the friction factor. American experience in this matter suggests that the speed for wet conditions is approximately 85 to 95 per cent of the design speed. Studies on motorways in this country, however, indicate that British drivers pay relatively little attention to wet weather conditions. For this reason it is probably safer to design on the assumption that vehicles will be travelling at the design speed of the highway.

Table 6.9 shows the safe stopping sight distances recommended by the Department of the Environment for use in the design of single- and dual-carriageway urban and rural roads in Britain.

TABLE 6.9. *Minimum sight distances used on urban and rural roads*

Design speed, km/h	*Stopping distance, m*		*Overtaking distance, m*	
	Urban	*Rural*	*Urban*	*Rural*
120	—	300	—	—
100	—	210	—	450
80	140	140	360	360
60	90	90	270	270
50	70	—	225	—
30	30	—	135	—

Commercial vehicles. The safe stopping sight distances shown in Table 6.9 and the values used in determining them are based on passenger car operation only. Commercial vehicles in general require longer braking distances than passenger cars travelling at the same initial speeds. In practice, however, it is usually not necessary to give special consideration to commercial vehicle stopping sight distance requirements. There are two main reasons for this.

Firstly, as indicated in Fig. 5.2, commercial vehicles travel at slower speeds than passenger cars at a given location, either because of regulation or by operator choice. Secondly, due to his higher position in the vehicle, the commercial driver is usually able to see a vertical obstruction on the roadway sooner than the private motorist. Thus, in the great majority of situations, these two additional considerations have the effect of compensating for any additional braking distances that might otherwise be needed.

The main exception where additional stopping distance should be provided for commercial vehicles is in the case of a horizontal sight restriction at the end of a steep hill. In locations where this occurs the commercial vehicle speeds may approach those of the passenger cars, while the additional height of eye of the commercial vehicle operator is of little value because of the sight restriction.

Passing sight distance

In order to achieve design capacity figures on the highway, sufficient sight distance must be provided for fast vehicles to overtake slower ones in safety. On 2- or 3-lane roads this can be a serious problem due to the need for the overtaking vehicles to travel in a lane used by vehicles travelling in the opposite direction. The minimum sight distance required by a vehicle to overtake safely on 2-lane or 3-lane roads is the visibility distance which will enable the overtaking driver to pass a slower vehicle without interfering with the speed of an oncoming vehicle travelling at the design speed of the highway.

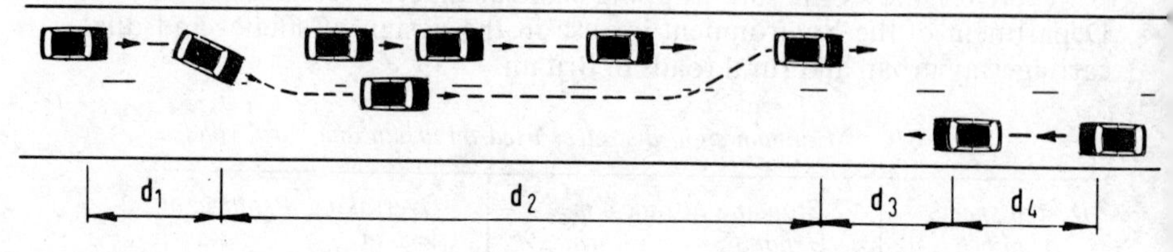

Fig. 6.10. Elements of total passing sight distance requirements on 2-lane roads

2-lane roads. As indicated by Fig. 6.10, there are four components of the minimum distance required for safe overtaking on 2-lane roads.

The dimension d_1 represents the time taken or distance travelled by a vehicle while its driver makes up his mind whether or not it is safe to pass the vehicle in front. This time period has been described as the hesitation time[15] and is in the order of 3·5 sec for comfortable overtaking conditions.

The dimension d_2 represents the time taken or distance travelled by the overtaking vehicle in carrying out the actual passing manoeuvre. Thus it begins the instant the overtaking driver turns the wheel and ends when he returns to his own lane. Measurements under controlled conditions regarding the length of this dimension are shown as solid lines in Fig. 6.11. The dotted lines shown in this graph provide estimates of the requirements at higher speeds.

The dimension d_3 has been called the safety dimension and is the time or distance between the overtaking vehicle and the oncoming vehicle at the instant the overtaking vehicle has returned to its own lane. From a safety aspect it is of course desirable that this distance should be as large as possible. Practical economic requirements on the other hand necessitate that it be as small as possible. One suggested value has been $1\frac{1}{2}$ sec. This means that if the combined relative speed is 160 km/h, then a safety margin of 67 m is available between the two vehicles.

The dimension d_4 represents the time taken or distance travelled by the opposing vehicle at the design speed of the road while the actual overtaking manoeuvre is taking place. Conservatively it should be the total distance travelled by the opposing vehicle during the time $d_1 + d_2$, but in practice this may be questioned as being too long. Examination of the overtaking vehicle's track in Fig. 6.10 shows that it can return freely to its own lane at any instant prior to drawing alongside the overtaken vehicle. If this initial time is not taken into account, then the opposing vehicle dimension d_4 can be taken as approximately equal to $\frac{2}{3}d_2$.

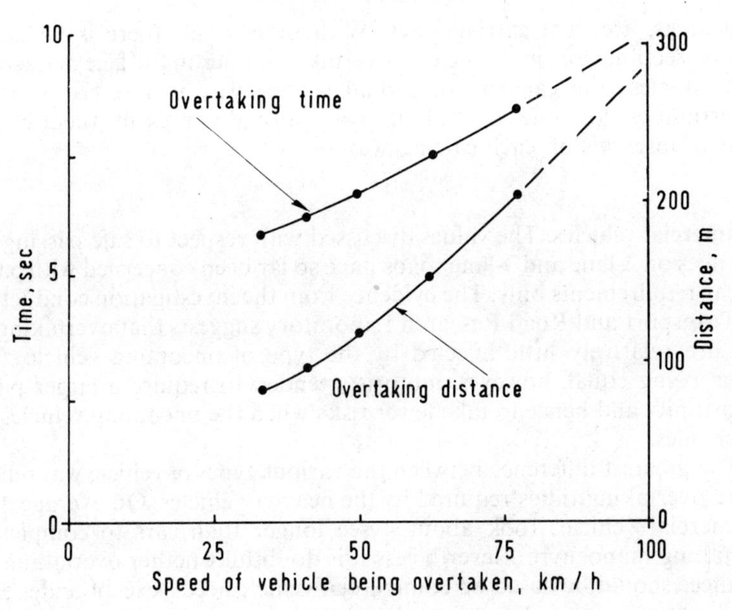

Fig. 6.11. Overtaking time and distance requirements

The recommended overtaking sight distances for use on single carriageway roads in Britain are given in Table 6.9. The urban overtaking distances in this table are based on a vehicle at design speed taking 8 sec to overtake one travelling 16 km/h slower.

3-lane roads. Adequate data are meagre relative to the distances required for safe overtaking on 3-lane roads. Ideally the minimum passing sight distance requirements for 3-lane roads should be the same as for 2-lane roads. From a practical aspect, however, this requirement can be considered rather stringent, since the centre lane makes it feasible for very many overtakings to be accomplished in the face of oncoming traffic, *provided* that the traffic on the highway does not exceed its capacity. The centre lane also makes the overtaking manoeuvre more flexible since either or both vehicles in the centre lane can more easily return to the proper lane if danger arises. For these reasons therefore the minimum passing sight distance requirements can be assumed to be composed of the same elements as for 2-lane roads except that the dimension d_4, the distance travelled by the oncoming vehicle, can be omitted.

As with 2-lane roads, it must be emphasized that values determined in this manner represent the minimum for safe overtaking. Where economically possible much longer sight distances should be designed into the highway.

Multi-lane roads. Most multi-lane roads in Britain, and especially those in rural areas, are dual carriageways. With these roads there is no need to provide sections for one vehicle to overtake another in the face of oncoming traffic. Unless the capacity of a dual carriageway is severely overtaxed, opportunities for safe overtaking will normally present themselves at frequent intervals on each carriageway.

Commercial vehicles. The values discussed with respect to safe passing sight distances on 2-lane and 3-lane roads have so far been concerned with passenger car requirements only. The evidence from the investigation conducted by the Transport and Road Research Laboratory suggests that overtaking vehicles are relatively little affected by the type of oncoming vehicle. Other things being equal, however, motorists tended to require a larger passing opportunity and hence to take fewer risks when the oncoming vehicles were larger ones.

The greatest difference between the various types of vehicle was reflected in the overtaking times required by the heavier vehicles. On average, heavy commercial vehicles took about 4 sec longer than cars to complete the overtaking manoeuvre. Nevertheless it is doubtful whether overtaking sight distances should be based on commercial vehicle needs, except under exceptional circumstances. Apart from the extra expense that this would involve, the commercial vehicle driver is able to see farther due to his greater eye-height, thereby compensating for any additional overtaking length that might be required.

Frequency of overtaking sections. Ideally all highways should be continuously safe for overtaking manoeuvres to take place at will. Usually, however, this is not possible so decisions regarding the frequency of sections have to be made on the basis of topography conditions, traffic volumes, design speed of the highway and economic considerations.

It is important to remember that high-speed roadways carrying near-capacity volumes require very frequent and long overtaking sections. A measure of the need for these sections can be illustrated by considering the effects of not having them on the capacity of a road—see Table 6.10.

TABLE 6.10. *Effect on rural road design capacity and speed where minimum overtaking sight distances are not provided*[1]

% of road length with sub-standard overtaking sight distance	*0*	*20*	*40*	*60*	*80*	*100*
% of standard design capacity	100	90	80	65	50	30
Est. reduction in average speed, km/h, of 2-lane road carrying 900 p.c.u./h	0	6·4	12·9	19·3	25·7	32·2

While they are most needed on roadways carrying heavy traffic volumes, overtaking sections are still most important safety adjuncts to highways carrying intermediate to low volumes of traffic. Long stretches of roadway with no overtaking opportunities will cause even the most careful motorist to attempt to overtake slower moving vehicles at the first possible opportunity, and inevitably accidents will occur. The 3-lane roadway is very noticeable in this respect since it automatically suggests safe passing opportunities irrespective of sight distance conditions. Therefore the use of 3-lane roads should be suspect and they are not to be recommended unless located on an alignment that results in nearly continuous safe overtaking sight distances.

HORIZONTAL CURVATURE

Horizontal curvature design is one of the most important features influencing the efficiency and safety of a highway. Improper design will result in lower speeds and a lowering of highway capacity. The importance of curve design upon safety is reflected by the accident statistics shown in Table 6.11, which shows that the sharper the curve the greater the tendency for accidents to occur.

The maximum comfortable speed on a horizontal curve is primarily dependent upon the radius of the curve and the superelevation of the carriageway. In addition, vehicle speeds and safety on high-speed roads are aided by the presence of such features as extra carriageway widths at the curves themselves and the insertion of transition curves between straights and curves.

TABLE 6.11. *Accident rates per 1·6 million veh-km for curves of various radii in Lancashire*, 1946–1947

Curve radius, m	*Total accident rate*	
	Urban	*Rural*
Straight to 610	4·70	1·47
610 to 305	3·42	2·46
305 to 150	4·20	4·01
150 to 60	7·20	3·72
<60	20·00	16·70

Properties of the circular curve

A circular curve joining two road tangents can be described either by its radius or its degree of curvature. In Britain the radius is usually utilized whereas in certain other countries, noticeably the United States, the degree of curve concept is preferred. The (metric) degree of curve is defined as the central angle which subtends a 100 m arc of the curve.

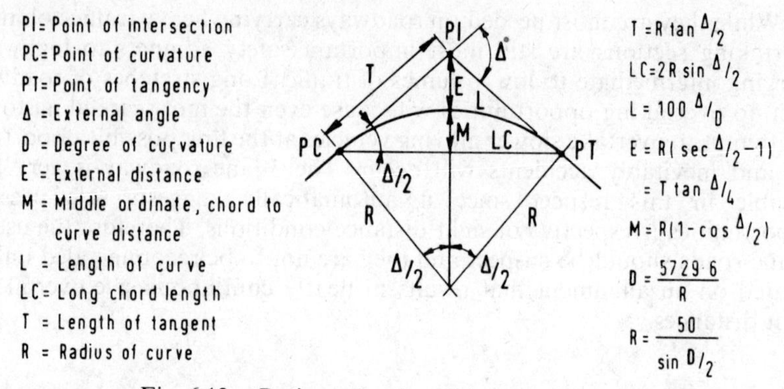

Fig. 6.12. Basic properties of the circular curve

Figure 6.12 shows the principal properties of a simple highway curve, together with some of the more important formulae. These formulae are given here without proof; their derivation can be found in any good textbook on surveying.

Curvature and centrifugal force

Any body moving rapidly along a curved path is subject to an outward reactive force called the centrifugal force. On highway curves this force tends to cause vehicles to overturn or to slide outward from the centre of road

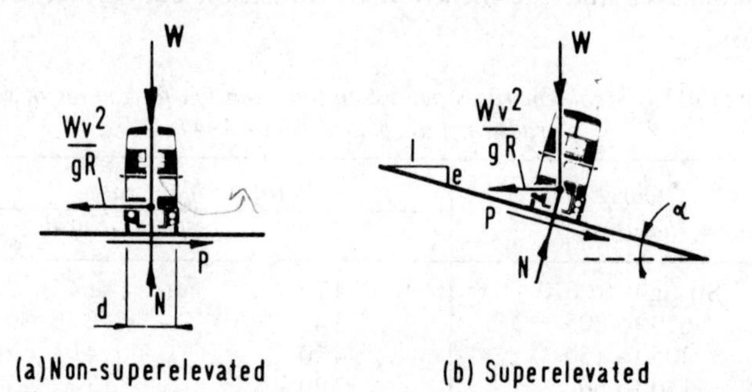

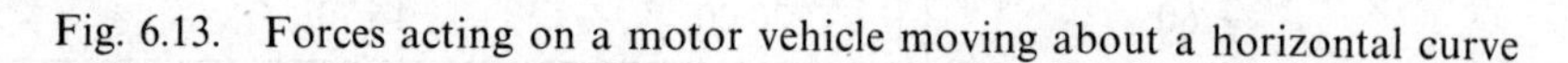

Fig. 6.13. Forces acting on a motor vehicle moving about a horizontal curve

curvature. Figure 6.13(a) illustrates the forces acting on a vehicle as it moves about a horizontal curve. Since the carriageway surface is flat and the forces are in equilibrium, then

$$P = \frac{Wv^2}{gR}$$

where W = weight of the vehicle, kg,

	v	= speed of the vehicle, m/sec,
	R	= radius of the curve, m,
	g	= acceleration due to gravity, m/sec^2,
and	P	= lateral frictional force resisting the centrifugal force, kg.
But	P	$= N\mu = W\mu = Wv^2/gR$
Therefore	μ	$= v^2/gR = P/W$
where	μ	= developed lateral coefficient of friction
and	P/W	$= v^2/gR$, known as the centrifugal ratio.

When the velocity is expressed in km/h and g is taken as 9·81 m/sec^2, then

$$\mu = \frac{V^2}{127R} \text{ (approx.)}$$

Thus, if the limiting value of μ is known, the minimum curve radius can be calculated for any given design speed.

The centrifugal force also tends to cause a vehicle to overturn. This force, acting through the centre of gravity of the vehicle, causes an overturning moment about the points of contact between the outer wheels of the vehicle and the carriageway surface. This overturning moment is resisted by a righting moment caused by the weight of the vehicle acting through its centre of gravity. Thus, for equilibrium conditions;

$$\frac{Wv^2}{gR} \times h = W \times \frac{d}{2}$$

and

$$h = d \times \frac{1}{2v^2/gR}$$

$$= d/2\mu$$

where d = lateral width between the wheels, m
and h = height of the centre of gravity above the carriageway, m.

The above equation illustrates, for instance, that if $\mu = 0{\cdot}5$, then the height to the centre of gravity must be greater than the lateral distance between the wheels before overturning will take place. Modern vehicles, however, and especially passenger cars, have low centres of gravity so that relatively high values of μ would have to be developed before overturning could take place. In practice the frictional value is usually sufficiently low for sliding to take place long before overturning. It is only with certain commercial vehicles having high centres of gravity that the problem of overturning may arise.

Superelevation. In order to resist the outward-acting centrifugal force it is customary to superelevate or slope the carriageway cross-section of curved sections of a modern highway in the manner shown in Fig. 6.13(b). For every combination of radius of curvature and highway design speed there is a particular rate of superelevation that exactly balances the centrifugal force. Where the superelevation is insufficient to balance the outward force, it is necessary for some frictional force to be developed between the tyres and the road surface in order to keep the vehicle from sliding laterally.

When the carriageway is superelevated, the forces acting on the vehicle are shown in Fig. 6.13(b). Since the forces are in equilibrium,

$$\frac{Wv^2}{gR}\cos\alpha = W\sin\alpha + P$$

$$= W\sin\alpha + \mu\left(W\cos\alpha + \frac{Wv^2}{gR}\sin\alpha\right)$$

Therefore

$$\frac{v^2}{gR} = \tan\alpha + \mu + \frac{\mu v^2}{gR}\tan\alpha$$

where α = angle of superelevation.
The quantity $(\mu v^2/gR)\tan\alpha$ is so small that it can be neglected. If $\tan\alpha$ is expressed in terms of the slope e, then

$$\frac{v^2}{gR} = e + \mu$$

This indicates clearly that the centrifugal force is resisted partly by the superelevation and partly by the lateral friction. If v m/sec is replaced by V km/h, and if $g = 9{\cdot}81$ m/sec^2, then the *minimum* radius equation becomes

$$\frac{V^2}{127R} = e + \mu$$

If $\mu = 0$ and the forces acting on the vehicle are in equilibrium, then the situation occurs where the centrifugal force is entirely counteracted by the superelevation. In this ideal situation, a moving vehicle will steer itself about the superelevated curve once its steering wheels have been set in the required track. When this occurs it must be remembered that, although no lateral friction is brought into use, sufficient friction must be developed in the longitudinal direction of travel to provide for traction of the wheels.

In practice high design speeds are only utilized in car-racing tracks, and cannot be used alone on normal highways because of the danger to slower moving vehicles. For instance, a carriageway surface may be relatively steeply superelevated in order to allow one vehicle travelling at high speed to safely traverse a curve without making use of the friction component. Another vehicle travelling at a slow speed about the same curve will develop a much smaller outward centrifugal force and the result may well be that the vehicle will tend to slide inward. In the extreme case of a vehicle at rest on the curved section, no centrifugal force at all is developed by the vehicle.

Therefore, in order to minimize the danger of sliding, the superelevation slope must never be greater than the minimum lateral coefficient that can be developed between the tyre and the carriageway under the design weather conditions.

Proper curve design does not normally take full advantage of the obtainable lateral coefficients of friction since obviously design should not be based on a condition of incipient sliding. Investigations have shown that the main factor controlling vehicle speed on a curve is the feeling of discomfort felt by the motorist as he negotiates the curve at a given speed. This sensation of the driver is related to the centrifugal ratio v^2/gR which, in turn, is related to the sliding resistance required to carry out the manoeuvres. Various observers have studied this problem and it appears that motorists feel uncomfortable and the vehicle becomes harder to hold on the road when the centrifugal ratio exceeds about 0·3. On the other hand, a value of 0·15 is a safe basis for curve design and it will eliminate most feelings of unease. This lateral ratio is equivalent to a lateral coefficient of friction of 0·15, which is considerably below friction values that are actually developable on the carriageway before slipping occurs. Since, however, highway curves should be designed to avoid slipping conditions, driver comfort represents a rational method of selecting a lateral coefficient of friction that can be used for design purposes.

Department of Environment recommendations. General practice in Britain with respect to superelevation design is not to think in terms of a self-steering vehicle travelling at the design speed of the road but rather of one negotiating a curve at about the average speed of the road; this will result in a gentler superelevation and, hence, there is less chance of slow-moving vehicles slipping sideways under icy conditions or, indeed, overturning when carrying exceptionally high loads. The 'average' speed tacidly accepted for normal usage is 63·2 per cent of the road design speed; in practice this means that the self-steering equation becomes

$$\frac{(0{\cdot}632V)^2}{(127R)} = \mathrm{e} = \frac{V^2}{314R}$$

Furthermore, it is recommended that the side slope should never exceed 0·07 (except on existing roads or (tight) interchange loops on motorways) and, in urban areas, preferably not exceed 0·04 on roads with at-grade intersections and little or no restriction of frontage access. The second and third columns of Table 6.12 show the radii which would be normally used on urban roads; they were determined using the above modified self-steering equation. The last two columns show the minimum radii that could be used; these assume that the vehicles travel at the design speed of the road, and utilize lateral friction factors of 0·18 up to and including 50 km/h and 0·15 at higher speeds. The recommended desirable and absolute minimum radii for rural roads are shown in columns 4 and 5 of Table 6.1.

TABLE 6.12. *Desirable and absolute minimum radii for use on urban roads*[7]

Design speed, km/h	Desirable min radius (m) for superelevation of		Absolute min radius (m) for superelevation of	
	0·04	0·07	0·04	0·07
80	500	300	260	230
60	275	170	150	130
50	200	120	90	80
30	75	50	35	30

Horizontal sight distance

In certain instances the radius of curvature determined by using limiting values of e and μ may not be adequate to ensure that, at the very least, the minimum safe stopping sight distance requirements are met. To provide the necessary horizontal sight distances it may be necessary to set back slopes of cuttings, fences, buildings or other such obstructions adjacent to the carriageway. If these obstructions are immovable, it may be necessary to redesign the road alignment in order to meet the safety requirements.

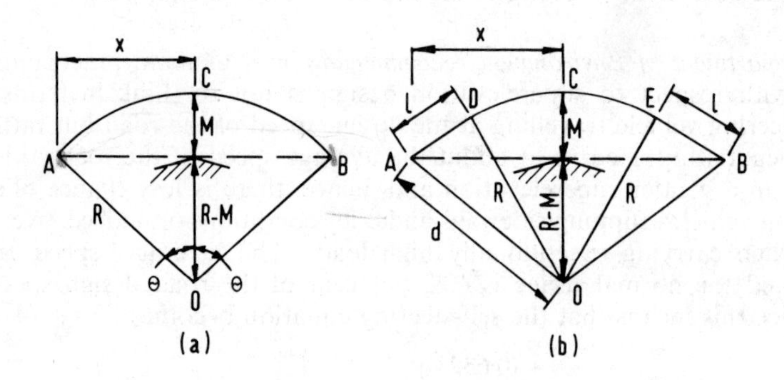

Fig. 6.14. Sight distance considerations on a horizontal curve

The most usual methods of calculating the offset distance necessary to secure the required horizontal sight distance are illustrated in Fig. 6.14. In both cases the driver's eye and the dangerous object are assumed to be at the centre of the nearside lane. The chord AB is the sight line and the curve ACB is the distance required to meet the sight distance criterion.

Figure 6.14(a) illustrates the situation where the required sight distance lies wholly within the length of the curved road section and ACB is equal to the required sight distance S. The minimum offset distance M from the centre-line to the obstruction can be estimated most simply by considering the track of the vehicle to be along the chords AC and CB rather than along the arc of the curve. By geometry,

$$R^2 = x^2 + (R-M)^2$$

and $$x^2 = \left(\frac{S}{2}\right)^2 - M^2$$

Therefore $$R^2 = S^2/4 - M^2 + R^2 - 2RM + M^2$$

Thus $$M = S^2/8R$$

Figure 6.14(b) illustrates the situation where the required sight distance is greater than the available length of curve. Thus the sight distance overlaps the highway curve on to the tangents for a distance l on either side. If the length of the curve is equal to L and the stopping distance is equal to S, then

$$S = L + 2l$$

Therefore $$l = \frac{S-L}{2}$$

By geometry,

$$\left(\frac{S}{2}\right)^2 = x^2 + M^2$$

and $$x^2 = d^2 - (R-M)^2$$

and $$d^2 = \left(\frac{S-L}{2}\right)^2 + R^2$$

Therefore $$\frac{S^2}{4} = \tfrac{1}{4}(S^2 - 2LS + L^2) + R^2 - (R^2 - 2RM + M^2) + M^2$$

Therefore $$M = \frac{L(2S-L)}{8R}$$

It should be noted that, when $S = L$, the above equation reduces to $M = S^2/8R$.

Widening of circular curves

Curve widening refers to the extra width of carriageway that is required on a curved section of a highway over and above that required on a tangent section. Curved sections of normal 2-lane roads are usually widened for two main reasons. First and most important, additional width is required because of the extra road space occupied by a vehicle turning about a curve. As shown in Fig. 6.15, the rear wheels of a vehicle follow a path of shorter radius than the front wheels and this has the effect of increasing the effective width of road space required by the vehicle. Thus, in order to provide the same clearance between opposing vehicles on curved sections as on tangent sections, there must be an extra width of carriageway available. This is especially so when a high percentage of commercial vehicles use the highway.

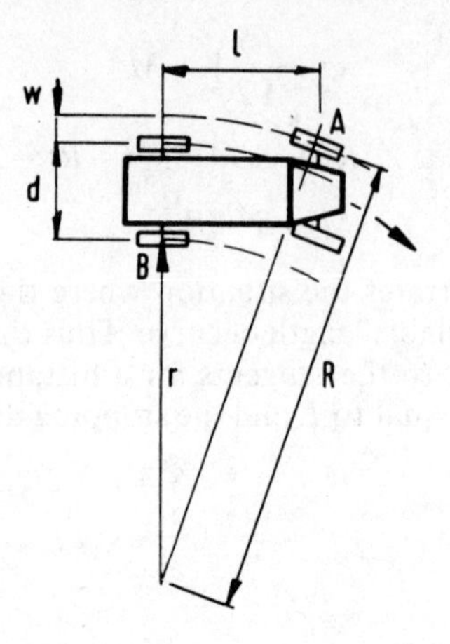

Fig. 6.15. Extra width requirements at circular curves

A second reason for providing the extra width is the natural tendency of drivers to shy away from the edges of the carriageway as they traverse a curve. This reduces the medial clearance between opposing vehicles and increases the accident potential of the curved section.

A measure of the extra width required can be obtained by considering Fig. 6.15. From this diagram,

$$w = R-(r+d)$$

and

$$(r+d) = \sqrt{R^2-l^2}$$

Therefore

$$w = R-\sqrt{R^2-l^2}$$

where w = extra width required,
R = distance OA which is the radius of the outer front wheel,
r = distance OB which is the radius of the inner rear wheel,
d = lateral width between wheels,
and l = length of wheel base.

To determine the extra width needed in any particular situation it is necessary to choose an appropriate design vehicle. This vehicle will normally be a heavy commercial vehicle, since its off-tracking is much greater than that of a passenger car. If the l-value of this design vehicle is taken as, say, 15 m, then

$$w = R-\sqrt{R^2-225}$$

$$\simeq R-R\left(1-\frac{112{\cdot}5}{R^2}\right)$$

$$\simeq \frac{112{\cdot}5}{R}$$

Thus the total extra width required on a normal 2-lane road curve is estimated by $2w = 225/R$, where R is the radius of the outer front wheel of the vehicle. R is usually also assumed to be the radius of the centre line of the road curve.

It is also good practice to provide the same amount of widening on curved sections of dual carriageways, as well as on 2-lane two-way roads. This ensures the high standard of geometric design which is compatible with these facilities. On 3-lane single carriageways the need for widening depends on whether the highway is to be operated as a 2-lane or 3-lane facility at the curved sections in question. If it is to be operated as a 3-lane road, widening is usually needed. If operation is to be on a 2-lane two-way basis, then extra widening is normally not necessary. The current governmental recommendations re widening on curves are:

Curve radius, m	*Increase per 10 m width of carriageway, m*
<150	1·6
150–300	1·2
300–400	0·8

Widening procedure. When a carriageway is to be widened at a curve, the extra width should be attained gradually over the approaches to the curve so that the 'natural' inclinations of vehicles entering or leaving the curved section are facilitated. The four main considerations to be taken into account are:

1. On simple circular curves the total extra width should be applied to the inside of the carriageway, while the outside edge and the centre line are both kept as concentric circular arcs. The result is much the same as providing a transition curve on the centre line. The main drawback is that the full width of the carriageway may not always be in use, since a fast-moving vehicle will tend to curve in from the outer edge at the beginning and end of the curved section as it makes its own transition.

2. Where transition curves are provided before and after a simple curve, the widening may be equally divided between the inside and outside of the curve, or it may be wholly applied to the inside edge of the carriageway. In either case, the final centre-line marking, if it is a 2-lane road, should be placed half-way between the edges of the widened carriageway.

3. The extra width should always be attained gradually, and never abruptly, to ensure that the entire carriageway is usable. When the curved sections utilize transition curves, they should be attained over the entire transition length so that there is zero extra width at the beginning of transition and the full extra width at the end. Where transition lengths do not occur, smooth alignment will be obtained if one-half to two-thirds of the extra width is distributed along the tangent just prior to the curve, and the balance along the curve itself.

4. From an aesthetic point of view, the edges of the carriageway should

at all times form smooth and graceful curves. Not only is this pleasing to the eye but it facilitates construction and provides reassurance for the motorist that the highway curve is properly laid out.

Transition curves

As a vehicle enters or leaves a circular curve the driver will, for his own comfort and safety, gradually turn the steering wheel from the normal position to that of the necessary deflection in order to combat the developing effect of the centrifugal force. By thus gradually changing direction, the motorist will cause the vehicle to trace out its own transition curve from the tangent to the curve.

The transition path traced out in this way will vary depending upon the speed of the vehicle, the radius of the curve, the superelevation of the carriageway and the steering action of the motorist. At moderate values of speed and curvature, the average driver is usually able to effect this path within the limits of the normal lane width. With higher speeds and sharper curves, vehicles in the inner lane tend to trace their transitions by crowding on to the adjoining lane, while those in the outside lane tend to cut the corner. This reduces the effective width of the roadway and can be a serious cause of accidents.

The primary purpose of transition curves is to enable vehicles moving at high speeds to make the change from the tangent section to the curved section to the tangent section of a road in a safe and comfortable fashion. The proper introduction of transition curves will provide a natural easy-to-follow path for motorists so that the centrifugal force increases and decreases gradually as the vehicle enters and leaves the circular curve. This minimizes the intrusion of vehicles on to the wrong lanes, tends to encourage uniformity in speed, and increases vehicle safety at the curve.

Spiral transition curves. The essential requirement of any transition curve is that its radius of curvature should decrease gradually from infinity at the tangent-spiral intersection to the radius of the circular curve at the spiral-circular curve intersection. Various forms of curves are used for this purpose, each having its own special advantages. The most common of these are the lemniscate, the spiral (or clothoid) and the cubic parabola. They are considered the 'natural' transition curves in contrast to others which constrain the motorist into a more definite path. In practice there is very little difference between the results obtained by all three procedures.

In recent years, the spiral transition curve has become widely accepted by highway engineers, so it will be briefly discussed here. Its acceptance is primarily based on the ease with which it can be set out in the field compared with the other two curves, and not because of any particular superiority as a transition curve.

Spiral properties. A circular curve joined to two tangents by spiral transition curves is shown in Fig. 6.16. The dotted lines illustrate the circular curve as it would appear if the transition curves were omitted. TS is the tangent-spiral intersection, SC is the junction of the spiral and circular curve, CS is the intersection of the circular arc and the second spiral and ST is the second tangent point. The spiral angle, i.e. the total angle subtended by the spiral, is θ_s.

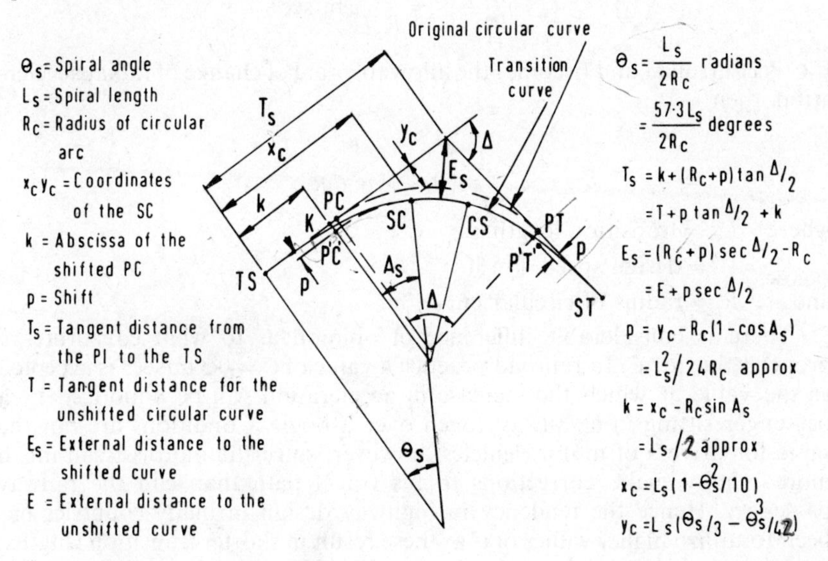

Fig. 6.16. Basic properties of the spiral transition curve

Also shown in Fig. 6.16 are some of the more important relationships affecting the use of the spiral curve in conjunction with the circular curve. Detailed derivation of these formulae can be found in surveying textbooks.

In Fig. 6.16 the amount by which either end of the circular curve is shifted inward from the tangent is indicated by the dimension K to P^1C^1. This distance p is called the *Shift*. It is very convenient to remember that the offset from point K on the tangent to the transition spiral is very nearly $P/2$, and that the line from K to P^1C^1 approximately bisects the spiral.

Length of transition. The three major factors governing transition curve design are the radius of curvature R_c, the external angle Δ, and the length of transition L_s. Of these, R_c and Δ are usually selected on the basis of conditions existing in the field. L_s, on the other hand, is selected on the basis of factors affecting the comfort and security of the motorist.

As a vehicle passes along the transition curve its centrifugal acceleration changes from zero at TS to v^2/R_c at SC. The transition length over which this change takes place is equal to the vehicle velocity v multiplied by the travel time t. Thus

$$t \text{ sec} = \frac{L_s \text{ m}}{v \text{ m/sec}}$$

If the rate of gain of radial acceleration is denoted by C, then

$$C = \frac{v^2}{R_c} \div \frac{L_s}{v} = \frac{v^3}{R_c L_s} \text{ m/sec}^3$$

If C is controlled and becomes the allowable rate of change of radial acceleration, then

$$L_s = \frac{v^3}{CR_c} = \frac{V^3}{3{\cdot}6^3 CR_c}$$

where L_s = transition length, m
V = design speed, km/h
and R_c = radius of circular curve, m

There is considerable difference of opinion as to what constitutes a proper value for C. In railroad practice a value of $C = 0{\cdot}3$ m/sec^3 is accepted as the value at which the increase in acceleration will be unnoticed by a passenger sitting in a railway coach over a bogie. Conditions are not the same for drivers of motor vehicles, however, since the motorist can much more readily make corrections to his travel path than can the railway passenger. Hence the tendency in highway design in many countries has been to utilize higher values of C as these result in shorter transition lengths; for instance, the C-values most often used in Great Britain(16) vary between 0·3 and 0·6 m/sec^3. In general, long spirals (low C-values) tend to be associated with large radii in order to accommodate the application of superelevation, or for aesthetic reasons.

Use of transition curves. Transition curves should be used on all highways whenever there is a significant change in road curvature. Obviously, however, there is less need for transitions as curves become flatter. At some point the difference between road curves with and without transition lengths is so small that the transition curve has little significance, except perhaps as a graceful method of changing from a cambered to a superelevated section. This point of no significance is usually reached when the shift is approximately 0·3 m.

Governmental recommendations regarding the minimum radii of rural road curves which require transitions are given in Table 6.1.

Application of superelevation. As discussed previously, superelevation should be introduced on curves with other than very large radii. In these cases it is most important that the superelevation of the carriageway be built up uniformly over as long a distance as possible so that the full superelevation rate is achieved at the circular arc part of the curve. This is not only for reasons of comfort and safety but also for aesthetic reasons. To avoid the carriageway edges having an unpleasing, distorted appearance, the change in cross-section from a camber-section to a fully superelevated one should be carried out, ideally, such that the slope of the outside edge of the carriageway does not exceed 1 in 200 when compared with the centre line. In other words, the difference in grade between the centre-line profile and the edge of a 2-lane road should not exceed 0·5 per cent.

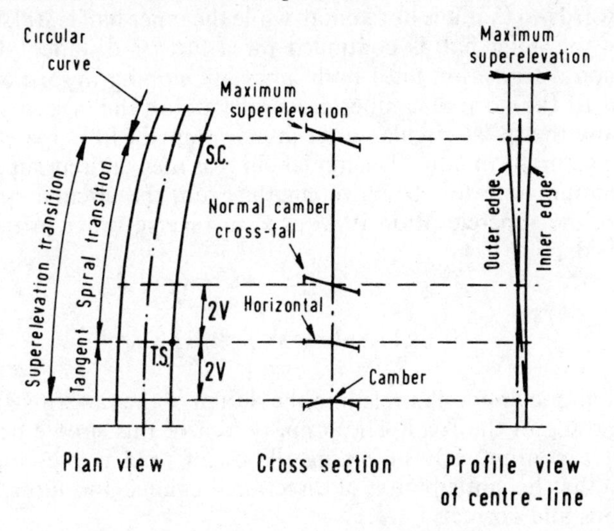

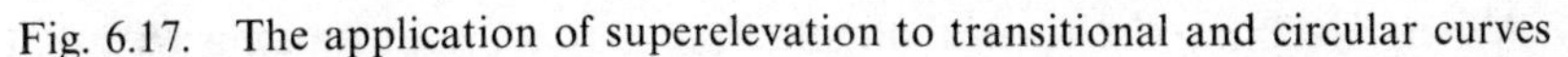

Fig. 6.17. The application of superelevation to transitional and circular curves

When a spiral transition curve is used in conjunction with a circular curve, it is usual for the superelevation run-off to be carried out over the total length of the spiral. The actual attainment of superelevation requires that the carriageway cross-section be tilted, the amount of movement varying with its location on the transition or circular curves. This tilting of the carriageway can be carried out by any one of the following three ways:

1. Rotation about the centre line of the carriageway.
2. Rotation about the inside edge of the carriageway.
3. Rotation about the outside edge of the carriageway.

In practice there is little to choose between the three methods, so that the choice in any particular situation should be that which provides the most pleasing and functional results. Regardless of which method is used, however, care should always be taken that the drainage requirements at the inside of the roadway are not detrimentally affected by the rotation procedure.

The method which this author prefers is that of rotation about the centre line. In the normal course of events, the longitudinal profile of the carriageway will have been decided upon before any superelevation calculations take place. More often than not therefore it is more convenient to keep the centre-line levels unchanged and to move both edges of the carriageway instead.

One method of attaining superelevation by rotating about the longitudinal centre line is illustrated in Fig. 6.17. With this method the superelevation transition is started on the tangent at a distance of $1{\cdot}4V$–$2V$ metres back from the beginning of the horizontal transition curve. In this case V is the design speed of the highway expressed in *metres per second*. By the time the tangent-to-the-spiral point is reached, the outer half of the carriageway is warped upwards so that it is horizontal, while the inner half is still cambered. Warping of the outer half is continued for a further distance of $1{\cdot}4V$–$2V$ along the spiral transition until both lanes are sloping inward at a cross-slope equal to the normal camber cross-fall rate on the tangents. Between this point and the spiral-circular curve intersection, the full superelevation is developed at a uniform rate. The circular curve is then entirely superelevated at the maximum rate for its entire length. From the circular-spiral curve intersection, the superelevation is reduced in reverse order to the manner just described.

VERTICAL ALIGNMENT

Vertical alignment design refers to the design of the tangents and curves along the profile of the road. The primary aim of this profile design is to ensure that a continuously unfolding ribbon of road is presented to the motorist so that his anticipation of directional change and future action is instantaneous and correct.

When designing the vertical alignment of a road, it must be ensured that, at the very least, minimum stopping sight distance requirements are met. Furthermore, on 2-lane and 3-lane roads the design must provide adequate stretches of highway to enable faster vehicles to overtake comfortably and safely. In addition, long stretches of steep gradients must be avoided because of their detrimental effect upon speeds, capacity and safety.

Gradients

Long steep uphill grades can have a considerable effect on vehicle speeds[17]. This is of the utmost importance on highways where commercial vehicles form a considerable portion of the traffic flow. Since restrictive sight distances are usually associated with steep uphill gradients, the speed of traffic is often controlled by the speeds of the slower commercial vehicles. As a result, not only are the operating costs of vehicles increased but the capacity of the road will have to be reduced in order to maintain the required level of service.

Uphill gradients can be a cause of accidents between vehicles in opposing traffic streams, since faster vehicles are tempted to overtake where normally they might not do so. Safety can also be a very important consideration on downhill gradients due to the possibility of relatively great increases in vehicle speeds. As a result the ability of motorists to stop or take other emergency measures can be seriously curtailed during inclement weather conditions. In particular, if vehicles have to be braked in order to traverse a curve at the bottom of a gradient, and the road is icy or wet, serious accidents can be caused by vehicles skidding out of control.

Maximum grades. Grades of up to about 7 per cent have relatively little effect on the speeds of passenger cars. On the other hand, the speeds of commercial vehicles are considerably reduced when long gradients, with grades in excess of about 2 per cent, are features of the highway profile.

Most highway authorities now accept a gradient of 4 per cent as being the maximum desirable on major highways; hence this criterion is usually a controlling feature of highway design. In certain instances, when the uphill sections are less than about 200 m long, grades of 5 to 6 per cent can be utilized. Vehicles usually accelerate upon entering an uphill section and this extra momentum is sufficient to overcome the effect of the extra increase in slope over these shorter distances.

Creeper lanes. The maximum grade is not in itself a complete design control. It is also necessary to consider the length of the gradient and its effect on desirable vehicle operation.

When the uphill climb is extremely long, an extra 'creeper' lane may have to be provided to allow ascending slower-moving vehicles to be removed from the main traffic stream, thus allowing the faster vehicles very much greater freedom of movement and action. It must be emphasized that this extra lane is not intended to convert the 2-lane highway into a 3-lane one. Rather it should be regarded as a temporary aid to traffic movement just as acceleration or deceleration lanes facilitate vehicle movement at intersections.

There are no definitive data available as to what should be considered the minimum tolerable vehicle speed or critical length of gradient to justify the introduction of creeper lanes, and the decision in many instances must be left to the judgement of the design engineer. In Britain, for example, it is recommended that a lowering of design standards on hills must be accepted, and that a creeper lane should only be provided when the predicted future traffic is greater than the capacity by 100 per cent in the case of 2-lane roads, and by 50 per cent in the case of dual carriageways; with 3-lane roads, it is customary to use one of the existing lanes (marked as a creeper lane by means of offset double white lines) when the lengths exceed the critical values given for the following gradients:

Gradient, %	3	4	5	6	7
Critical length, m	488	335	244	207	168

The point at which the creeper lane should be initiated depends on the speeds at which the commercial vehicles begin to ascend the hill. When there are no restrictive sight distances or other features that might result in low vehicle approach speeds, the extra lane should be introduced some distance uphill from the beginning of the slope, ideally at the point where vehicle speeds have been reduced to the speed for capacity. When restrictions cause lower approach speeds, then the creeper lane will obviously have to be initiated closer to the bottom of the hill.

For safety reasons, the extra lane should end at a point well beyond the crest of the hill so that the slower vehicles can return to their normal lane without hazard. In addition, it should not be ended abruptly but should be tapered in a manner similar to an acceleration lane, so that the vehicles can make the transition to the normal lane efficiently as well as safely.

Minimum grades. These are of interest only at locations where surface drainage is of particular importance. Normally the camber of the road is sufficient to take care of the lateral carriageway surface drainage. In cut sections, however, it may be necessary to introduce a slight longitudinal gradient into the road surface in order to achieve longitudinal drainage in the side ditches; a similar situation arises when the road is kerbed. A minimum grade of 0·5 per cent (1 in 200) is desirable in both of these instances. Satisfactory drainage has, however, been obtained with grades in the order of 0·3 per cent.

Vertical curves

Just as a circular curve is used to connect horizontal straight stretches of road, a parabolic curve is usually used to connect gradients in the profile alignment. These curves are convex when two grades meet at a 'summit' and concave when they meet at a 'sag'.

While other curve forms can be used with satisfactory results, the tendency has been to utilize the parabolic curve in profile alignment design. This is primarily because of the ease with which it can be laid out as well as enabling the comfortable transition from one grade to another. Normally vertical curves of this type are not considered necessary when the total grade change from one tangent to the other does not exceed 0·5 per cent.

Parabolic curve properties. The form of parabolic curve most often used is the simple parabola, an example of which is shown in Fig. 6.18. If Y in this figure is considered to be the elevation of the curve at a point x along the

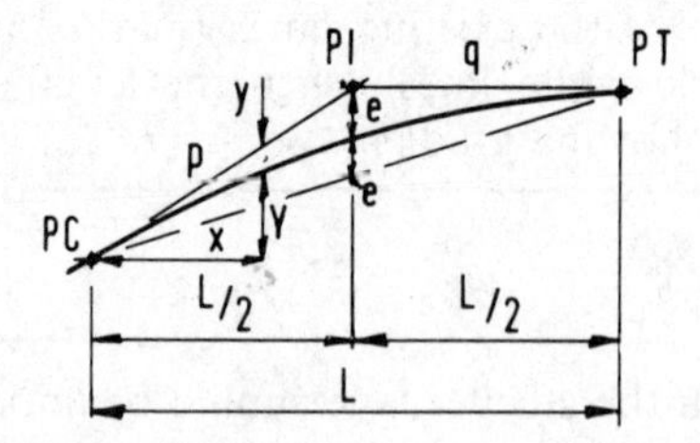

Fig. 6.18. A simple symmetrical parabolic curve

curve, and since the rate of change of slope along a parabola is a constant, then

$$\frac{d^2Y}{dx^2} = k, \text{ a constant}$$

Integrating, $$\frac{dY}{dx} = kx + C$$

When $x = 0$, $$\frac{dY}{dx} = p, \text{ the slope of the first tangent.}$$

When $x = L$, $$\frac{dY}{dx} = q, \text{ the slope of the second tangent.}$$

Therefore $$C = p \text{ and } k = \frac{q-p}{L}$$

and $$\frac{dY}{dx} = \left(\frac{q-p}{L}\right)x + p$$

Integrating, $$Y = \left(\frac{q-p}{L}\right)\frac{x^2}{2} + px + C_1$$

When $x = 0$, $$Y = 0 \text{ and } C_1 = 0$$

Therefore $$Y = \left(\frac{q-p}{L}\right)\frac{x^2}{2} + px$$

From geometry $$\frac{y+Y}{x} = \frac{p}{1}$$

Therefore $$y = -\left(\frac{q-p}{2L}\right)x^2$$

y, measured downward from the tangent, gives the vertical offset at any point along this curve. These vertical offsets from the tangents are used to lay out the curve.

It is often necessary to calculate the highest (or lowest) point on the curve to ensure that minimum sight distance (or drainage) requirements are met. The location of this point is given by

$$x = \frac{Lp}{p-q} \text{ and } y = \frac{Lp^2}{2(p-q)}$$

As these equations illustrate, the high or low point of a symmetrical curve is not necessarily directly below or above the point of intersection of the tangents, but may in fact be located on either side of this point.

Figure 6.19 illustrates in exaggeration some varying forms of highway vertical curves. The algebraic signs given to the slopes are quite important;

the plus sign is used for ascending slopes from the point of curvature while minus signs are used for descending ones. Proper use of these algebraic signs indicates automatically whether the elevations along the vertical curve are obtained by addition or subtraction of offsets from the tangent elevations. Negative answers are measured downwards from the tangents for all summit curves, while positive answers are measured upwards for sag curves.

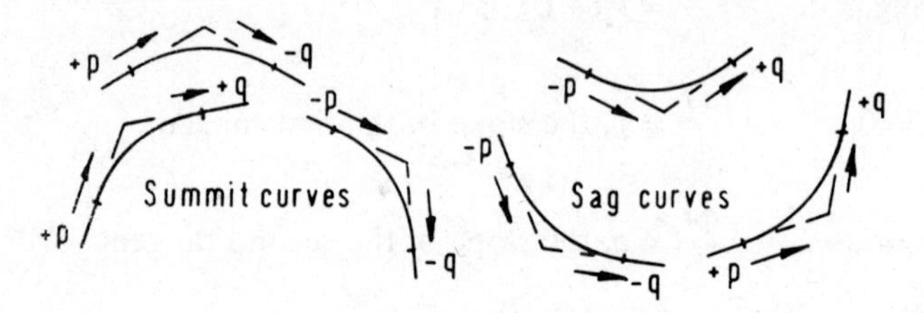

Fig. 6.19. Typical vertical curve forms

EXAMPLE. A -2% grade is being joined to a -4% grade by means of a parabolic curve of length 1000 m. Calculate the vertical offset at the point of intersection of the tangents.

Solution. Assume that the length of the curve is equal to its horizontal projection. The actual difference between the two can be neglected for all practical purposes. Thus $L = 1000$ m.

The vertical offset y at any point x along the curve is given by

$$y = \left(\frac{q-p}{2L}\right)x^2$$

When $x = \frac{L}{2}$, $y = e$, the vertical offset at the P.I.

Therefore
$$e = \left(\frac{q-p}{2L}\right)\frac{L^2}{4} = \frac{L}{8}(q-p)$$
$$= \frac{1000}{8}\left[-\frac{4}{100} - \left(-\frac{2}{100}\right)\right]$$
$$= -2{\cdot}5 \text{ m.}$$

Sight distance requirements. In determining the lengths of vertical curves, the controlling factors are the security and comfort of the motorists, and the appearance of the profile alignment. Of these, the sight distance requirements for safety are by far the most important on summit curves. With sag curves, on the other hand, safety is of less importance and more consideration can be given to the other factors.

Summit curves. When deriving formulae for the *minimum* lengths of summit curves, there are two design conditions that have to be considered.

Figure 6.20(a) illustrates the first condition, where the required sight distance is contained entirely within the length of the vertical curve, i.e. where $S<L$. In this figure h_1 and h_2 are the height of the driver's eye and the height of the dangerous object on the roadway. Since the curve is a parabola, the

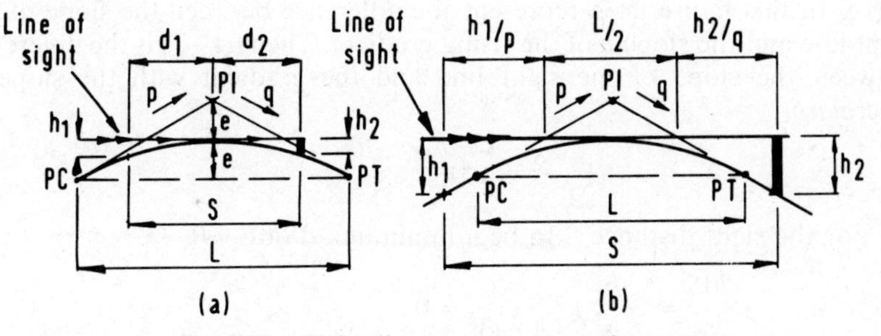

Fig. 6.20. Sight distances over summit curves

offsets from the line of sight are proportional to the square of the distance from the point where the curve is tangent to the line of sight, Thus,

$$h_1 = kd_1^2 \text{ and } h_2 = kd_2^2$$

But $$e = k\left(\frac{L}{2}\right)^2$$

Therefore $$\frac{h_1+h_2}{e} = \frac{4d_1+4d_2^2}{L^2}$$

and $$d_1+d_2 = \sqrt{\frac{h_1 L^2}{4e}} + \sqrt{\frac{h_2 L^2}{4e}}$$

But $e = LA/8$, where A = algebraic difference in slopes expressed in absolute values; it is always expressed in decimal form.

Substituting, $$d_1 = \sqrt{\frac{2h_1 L}{A}} \text{ and } d_2 = \sqrt{\frac{2h_2 L}{A}}$$

Therefore $$L = \frac{A(d_1+d_2)^2}{(\sqrt{2h_1}+\sqrt{2h_2})^2}$$

But $d_1+d_2 = S$, the sight distance.

Therefore $$L = \frac{AS^2}{(\sqrt{2h_1}+\sqrt{2h_2})^2}$$

If the dangerous object is assumed to be at carriageway level, then

$$L = \frac{AS^2}{2h_1}$$

If $h_1 = h_2$, then

$$L = \frac{AS^2}{8h_1}$$

Figure 6.20(b) illustrates the second condition, where the required sight distance overlaps on to the tangent sections on either side of the parabolic curve. In this figure let g represent the difference between the slope of the sight line and the slope p of the rising gradient. Then $(A-g)$ is the difference between the slope of the sight line and the gradient with the slope q. Therefore,

$$S = \frac{L}{2} + \frac{h_1}{g} + \frac{h_2}{A-g}$$

For the sight distance S to be a minimum, $dS/dg = 0$

Therefore $$\frac{dS}{dg} = -\frac{h_1}{g^2} + \frac{h_2}{(A-g)^2} = 0$$

Solving, $$g = \frac{A\sqrt{h_1 h_2} - h_1 A}{h_2 - h_1}$$

Substituting, $$S = \frac{L}{2} + \frac{h_1}{\left(\dfrac{A\sqrt{h_1 h_2} - h_1 A}{h_2 - h_1}\right)} + \frac{h_2}{\left(A - \dfrac{A\sqrt{h_1 h_2} - h_1 A}{h_2 - h_1}\right)}$$

Therefore $$L = 2S - \frac{2(\sqrt{h_1} + \sqrt{h_2})^2}{A}$$

If $h_2 = 0$, then $L = 2S - \dfrac{2h_1}{A}$

If $h_1 = h_2$, then $L = 2S - \dfrac{8h_1}{A}$

The decision as to which condition should be used at a particular site can be made by solving either of the equations

$$e = \frac{(q-p)L}{8} \text{ or } e = \frac{(q-p)S}{8}$$

depending upon whether L or S is the known value. In either case, if e is found to be greater than h_1, then the equation for the first condition, i.e. when L is greater than S, should be used. If it is found that e is less than h_1 then the equation where L is less than S should be used.

EXAMPLE. A vertical curve is to be constructed between an ascending $3\frac{1}{2}\%$ grade and a descending 4% grade. The required safe stopping sight distance is 300 m, the dangerous object is considered to be at the carriageway level, and the motorist's eye height is 1·05 m. Determine the minimum length of vertical curve that will satisfy this sight distance requirement.

Solution. $q = -\frac{4}{100}, p = +\frac{3 \cdot 5}{100}, S = 300\,\text{m}$

Therefore, $$e = \frac{(q-p)S}{8} = \frac{\left(-\frac{4}{100} - \frac{3 \cdot 5}{100}\right)300}{8}$$

$$= \frac{-7 \cdot 5 \times 300}{8 \times 100}$$

$$= -2 \cdot 82$$

But h_1, the eye height, is equal to 1·05 m, therefore the equation for $L > S$ will be used.

When $L > S$, the length of the vertical curve is given by

$$L = \frac{AS^2}{(\sqrt{2h_1} + \sqrt{2h_2})^2}$$

But $h_1 = 1 \cdot 05\,\text{m}$, $h_2 = 0$, $A = 7 \cdot 5/100$ and $S = 300\,\text{m}$

Therefore, $$L = \frac{\frac{7 \cdot 5}{100} \times 300 \times 300}{(\sqrt{2 \times 1 \cdot 05} + 0)^2}$$

$$= 3214\,\text{m}$$

Department of Environment recommendations re vertical curve lengths are summarized in Table 6.13. For both urban and rural situations, the curve length is determined from the formula $L = KA$ metres, where A is the algebraic difference in gradients (expressed as a percentage) and K has a value selected from the table for the design speed of the road. It is important to appreciate that the curve lengths determined in this way for *both* safe stopping and safe overtaking sight distance are based on a driver eye height of 1·05 m and an object height of 1·05 m. This recommendation may be queried with respect to safe stopping sight distance; while it is very economical in terms of earthwork requirements (e.g. compare the appropriate L-value determined from Table 6.13 with that calculated in the example given above) because of the shorter curve length required, it also means that inadequate sight distance will be available should, for example, an object less than 1·05 m high be lying on the carriageway.

TABLE 6.13. *Recommended minimum vertical curve lengths*[1, 7]

Design speed, km/h	Urban road			Rural road		
	Min K-value for:		*Min vert. curve length	Min K-value for:		
	Stopping	Overtaking		Stopping (Crests)	Stopping (Sags)	Overtaking (Crests)
120	—	—	—	105 (175 for motorways)	75	—
100	—	—	—	50	50	240
80	20	—	50	20	30	150
60	10	—	40	10	20	90
50	5	60	30	—	—	—
30	1	20	20	—	—	—

* Use only if greater than KA

Sag curves. Whereas with summit curves the most important factor is the length of curve necessary for safety, there are at least four widely accepted criteria for determining the minimum lengths of sag curves. These are the vehicle headlight sight distance (British practice), motorist comfort, drainage control and general aesthetic considerations. In addition, since sag curves are often associated with highways underpassing structures such as bridges, in certain instances the curve length may be chosen to ensure the necessary vertical clearance and to maintain adequate sight distance.

There is still a considerable difference of opinion as to what value of radial acceleration should be used on vertical curves for comfort design purposes. The most commonly quoted values are between 0·30 and 0·46 m/sec².

If the vertical radial acceleration is assumed to be equal to a m/sec²,

then
$$a = \frac{v^2}{R} = 3{\cdot}6^2\left(\frac{V^2}{R}\right)$$

and
$$R = \frac{V^2}{13a}$$

where R = radius of the circle equivalent to the parabolic curve,
v = vehicle speed, m/sec,
and V = vehicle speed, km/h

Since the central angle Δ of the equivalent circular curve is very small, and the circle practically coincides with the parabola,

$$L = R\Delta = \frac{V^2\Delta}{13a}$$

But
$$\Delta = A$$

Therefore
$$L = \frac{V^2 A}{13a}$$

where L = length of the sag curve, m

and A = algebraic difference in slopes, expressed as a decimal.

EXAMPLE. Determine the minimum length of curve required to connect a descending 4% grade to an ascending 3% grade. The design speed of the road is 100 km/h and the acceptable radial acceleration is 0·3 m/sec².

Solution. $a = 0{\cdot}3\ \text{m/sec}^2$, $V = 100\ \text{km/h}$, $A = 7/100$

Therefore, $$L = \frac{V^2A}{13a} = \frac{(100)^2(0{\cdot}07)}{13(0{\cdot}3)}$$

$$= 180\ \text{m}$$

When a highway passes underneath a structure such as a bridge, the motorist's line of sight may be obstructed by the edge of the bridge. In such cases the controlling factor should be the sight distance requirement for safety. Again, when calculating the required length of curve, two considerations have to be taken into account:

1. When the required sight distance is less than the length of the sag curve, then

$$L = \frac{S^2A}{8\left(C - \dfrac{h_1+h_2}{2}\right)}$$

2. When the sight distance is greater than the length of the sag curve, then

$$L = 2S - \frac{8\left(C - \dfrac{h_1+h_2}{2}\right)}{A}$$

where L = length of the sag curve, m
S = sight distance, m
A = algebraic differences in tangent slopes, expressed in decimal form,
C = vertical clearance to the critical edge of the structure, m
h_1 = vertical height of eye, m
and h_2 = vertical height of hazardous object on the carriageway, m

In both of these equations, the critical edge of the structure is assumed to be directly over the point of intersection of the tangents. In practice, both equations can be considered valid provided that the critical edge is not more than about 60 m from the point of intersection.

EXAMPLE. Determine the minimum length of valley curve required to connect a descending 4 per cent grade to an ascending 3 per cent grade. The vertical clearance is to be 5·1 m (British practice) and the required sight distance is 300 m. The height of eye for a commercial vehicle is 1·83 m and the hazardous object has a vertical height of 0·46 m. (*Note.* The h_1 and h_2 values are those used in sag curve design at underpasses in the United States.)

Solution. Assuming that the sight distance is greater than the required length of curve, then

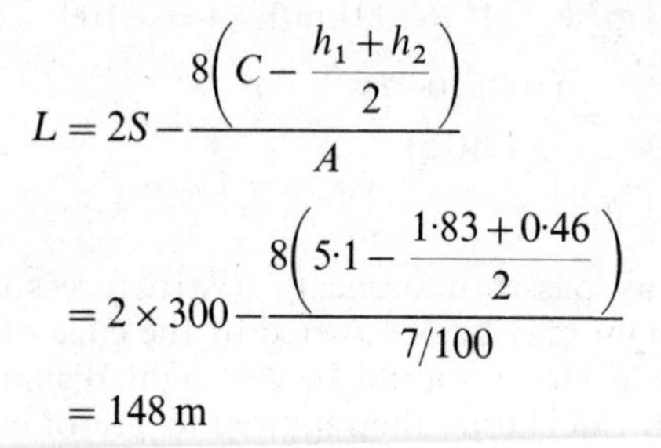

$$L = 2S - \frac{8\left(C - \dfrac{h_1 + h_2}{2}\right)}{A}$$

$$= 2 \times 300 - \frac{8\left(5{\cdot}1 - \dfrac{1{\cdot}83 + 0{\cdot}46}{2}\right)}{7/100}$$

$$= 148 \text{ m}$$

General considerations

Proper design of the vertical alignment requires that considerations other than safety and comfort also be taken into account. For instance, a smooth grade line with gradual changes should always be in preference to one with numerous breaks and short lengths of grade. Roller-coaster types of profile should be avoided as they are dangerous as well as aesthetically unpleasing. Broken-back grade lines, i.e. a section composed of two vertical curves in the same direction separated by a short tangent length, should also be avoided. This is particularly noticeable in valley topography where the full view of both vertical curves is not at all pleasing.

Where single-level intersections occur on highway sections with moderate to steep gradients, the slope at the intersection itself should always be reduced. This will considerably help vehicles performing turning movements and may well serve to reduce potential accidents.

On long gradients, it may be desirable in certain instances to have a steeper slope near the bottom of the hill and lighten the slope near the top instead of using a uniform sustained grade that may be only just below the maximum allowable. This procedure is particularly applicable to gradients on low-speed roads, where the approaching vehicles can accelerate into the rising sections, and on high-speed roads when a rising gradient occurs just after a falling one.

Vertical curvature superimposed on horizontal curvature, or vice versa, generally has a very aesthetically pleasing result. It should, however, always be carefully analysed for possible effects upon traffic. In particular, sharp horizontal curvature should never be introduced at or near the top of a pronounced vertical curve. At night-time this can be especially dangerous since drivers may not notice the horizontal change in direction and severe accidents may occur. For similar reasons, sharp horizontal curves should never be introduced near the bottom of steep gradients because of the high vehicle speeds that can be expected. Not only is there the danger of vehicles overshooting such curves at night-time, but accidents involving skidding can be expected during inclement weather as drivers attempt to slow down to negotiate the sharp bends.

When traffic conditions justify the provision of a dual carriageway, consideration should always be given to the feasibility of varying the width of the central reservation and possibly using two completely separate horizontal and vertical alignments. In this way a superior design making the maximum possible use of the one-way feature of dual carriageways may be obtained at little additional cost.

CROSS-SECTION ELEMENTS

By cross-section elements are meant those features of the highway which form its effective width and which affect vehicle movement. The constituent parts of primary interest are the number and width of traffic lanes, the central reservation, shoulders, camber of the carriageway and, where necessary, the side slopes of cuttings or embankments.

Figure 6.21 shows the various elements of some simplified highway cross-sections in urban and rural areas. Specific dimensions are not shown since, as will now be discussed, recommended dimensions vary considerably throughout the world because of economic and other considerations.

Traffic lanes. The number of traffic lanes to be used in a specific situation is dependent on the volume and type of traffic that has to be handled. Normally, however, the minimum number is two, single-lane roads being rarely constructed in developed countries today. Even though traffic volumes may be light, safety considerations and ease of traffic operation require two lanes.

While 2-lane roads constitute the predominant part of the British highway system, there are also a substantial number of 3-lane roads. These are constructed when the design volume exceeds the capacity of a 2-lane road, but is not sufficient to justify a 4-lane facility. Because of safety considerations 3-lane roads should only be constructed in rural areas where it is practical to provide nearly continuous overtaking sight distances. In urban or suburban areas, 3-lane two-way facilities should never be constructed because of their high accident potential, i.e. in heavy traffic drivers often take the view that they are as entitled to the middle lane as the opposing traffic and attempt to overtake regardless of the hazardous conditions.

If it is expected that long-term traffic needs will eventually justify a 4-lane road, but it is not required at this time, an economical procedure is to construct a 3-lane road composed of two outer lanes of high quality construction and a middle lane of lower quality and cost. Additional lanes of high-type construction can be then added on the outside at a later time and the middle lane converted into a central reservation.

Four or more lanes are needed to enable vehicles to overtake on lanes not used by opposing traffic. Normally these highways have dual carriageways separated by a central reservation. At least one urban motorway in the United States, the Southside Expressway in Chicago, has a total of fourteen traffic lanes on four carriageways. These are composed of four through lanes on one carriageway plus three service lanes on another carriageway in each direction of travel.

British governmental practice is not to assign a certain width to a lane, but rather to specify the carriageway width in relation to the traffic needs at a given location and then to assign a given number of lanes to that carriageway (see Table 6.14).

TABLE 6.14. *Recommended carriageway widths*[1, 7]

Road type	*Description of carriageway(s)*	*Carriageway width, m*
Primary distributor (Urban)	Dual 4-lane	14·60
	Overall width for 4-lane divided carriageway with central refuges	14·60
	Single 4-lane, no refuges	13·50
	Dual 3-lane	11·00
	Single 3-lane (for tidal flows only)	9·00
	Dual 2-lane (normal)	7·30
District distributor (Urban)	Single 2-lane (normal)	7·30
	Dual 2-lane (normal)	7·30
	Dual 2-lane (if the proportion of heavy commercial traffic is fairly low)	6·75
Local distributor (Urban)	Single 2-lane, in industrial districts	7·30
	Single 2-lane, in principal business districts	6·75
	Single 2-lane, in residential districts used by heavy vehicles (min)	6·00
Access roads (Urban)	*In industrial and principal business districts use the values given above for Local Distributors*	
	Single 2-lane, in residential districts (min)	5·50
	2-lane back or service roads, used occasionally by heavy vehicles	5·00
	2-lane back roads in residential districts, used only by cars (min)	4·00
Rural	Single lane (Scotland and Wales mostly)	3·50
	Min width in rural junctions	4·50
	Single 2-lane (min)	5·50
	Motorway slip road	6·00
	Single or dual 2-lane (normal)	7·30
	Single 3-lane	10·00
	Dual 3-lane	11·00
	Dual 4-lane	14·60

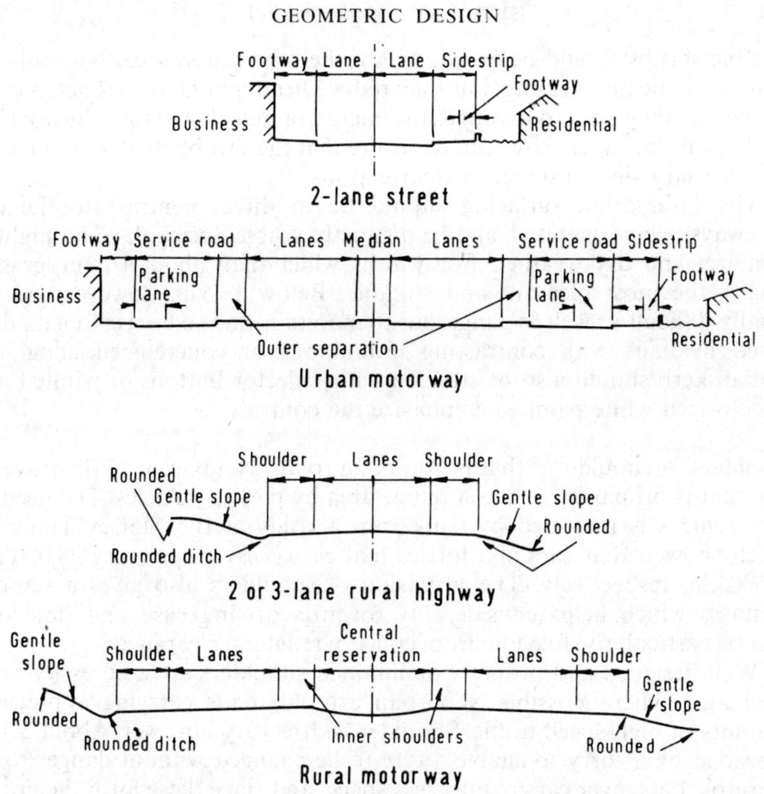

Fig. 6.21. Typical highway cross-sections

Central reservations. Most dual carriageways have at least two traffic lanes in each direction divided by a central reservation at least 1·75 m wide, even in urban areas where space is restricted.

There is considerable difference of opinion between major highway authorities as to the *desirable* width of the central reservation. For instance, in Britain 4·5 m is recommended as the normal minimum width of central reservations in rural areas.[(1)] In the United States the equivalent A.A.S.H.O. recommended figure is 18·3 m.

While it has been proved that the wider the central reservation the greater the reduction in head-on collisions, it has not been possible to establish any overall relationship between the total number of accidents and the reservation width. It has been found, however, that an almost straight-line relationship exists between the median width and the percentage of vehicles involved in median accidents which actually crossed the central space.[(23)] It would appear that a central reservation width of 12–15 m is needed to bring an encroaching high-speed vehicle under control to avoid the possibility of becoming involved in a head-on collision on the other carriageway.

Central reservations need not be of constant width. In fact, in rural areas, consideration should always be given to varying the width in order to obtain a safe, pleasing and economical design that fits the topography. Where possible shrubs should be grown within the reservation area. Not only are these aesthetically pleasing, but they reduce headlight glare and act as crash barriers as they help to dissipate the energy of out-of-control vehicles. Care must always be taken, however, to ensure that the shrubbery does not reduce the necessary sight distance at intersections.

The reservation surfacing should be in direct contrast to the carriageways being separated, and be distinctly visible during day and night, in both wet and dry weather. For widths wider than about 1·75 m, grass is usually the most pleasant and suitable. Below 1·75 m, however, grass is usually difficult as well as dangerous to maintain, and so use is often made of raised medians with contrasting bituminous or concrete surfacing; the median kerb should also be studded with reflector buttons or painted with reflectorized white paint to emphasize the contrast.

Shoulders. A shoulder is that portion of a roadway adjacent to the travelled way that is primarily used as a refuge area by parked vehicles. The need for such refuges is indicated by data from a study at the Mersey Tunnel[18] which showed that cars and lorries had *emergency* stops every 28,000 and 15,000 km, respectively. The provision of shoulders also gives a sense of openness which helps considerably towards driving ease and (for lorry drivers, particularly) freedom from concern re lateral clearances.

Well-designed and properly maintained shoulders are a necessity on all rural and, where possible, suburban arterial roads carrying appreciable amounts of high-speed traffic. Shoulder widths vary however. About 3·35 m is needed by a lorry to enable a tyre to be changed without danger to the operator. Passenger cars require less space, and since these form the greater part of the total number of stoppages shoulder widths of 3 m are usually recommended for major highway design purposes. This normally allows a 1 m gap between the parked car and the edge of the carriageway, which is ample from the point of view of safety. Shoulder (or verge) widths as low as 0·6 m are used on less important roads. A shoulder should be capable of supporting vehicles under all conditions of weather, without rutting or shoving of the surface. If a motorist becomes bogged down while parked, it is doubtful whether he will make full use of these facilities again. Also skidding and overturning may occur if vehicles drive on to soft shoulders at high speeds and then attempt to decelerate.

The best but unfortunately also the most expensive way of providing a stabilized shoulder is to extend the roadbase beyond the edge of the carriageway. Not only is the shoulder made stable in this manner, but also added structural strength is given to the carriageway pavement. Furthermore, the shoulder can then be safely used as a slow traffic lane should maintenance work have to be carried out on the regular carriageway.

The shoulder surfacing must be distinctly different from the carriageway, otherwise motorists will use it as a regular traffic lane. Surfacings of grass are the most clearly delineated and aesthetically pleasing, but drivers are often afraid to use them for fear of being bogged-down. Probably the most effective type is a bituminous surfacing with either an added pigment material or different coloured stone chippings. Normally, the surfacing can be composed of less expensive materials than is used on the carriageway.

Lay-bys. When economic considerations do not allow the use of full-length shoulders then lay-bys should be built at favourable locations along the highway. Great care must be taken, however, *not* to construct lay-bys where the sight distances from their exit and entry points are inadequate for safety.

British lay-bys are not less than 30 m long and 3 m wide. Obviously however they should be longer where possible. If located on high-speed roads they should be provided with adequate acceleration and deceleration lanes on either end. Their spacing along the highway should be related to the volume of traffic; British practice is that there should be one for every 1·6 km on each side of a highway carrying 6000 p.c.u./day, and one at 4·8–8 km intervals on roads carrying 3000–6000 p.c.u./day.

Camber. The term camber is used in highway engineering to describe the convexity of the carriageway cross-section. The main object of cambering is to drain water and avoid ponding on the road surface.

Early road builders used much greater cambers than are used today, e.g. Telford used a camber with a side-slope of 1 in 30. These early roads had rough open surfaces and so severe cross-slopes were needed in order to remove the water as quickly as possible. Nowadays, however, these types of surfacing have been replaced by relatively impermeable ones, so that it has been possible to reduce the amount of camber very considerably and so increase the ease of driving. Today high-quality roads in Britain have average cross-slopes of between 1 in 40 and 1 in 48, and future years could see them with slopes as flat as 1 in 60, e.g. one study[(19)] showed that the major benefit resulting from changing the cross-slope from 1 in 60 to 1 in 30 was a reduction in the amount of water which can pond in deformations of the pavement (the depth of water flowing across the road was little affected).

Modern 2-lane roads desirably have either parabolic or circular cross-sections. These cross-sections have the advantage that the swaying of commercial vehicles is kept to a minimum as they cross and recross the crown of the road during an overtaking manoeuvre. On carriageways with three or more lanes, greater care has to be taken in deciding the manner in which the camber should be applied. If, for instance, the application of the camber takes the shape of a parabola, then the outer lanes may have an undesirably steep cross-slope which could seriously interfere with safe traffic operation. The equation of the parabola is such that the centre of the road will be very flat while the desired cross-slope will be exceeded towards the outside. Common practice in this situation is to use a curved crown section for the central lane or lanes and have a tangent plane section on each of the outer lanes. The cross-slope on the tangent planes is made the same or slightly steeper than that at the end of the curved section so that the accumulated water is more easily removed.

On dual carriageways it is desirable for each carriageway to be cambered. Not only does this minimize the sheeting of water during rainstorms, but also the difference between the low and high points in the carriageway cross-section is kept to a minimum. This latter advantage is a result of the smaller width which is sloped in a given direction and the avoidance of the higher rate of cross-slope which is necessary to get rid of the accumulated water when sloping a wider carriageway in one direction only. Changes from normal to superelevated sections are also easily made. The disadvantage of using cambered sections on both carriageways is that more inlets and underground drainage lines are required, with pick-up facilities near both edges of each carriageway.

Where carriageways are sloped in one direction to drain from the median space to the outside (i.e. British practice), savings are effected in drainage structures and treatment of intersecting roadways is easier. In areas subject to heavy rains or where the central reserve is used to store cleared snow, this procedure is most undesirable however.

Another possible arrangement often suggested for dual carriageways is that each carriageway should have a one-way cross-slope draining towards the central reservation. This has the advantage that the outer lanes, which are most used by commercial traffic, are more free of surface water. In addition there is the economical advantage that all surface water can usually be collected in a single drainage conduit within the reservation. However, a very serious objection to this procedure is that all the drainage must pass over the inner high-speed lanes. This can result in annoying and dangerous splashing on the windscreens of vehicles. Hence cross-sections with drainage concentrated in the central reservation are not generally used except on long bridge structures.

Side-slopes. Soil mechanics procedures now make it possible to determine accurately the maximum slopes at which earth embankments or cuts can safely stand. In practice, however, these maximum values are not always used, flatter sections being preferred for reasons of safety and ease of maintenance.

From the traffic safety aspect, the flatter the side-slopes the better, since vehicles driving on to them can more easily be brought under control. In cut sections also the horizontal sight distance at curves is considerably increased by the use of flatter side-slopes.

Although the tendency is also to construct side-slopes that are as steep as possible for reasons of economy, this can be a short-sighted policy if future maintenance problems are not taken into account. Modern highway grading equipment can perform well on most slopes when cutting or filling. However, when the slope surfaces are being stabilized by spreading topsoil, and when they are being generally finished and rounded, it is virtually impossible to operate the necessary equipment efficiently on slopes greater than about 2 horizontal to 1 vertical. The flatter the side-slopes the easier it is to grow grass on them and the less chance there is of erosion. If the slopes are to be properly maintained and kept in pleasing appearance the grading equipment necessary to do this can work most efficiently on slopes of 3 to 1 or flatter.

INTERSECTIONS

Particular attention must be paid by the highway engineer to the design of intersections because of the very considerable influence that they have on vehicle safety and efficiency of movement. Their effect on speed and flow of vehicles has already been discussed. A measure of their influence on safety is given by American national accident statistics for the year 1968. It is reported[20] that 41 and 27 per cent of total accidents, and 39 and 17 per cent of fatal accidents, in urban and rural areas respectively were intersectional.

Highway intersections can be divided into two main classifications: at-grade (or single-level) intersections and grade-separated or 'fly-over' intersections. The main factors influencing the choice of a particular type of intersection, as well as its specific layout, are cost, capacity, delay to vehicles, safety and aesthetics. The following discussion is primarily concerned with the layout of intersections and the manner in which they are related to some of the above considerations.

Intersections at-grade

At-grade intersections occur in a multiplicity of shapes. They can, however, be divided into the seven basic forms shown in Fig. 6.22. It is not possible in the space available here to discuss all the forms in detail. One point that should be mentioned, however, is that in general staggered crossroads tend to be much safer than straight-over ones. Furthermore the simple non-skew right/left stagger shown in Fig. 6.22 is safer than an equivalent left/right stagger, in the absence of a carriageway area for (in Britain) right-turning vehicles to wait. A minimum stagger length of 36·5 m is commonly used to design staggered intersections in Britain.

Non-channelized intersections. This type of intersection can be either plain or flared. The plain type is the simplest as well as being the most dangerous and inefficient; since no special provision is made for turning or straight-through traffic these are also the most economical form of intersection. Flared intersections have wider than normal carriageway widths at the junction of the intersecting roadways. The full extra widths are usually in units of lane width and located on either side of the intersection carriageway area.

Flared intersections have two particular advantages which more than justify the extra expense involved in their construction. First, and most important, the number of accidents is reduced as the addition of a third lane allows through vehicles to proceed relatively unhindered by turning vehicles. Secondly, the extra carriageway widths serve to reduce accident severity by enabling turning vehicles to merge and diverge from the main traffic streams at lower relative speeds. In urban areas the roadway also may be flared at an intersection to provide storage space for vehicles waiting to turn.

Speed-change lanes. As illustrated in Figs. 6.23 and 6.24, speed-change lanes are additional lanes designed and used exclusively for the acceleration

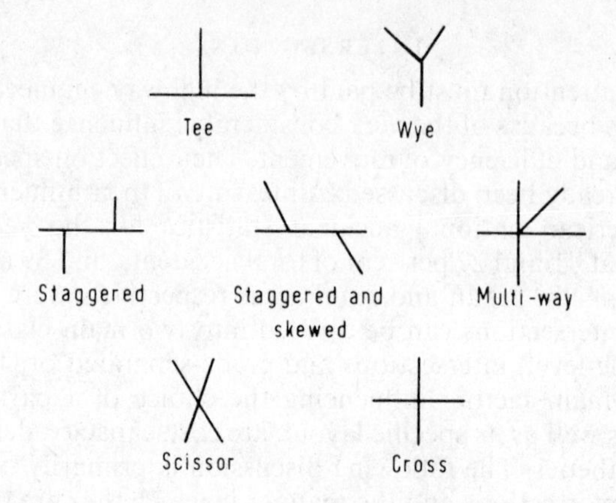

Fig. 6.22. Basic forms of intersections at-grade

and deceleration of vehicles entering or leaving the through-traffic lanes at flared intersections. Speed-change lanes are required at all at-grade intersections on high-speed high-volume roads, as well as being automatic at all motorway interchanges. They are desirable at intersections on all other roads.

Deceleration lanes have a priority of construction over acceleration lanes since, without them, vehicles leaving the through carriageway have to slow down within a high-speed traffic lane and this is a major cause of rear-end collisions. Three forms of deceleration lane are shown in Fig. 6.23. Studies in

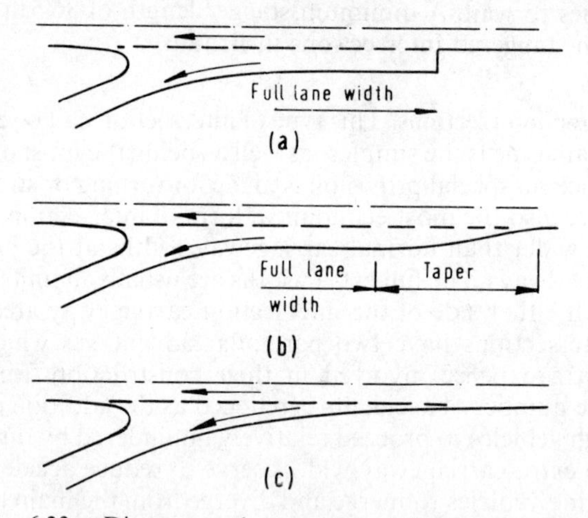

Fig. 6.23. Diagrammatic examples of deceleration lanes

the United States have shown that the type most effectively used by drivers is that in Fig. 6.23(c). This type is particularly adapted to high-speed roads with large volumes of turning movements. Its length is dependent on the speed at which vehicles can manoeuvre on to the auxiliary lane, the rate of deceleration of the vehicle, and the speed at which vehicles turn after traversing the lane.

Acceleration lanes permit entering vehicles to increase speed in order to enter the major road at the speed of the traffic on it. If the through traffic is very heavy, then the entering traffic is provided with space to manoeuvre as it awaits openings in the main stream of vehicles.

As much as possible of the acceleration lane should be adjacent to and flush with the through carriageway. At the end of the acceleration lane there should be no vertical kerb between the lane and the shoulder, so that a vehicle that cannot find a gap in the traffic stream is provided with a safety outlet and can over-run on to the shoulder if necessary.

The type of acceleration lane most effectively used by motorists is that of Fig. 6.24(b). Again its length is dependent on the design speed of the through highway, the rate of acceleration, the design speed of the ramp and the volumes of through and entering traffic.

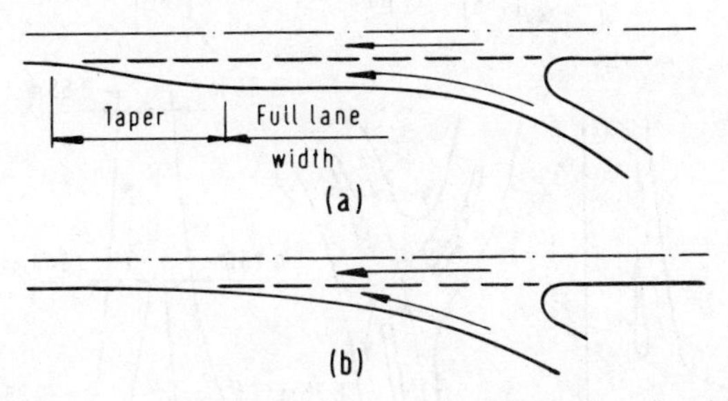

Fig. 6.24. Diagrammatic examples of acceleration lanes

Channelized intersections. The term channelization is used in highway engineering when discussing highway facilities at which two or more traffic streams are separated, each stream being confined to a single roadway channel. In this way, the safe movement of traffic is facilitated, vehicle conflict points are reduced and traffic friction points minimized.

At channelized intersections directional islands are used to divert the traffic into definite travel paths. These islands are each usually greater than about 4·5 m^2 in area, and they do not necessarily have distinctly defined shapes or forms. Instead they may vary from carriageway areas marked by paint to distinct and aesthetically pleasing surfaces defined by raised curbs.

Channelization of an at-grade intersection is carried out to achieve one or more of the following basic purposes:

1. To diminish the numbers of possible vehicle conflicts by reducing the possible carriageway area of conflict. This principle is illustrated in Fig. 6.25(a) which suggests the large area in which possible conflicts can take place at a conventional wYe intersection. When elongated islands are used in the intersection, the size of the trouble area is significantly reduced and there is less chance of confusion and resultant uncontrolled vehicle movements.

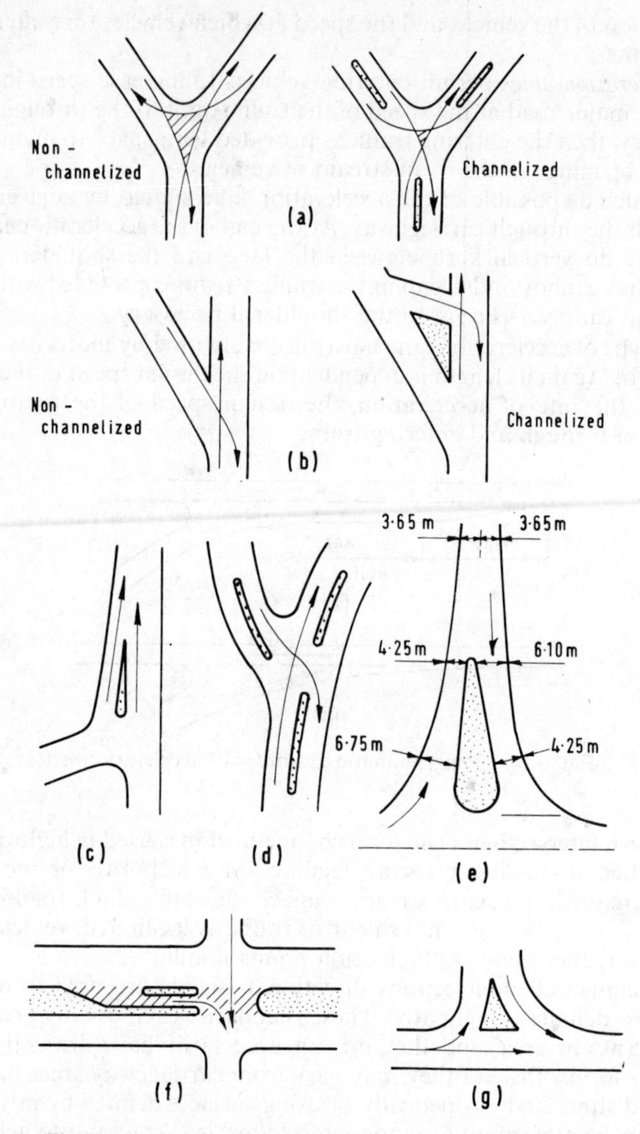

Fig. 6.25. Techniques of channelization at intersections

2. To control the angle of vehicle conflict. Figure 6.25(b) illustrates the very dangerous situation where two opposing traffic streams merge without weaving at about a 30 deg angle. If the intersection is channelized so that the two traffic streams are caused to merge at or near a right-angle, then the conflict zone is again made smaller. Assuming that proper sight conditions are available, the motorist is also provided with favourable conditions

for estimating the speed and position of opposing traffic, while the time required by each vehicle to cross the conflicting traffic stream is reduced. Also reduced is the combined energy of impact if an accident does occur.

In a similar fashion, Fig. 6.25(c) shows how an intersection can be modified by flaring and an island in order to promote the merging of traffic streams at angles of about 10 to 15 degrees. This procedure enables the merging vehicles to run parallel to the main traffic stream until a safe gap appears in the main stream of traffic. At angles greater than about 15 deg this manoeuvre is more difficult, longer gaps are required, and stop-sign control is desirable.

3. To reduce the speed of traffic entering the intersection. Figure 6.25(d) shows how the speed of the entering *minor* traffic stream can be reduced by causing it to bend substantially. Figure 6.25(e) illustrates how vehicle speeds can be similarly reduced by funnelling procedures.

4. To provide protection for vehicles leaving or crossing the main traffic stream. The shadowed area in Fig. 6.25(f) provides storage space for vehicles wishing to turn to the right or cross the opposing traffic stream. This layout has the added safety feature that a motorist attempting to cross the main roadway has only to be concerned with obtaining one safe gap in each traffic stream at a time.

Large volumes of traffic having complex turning movements can often be handled very efficiently by a combination of traffic signal control and island channelization. If vehicle-actuated detectors are placed in the channels and the signals are located on the islands, then vehicle conflicts and delays can be reduced substantially.

5. To regulate traffic movement by island location and shape. Figure 6.25(g) illustrates just one example of this principle. A triangular-shaped island is used to encourage drivers entering a one-way traffic stream to follow the correct direction of travel.

6. To provide a refuge for pedestrians wishing to cross the road. All of the islands shown in Fig. 6.25 also, of course, serve this purpose.

Use of channelization techniques. There are no uniformly accepted criteria which indicate whether or not particular intersections should be channelized. Hence every intersection must be considered on its own merits when determining whether the use of these techniques would be advantageous. There can be no doubt that the proper use of any of these procedures, or combinations of them, will result in improved traffic operation which will more than justify the relatively minor extra expense involved. Comprehensive examples of nearly every form of channelized intersection, and detailed comments on them, are readily available in the literature.[24] These can be used as guides to the feasibility of the use of channelization methods in any particular situation.

Roundabout intersections

In one sense roundabouts can be considered as a form of channelized intersection wherein vehicles are guided on to a one-way roadway and required to move in a clockwise direction (in Britain) about a central island. At one time the roundabout intersection was considered to be the answer to the problems associated with intersections on highways. This is evidenced by the multitude of roundabouts to be found on British highways at this time. In fact, roundabout intersections have particular advantages and disadvantages, and the decision as to whether a roundabout should be used at any individual location requires an understanding of these (see also p. 237).

Where roundabouts are properly used and designed the efficient flow of traffic is promoted by the orderly movement of vehicles about the central island. There is little delay to traffic due to speed reductions and no delay at all due to stopping. Furthermore, the possibility of having vehicles conflict is considerably reduced since all traffic streams merge and diverge at small angles. When accidents do occur they rarely have fatal consequences, damage being usually confined to vehicles only.

The roundabout design is most suited to intersections with five or more approach roads and/or where there are very heavy right-turning movements.

Roundabout intersections usually require greater land area and cost more than other at-grade intersections capable of handling the same traffic flow. Furthermore, they are not adaptable to locations with difficult topographic conditions, since they require large areas of relatively flat land.

Because of the long weaving lengths required to handle large traffic volumes, conventional roundabouts normally cannot be properly used in built-up areas. In practice, of course, an existing two-way street system in an urban area is often converted into a one-way system, thus in essence creating a roundabout (gyratory) system with central islands of blocks of buildings.

Roundabout intersections on high-speed roads require extremely long weaving lengths to ensure low relative speed differences between vehicles within and entering the intersection. If excessive area requirements are to be avoided and large volumes of weaving traffic still accommodated, then a substantial reduction in speed must be expected and accepted. This not only causes an increase in vehicle delays but can also lead to an increase in accidents.

For safe and efficient operation, there is need for many warning and directional signs at roundabouts. At night-time the central island and the entrances and exits must be well lighted. In addition, aesthetically pleasing landscaping of the central island is required.

Roundabout intersections are not as readily adaptable as traffic signal-controlled intersections to the long-term stage development of a highway. If constructed to meet the long-term needs, they usually result in over-design when compared with the immediate traffic requirements.

Grade-separated intersections

When at-grade highways cross, the number of vehicles that can pass through the intersection is controlled by the characteristics of the junction rather than of the highways themselves. Not only do they provide many opportunities for vehicle conflicts, with the resultant expected accidents, but at-grade intersections also reduce vehicle speeds and increase operating costs. It is when these difficulties become unduly great that grade-separated intersections become most advantageous.

Grade-separated structures generally have very large initial costs when compared with single-level intersections. The main situations which justify the very considerable extra expenditure can be summarized as follows:

1. *At intersections on motorways.* The construction of a highway with complete control of access automatically justifies the use of grade-separated structures in order to ensure the free movement of high-speed traffic.

2. *To eliminate existing traffic bottlenecks.* The inability of an important at-grade intersection to provide the necessary capacity is in itself a justification for a grade-separation on a major highway.

3. *Safety considerations.* Some at-grade intersections are accident prone, regardless of the traffic volumes they carry. For instance, many lightly travelled rural roads having high vehicle operating speeds have relatively large numbers of accidents at certain intersections. In these locations land is relatively cheap and so it may be possible to construct fairly low-cost grade-separations and so eliminate these accidents.

4. *Economic considerations.* At at-grade major road junctions very considerable economic losses can be incurred due to intersectional frictions and the resultant delays to traffic. These are usually in the form of costs for fuel, tyres, oil, repairs and accidents, as well as the time costs of the road-users. If these intersections are converted to grade-separated ones the very considerable long-term economic gain to the community may by far outweigh the burden of the large initial costs.

5. *Topographic difficulties.* At certain sites the nature of the topography or the cost of land may be such that the construction of an at-grade intersection is more expensive.

Grade-separation without slip roads. In essence this type of structure is simply a bridge or series of bridges which enable the traffic streams on the intersecting highways to cross over each other without any vehicle conflicts taking place. This type of intersection is most often constructed in rural areas where a minor road crosses a major road and the turning movements are not sufficient to justify expenditure on interconnecting ramps.

In urban and suburban areas grade-separations of this type are often used to cut down on the total number of intersections on major streets. In this way overall traffic safety and efficiency of movement is increased by concentrating the turning traffic at a few locations where adequate ramp facilities can be built. They may also be constructed at locations in urban or rural areas where the site conditions are so difficult that it is not economically feasible to construct ramps connecting the roadways.

Interchanges. An interchange is a system of interconnecting roadways or sliproads in conjunction with a grade-separation or grade-separations which provide for the interchange of traffic between two or more roadways on different levels. As with the at-grade intersections, there are very many types of interchange which are used in varied situations. Since, however, the basic purpose of an interchange is to provide an easy and safe means by which vehicles may transfer from one roadway to another, it is possible to classify the many types according to the manner in which they perform this function. Once this is done it is easier to discuss the basic forms involved and their particular usage.

The simplest way of dividing interchanges is to classify them according to the number of approach roads or intersection legs they serve. Thus interchanges can be considered as 3-way, 4-way and multi-way. Figure 6.26 illustrates several forms of interchange falling within these general classifications.

3-way interchanges. If one of the intersecting legs of a 3-way intersection is an approximate prolongation of the direction of approach of another, and if the third leg intersects this prolongation at an angle of between 75 and 105 degrees, the intersection is called a Tee-intersection. The equivalent interchange is called either a *Tee* or a *Trumpet* interchange. If one leg of the intersection is a prolongation of the approach of another, and the third leg intersects this prolongation at an angle less than 75 or greater than 105 degrees, it is called a *wYe* intersection. The equivalent interchange is also called a wYe interchange.

Two examples of traffic movement at Tee and wYe interchanges are shown in Fig. 6.26. Both utilize a single bridge structure and illustrate the situation where the greater volume of interchange traffic is given preferential turning treatment at the expense of the lower turning volume which has to use the semi-directional loop. These designs are most suitable for connecting two major roads, or a major road to a motorway, provided that the loop movement is relatively small. If the loop movement is heavy then extra bridge structures may have to be constructed so that both turning movements are favoured equally.

4-way interchanges. The simplest type of 4-way interchange is the *Diamond.* Consisting of a single bridge and four one-way sliproads, it has the particular advantage that it can be located within a relatively narrow land area, since it needs little extra width beyond that required for the major road itself. As Fig. 6.26 illustrates, this interchange has direct high-speed entrance and exit slip roads on the main road and at-grade terminals on the minor road. Thus its use is confined to intersections of major and minor highways. Because of its narrow width it can be easily used in urban areas where the high cost of land might make it impractical to use other interchange forms. The junctions of the slip roads and the minor road may need signalization if the minor road carries fairly heavy volumes of traffic, and the exits of the normally single-lane slip roads may have to be widened to allow for temporary vehicle storage.

The *Cloverleaf* is the only 4-way single-structure interchange having no terminal right turns at-grade. Internationally, it is probably the most common form of interchange and is often regarded by motorists as the ultimate answer to intersection problems. It has the great advantage of being very uncomplicated to use, while the eight turning movements are accomplished with no direct vehicle conflicts. Nevertheless there are a number of features about this type of interchange which limit its usefulness.

The first and most important is that if a Cloverleaf is used at the junction of two high-speed heavy-volume highways an excessively large area of land may be required to enable the loops to handle the traffic at relatively low speed differentials. Experience with this type of interchange indicates that loop design speeds in excess of 50 km/h and loop radii less than 61 m are rarely justified. At greater speeds, extra time is required to travel the necessarily longer loop distances. Shorter radii can be dangerous(22) and yet not reduce the land area requirement by very much, i.e. the length required for a 61 m radius is generally about the minimum required to comfortably secure the necessary difference in elevation between the two roadways.

A second undesirable feature of the Cloverleaf is that vehicles wishing to make right-hand turning movements must negotiate a 270 degree semi-direct turn. Not only can this represent a relatively difficult design problem but, in addition, as vehicles leave a particular loop upon entering the main highway, it is necessary for them to weave their way through other vehicles attempting to enter the adjacent exit loop. Therefore, when traffic volumes are heavy, the area required for the cloverleaf may have to be considerably enlarged unless ancillary weaving lanes are provided between the loops.

The *Turban*, the *Braided* and the *All-directional* interchanges are particular types with all right-turning, as well as left-turning, movements made directly. Thus they fall into the category known as 'directional' interchanges, whereas the Cloverleaf would be categorized as 'semi-directional'. Directional interchanges are the highest type of interchange and are most suitable at the intersections of motorways carrying high volumes of traffic. When compared to the semi-direct loop types of interchange they recuce vehicle travel distance, increase the speed of traffic operation, have greater capacity and, more often than not, eliminate any weaving problems.

Directional interchanges are very expensive to construct because of the type and number of bridge structures required. On the other hand, they can need relatively small land areas and thus certain types can be used in urban locations—provided they are environmentally acceptable. In rural areas traffic volumes are rarely sufficiently heavy to justify direct connections in all quadrants of an interchange; direct connections are often made in one or two quadrants, and the remaining turning movements are handled by loops.

Multi-way interchanges. When an intersection has more than four approach arms and an interchange is required, a roundabout interchange can be particularly advantageous. With this interchange it is usual to have the main highway underpassing or overpassing an at-grade roundabout

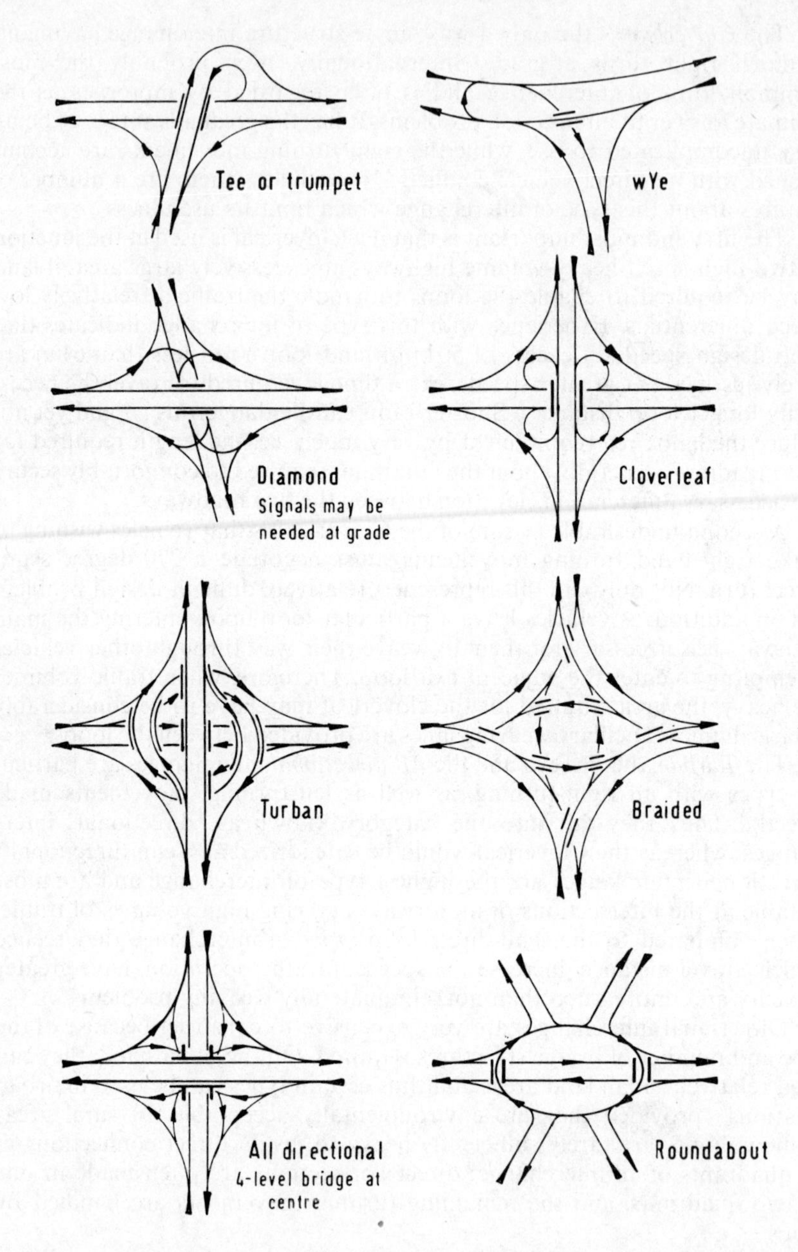

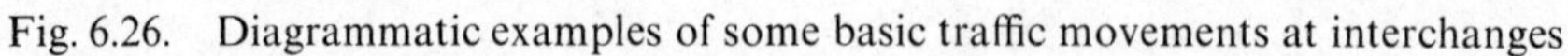

Fig. 6.26. Diagrammatic examples of some basic traffic movements at interchanges

intersection. Vehicles enter and leave the main road on diagonal ramps in a manner similar to that illustrated for Diamond interchanges.

The Roundabout interchange has the great advantage that it can be adapted from an existing at-grade roundabout that is overloaded. The interchange frees the movement of the major through-traffic stream while the minor and interchange traffic is confined to the roundabout. Because of the great number of existing at-grade roundabout intersections on British highways, it is very probable that very many of these interchanges will be constructed in this country in the future. While roundabout interchanges can require relatively small overall areas, and require less carriageway area than other types of interchange, e.g. cloverleaf, the fact is that the capacity of the junction as a whole is limited by the capacity of the roundabout itself.

SELECTED BIBLIOGRAPHY

1. Ministry of Transport. *Layout of Roads in Rural Areas.* London, H.M.S.O., 1968. (And as amended in Ministry of Transport Technical Memorandum No. T8/68.)
2. Almond, J. Speed measurements at rural census points, *Traffic Engineering and Control*, 1963, **5,** No. 5, 290–294.
3. Greenshields, B. D. A study of traffic capacity, *Proc. Highway Research Board*, 1934, **14,** 448–474.
4. Lighthill, M. J. and G. B. Whitham. On kinematic waves, II: A theory of traffic flow on long crowded roads, *Proc. Roy. Soc.*, 1955, **A339,** No. 1178, 317–345.
5. Webster, F. V. and J. G. Wardrop. Capacity of urban intersections at-grade, *Proc. Sixth International Study Week in Traffic Engineering, Salzburg*, 1962.
6. Ministry of Transport. *Urban Traffic Engineering Techniques.* London, H.M.S.O., 1965.
7. Ministry of Transport. *Roads in Urban Areas.* London, H.M.S.O., 1966. (And as amended in Ministry of Transport Technical Memorandum No. T8/68.)
8. Blackmore, F. C. Priority at roundabouts. *Traffic Engineering and Control*, 1963, **5,** No. 2, 104–106.
9. Blackmore, F. C. Capacity of single-level intersections. *RRL Report* LR 356, Crowthorne, Berks., The Road Research Laboratory, 1970.
10. *Technical Memorandum No. H7/71.* London, The Department of the Environment, 1971.
11. Poole, D. *Notes on the Design and Performance of Mini-Roundabouts.* The City Engineers Department, Newcastle-Upon-Tyne, Sept 1972. (Unpublished).
12. Webster, F. V. and R. F. Newby. Research into relative merits of roundabouts and traffic signal controlled intersections, *Proc. Inst. Civ. Engrs*, 1964, **27,** 47–75.
13. Millard, R. S. Roundabouts and signals, *Traffic Engineering and Control*, 1971, **13,** No. 1, 13–15.
14. A.A.S.H.O. *A Policy on the Geometric Design of Rural Highways.* Washington, D.C., American Association of State Highway Officials, 1965.
15. Crawford, A. The overtaking driver, *Ergonomics*, 1963, **6,** No. 2, 153–170.
16. County Surveyor's Society. Highway Transition Tables. London, Carriers, 1969.
17. Ackroyd, L. W. and Bettison, M. Effect of maximum motorway gradients on the speeds of goods vehicles, *Traffic Engineering and Control*, 1971, **12,** No. 10, 530–531.

18. BARTLETT, R. S. and CHHOTU, S. R. An Analysis of Vehicle Breakdowns in the Mersey Tunnel. *TRRL Report* LR 484, Crowthorne, Berks., The Transport and Road Research Laboratory, 1972.
19. ROSS, N. F. and RUSSAM, K. The Depth of Rain Water on Road Surfaces. *RRL Report* LR 236, Crowthorne, Berks., The Road Research Laboratory, 1968.
20. BOX, P. C. and ASSOCIATES. Ch. 4: Intersections. In MAYER, P. A. Traffic Control and Roadway Elements—Their Relationship to Highway Safety. Washington, D.C., Highway Users Federation for Safety and Mobility, 1970.
21. BENTLEY, J. B. A review of highway curve design, *Journal of the Institution of Highway Engineers*, 1971, **18,** No. 11, 7–14.
22. HEWITT, R. H. Road transition curves for an accelerating vehicle, *Journal of the Institution of Highway Engineers*, 1971, **18,** No. 3, 7–16.
23. HURD, F. W. Accident experiences with transversable medians of different widths, *Highway Research Board Bull.* **137,** 1956, 18–26.
24. COMMITTEE ON CHANNELIZATION. Channelization: The design of highway intersections at-grade, *Highway Research Board Special Report* No. 74, 1962.

7 Traffic management

Traffic management is concerned with short-term measures to improve the efficient and safe movement of both pedestrian and vehicular traffic on the existing road network. The cost of traffic management work is normally fairly low.

In order to obtain the maximum return from the management investment it is necessary for the traffic engineer to have a basic understanding of the factors underlying the use of any particular technique. The function of this chapter is to present some of the more important of these techniques and to discuss in detail the basic factors pertinent to their usage. (Measures dealing with the control of standing vehicles are dealt with separately in the chapter on Parking.)

HIGHWAY SAFETY

The function of a street or highway is to serve the travelling public. To fulfil this function the roadway should be designed so that the motorist, the cyclist and the pedestrian can all travel quickly, economically and, most important of all, safely to their desired destinations. Unfortunately these ideal travel arrangements have yet to be achieved, as is evidenced by the ever-mounting toll of tragedies on the roads.

Accident statistics

Figure 7.1 expresses in graphical form the annual road casualty figures since 1928. It is of interest to consider briefly some of the more interesting features which these data reflect. First, and perhaps most important, is that, while the total number of casualties now stands at an all time high, the number of fatalities has actually *decreased* when compared to the figures before the war. This is in spite of the fact that the number of motor vehicles has more than trebled since the pre-war peak. This should not be viewed with complacency, however, as, although some of this decrease in fatalities is undoubtedly due to higher standards of roadway design and better traffic management, much of it is also due to higher standards of medical care and to significant reductions in the numbers of pedal cycles, public service vehicles, and (since 1961) motor cycles on the road.

A second most important factor is that, consistently, nearly one-half of the people killed on the roads in Britain are pedestrians. Nearly two-thirds of the pedestrian fatalities and serious injuries involve children under fifteen and adults over 60 years of age. About 90 per cent of these accidents occur on the streets in urban areas, indicating that this safety problem is primarily one of helping pedestrians to cross the road.

A third interesting feature is that, not only has the number of pedestrian fatalities decreased, but the total number of these casualties has also decreased when compared to *pre-war* figures. This can also be attributed to

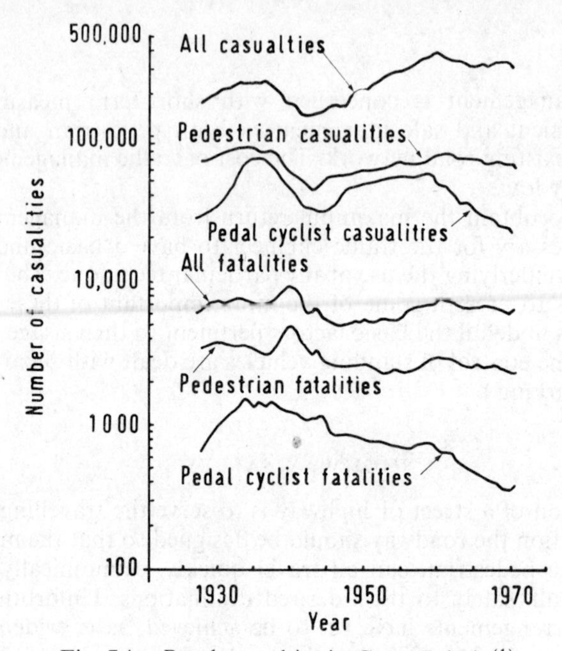

Fig. 7.1. Road casualties in Great Britain[1]

improved medical care and better traffic management, a 'natural' lessening of the number of accidents involving pedestrians and pedal cycles/motor cycles/public service vehicles, as well as to an increasing awareness by the pedestrian to the dangers on the roadway.

Another factor illustrated by Fig. 7.1 is that both the number of pedal-cyclist fatalities and the total number of these accidents is getting less. Much of the reason for this is undoubtedly the fall in the use of the bicycle as a means of transport. Coincident with this fall in the number of casualties, however, is a rise in the *risks* to which the pedal cyclist is exposed as a result of the increasing motor traffic. This is reflected in the statistics for 1959 and 1969 which show that the number of cyclist casualties per 100 million kilometres travelled rose from 383 to 539, an increase of over 40 per cent. (In 1969 also, 40·5 per cent of pedal cyclist casualties involved children under 15 years of age.) In other words, notwithstanding the decrease in the numbers of cyclists and cyclist accidents, it is more risky to ride a bicycle on the highway today.

Not only do the highway accidents statistics for this country reflect an appalling cost in terms of human suffering, but they also represent a very serious economic loss to the nation as a whole. Table 7.1 shows some estimates of the resource cost of accidents in Great Britain in 1968, and clearly points up the fact that accidents on the highway are a most significant 'invisible' economic loss to this country.

TABLE 7.1. *Cost of accidents to the community*[2]

Costs leading to a diversion of current resources	
Medical treatment, ambulances and funeral costs	15£ million
Cost of damage to vehicles and other property	173
Administration costs	28
Gross losses of future output	84
Total cost	300

International comparisons. When the numbers of highway fatalities, motor vehicles and populations of different countries are compared, the following common features emerge:

1. Tendencies are not very different in different countries, with the result that the relationship between road fatalities, licensed motor vehicles and population can be generally expressed by the empirical formula

$$\text{Fatalities} = 0{\cdot}0003[(\text{Vehicles})(\text{Population})^2]^{1/3}$$

This formula, which is based on results[3] from 68 different countries, can be used to predict the number of fatalities expected in any future year for which the number of vehicles per head of population can be predicted. Most interesting perhaps, is that it suggests that each country has an 'inbuilt' toleration level with regard to what is acceptable regarding road accidents, particularly fatalities. In other words, road-users become more careful and authorities more stringent as the dangers on the road increase.

2. As the number of vehicles increases, the numbers of motorist fatalities and casualties increase much faster than the number of pedestrian casualties.

3. The ratio of one-vehicle accidents to total accidents decreases as the motor vehicle population becomes greater.

4. The ratio of slight to total casualties increases when the number of registered vehicles increases.

In general, absolute comparisons of accident and casualty rates in different countries must be treated with care, as they may contain results arising from such factors as differing traffic compositions, variations regarding the proportion of travel occurring in built-up areas, different qualities of street lighting, road standards, vehicle legislation, etc. Even so, Table 7.2 which compares death rates per vehicle-kilometre in Britain (which makes allowance for the different usage of the roads) with that in the United States and in three continental European countries of a similar population and degree of motorization to Britain, show that conditions in this country are, on the whole, quite good as compared with others. The rates for both France and Italy are actually underestimates as France only counts deaths occurring within 6 days of the accident, and Italy those within 7 days, as compared with the internationally agreed time of 30 days (as in Britain). The United States counts fatalities which occur within 1 year, thus slightly overestimating their rate. The British figure is higher than the American one, primarily because of higher components from pedestrians, pedal cyclists and motorcyclists.

TABLE 7.2. Motor vehicle fatalities per 10^8 vehicle-kilometres[4]

Year	*Great Britain*	*U.S.A.*	*France*	*West Germany*	*Italy*
1969	3·8	3·3	8·6	7·0	6·6

Types of highway. The accident statistics on different types of road have been studied and some of the results of these studies are summarized in Table 5.11. From these data it is seen that one is much more likely to be involved in an accident when travelling on a town street as on, say, a motorway; this can be attributed to the greater number of decisions to which the urban road driver is continually being subjected, e.g. at intersections, from pedestrians, etc. On the other hand, it should be noted that the fatal accident rate on roads in built-up areas is only 3·7 per 10^6 vehicle-kilometres as compared with 3·0 on roads in rural areas,[4] thereby indicating the greater severity of accidents (primarily because of higher speeds) which occur on rural roads. The implications of the data in terms of the safety features built into the design are obvious.

Intersections. One of the most fruitful applications of traffic management lies in the improvement of highway intersections. Very often minor improvements can be carried out which will reduce accidents and improve highway safety beyond all proportion to their costs.

A measure of the importance of the role played by intersections in highway accidents can be gathered from the British accident statistics for the year 1966. In that year 64 per cent of the fatal and serious accidents in daylight in urban areas, and 64 per cent of those at night, occurred at (including within 6·1 m of) a junction; the equivalent figures for non-built-up areas were 36 per cent in daylight and 28 per cent in darkness. In this respect it should be remembered that there are some 5·77 and 1·19 intersections per kilometre of urban and rural road, respectively, in Great Britain.[5]

3-*way intersections.* Intensive analyses[6,7] have been carried out on accidents at rural 3-way intersections and some of the results are shown in Fig. 7.2. The main features to be gathered from these studies are as follows:

1. Right-turn movements are much more serious than left-turn movements.

Twenty-five per cent of the accidents were the result of collisions between vehicles turning right from the main road and vehicles travelling in the same direction on the main road. Seventeen per cent involved vehicles entering the main highway, in order to turn right, and coming into collision

Left turns	Entry 14%	4%	6%	4%
	Exit 5%	2%	1%	2%
Right turns	Entry 28%	8%	17%	3%
	Exit 40%	25%	12%	3%
Two turning vehicles		13%		

Fig. 7.2. Accidents at rural 3-way intersections[6]

with main road vehicles travelling in the opposite direction. This emphasizes the need to segregate vehicles entering and leaving the main through highway.

2. A right-hand splayed junction (see Fig. 7.3) is preferable to either a left-hand splay or a square layout.[7]

The likely reasons for this are that a right-hand splay layout (a) enables a right-turn to be made *from* the main road without the vehicle necessarily having to slow down, and (b) forces drivers about to turn right *into* the main road to slow down on the minor road before doing so, and they do so while facing at such an angle that they can see along the major road in both directions.

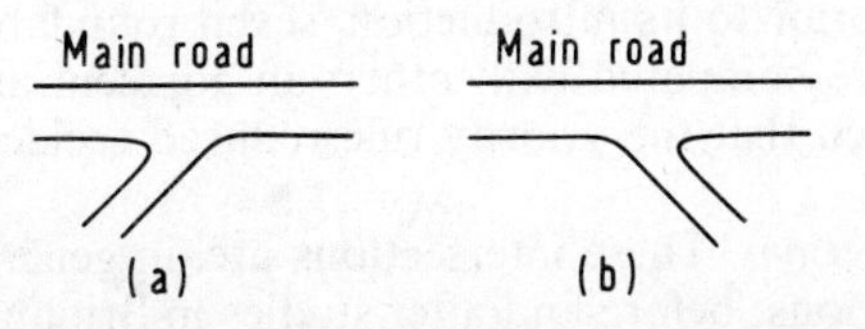

Fig. 7.3. Splay junctions (a) left-hand, (b) right-hand

3. The frequency of accidents is proportional to the *square root* of the product of the flows, and not, as might perhaps be expected, to either the addition or product of the flows.

This would appear to bear out what highway engineers have intuitively surmised for some time, that motorists are much more careful at busy intersections. This deduction can be also used in certain instances to justify the restriction of access on to a major road. To illustrate its use consider the case of two 3-way intersections located close to each other on a major highway. If one of the minor roads is refused access and its traffic diverted to the other minor road, then the number of accidents at the remaining 3-way intersection on the major road should be less than that previously experienced.

TABLE 7.3. *Accidents at roundabouts*[8]

Type and location	*Percentage of accidents*
One-vehicle accidents:	
Entering roundabout	22
On roundabout	20
Leaving	7
Two-vehicle accidents:	
Both entering roundabout	8
One entering, one on	17
One entering, one leaving	3
Both on	10
One on, one leaving	11
Both leaving	2

Roundabouts. A study of nearly 500 accidents at conventional roundabouts in London resulted in the data shown in Table 7.3. The most noteworthy feature of this table is that so many of the accidents involved only one vehicle. Two-thirds of the one-vehicle accidents occurring while entering the roundabout happened at night-time; this emphasizes the need to light intersections, especially central islands. Over half of the one-vehicle accidents involved motor-cycles—most of which skidded—and one-seventh involved pedestrians.

In order to evaluate the effect of the priority rule 'Give way to the vehicle from the right' prior to its introduction, seven roundabouts where the rule was in force were compared with others in adjacent areas. The statistical evaluation showed that the priority rule reduced accidents by about 40 per cent.

4-*way intersections.* These intersections are, in general, more dangerous than 3-way junctions; before-and-after studies in Britain have indicated that very significant accident reductions can be obtained by changing a straight-over four-way intersection into two three-way staggered junctions.

Interesting results have been obtained in the United States[9] regarding the use of 'GIVE WAY' and 'STOP' signs at four-way junctions. These studies have shown that 'YIELD' signs can be an effective way of reducing accidents at low volume intersections, i.e. where one roadway is given priority over another. It was found for two-way 'STOP' signs that accident rates at junctions are much more sensitive to changes in minor road volume than in major road volume, and that low cutting volume intersections have higher accident rates per crossroad vehicle than do higher cutting volume junctions. Changing from two-way to four-way STOP signs was found to result in accident reductions provided that the new installation was warranted by accident frequency and volumes were moderate and balanced.

Seasonal and day-of-week effects. A statistical analysis of casualty totals and daily figures of motor vehicle-km travelled in 1959 showed that the casualties per million vehicle-km tend to be higher in Winter and on Saturdays. It is of interest to note, however, that in the same year the number of fatalities per day varied from 2 to 73; the total number of casualties varied from 408 on Jan. 18 to 1618 on July 4 and 2553 on Dec. 24.

Some idea of the influence of the weather on the safety characteristics of the road can be gained from the fact that the resurfacing of a number of roads which were slippery in wet weather reduced wet road accidents by 80 per cent and all road accidents by 45 per cent. Wetness of the carriageway is also associated with an increase in adult pedestrian casualties.

Thick fog (visibility <200 m) occurs in Britain relatively infrequently. It occurs least in coastal areas and increases to maximum frequency in central England—where, on the whole, it is relatively infrequent, patchy and rarely widespread, e.g. there were only four whole days of widespread thick fog between 1963 and 1966. During these four days of thick fog it was found[10] that (*a*) traffic flow was reduced by about 20 per cent, (*b*) accidents in darkness and those involving pedestrians were reduced slightly, and (*c*) overall, there was no change in the fatal and serious accident rate per unit of traffic whereas the slight injury and total accident rates increased by about 70 and 50 per cent, respectively.

Skidding. The skidding rate, i.e. the percentage of accidents involving skidding is, as might be expected, higher on wet roads than on dry, e.g. one study[11] showed that 1 in 3 accidents occurring on wet roads involved skidding, which is almost twice the rate for accidents in the dry. Furthermore the skidding rate at high speeds is greater than at lower. Over the years 1959–1966, and notwithstanding the fact that the performance of cars increased, there was an improvement in skidding resistance; this may be attributed partly to improvements in road surfaces and partly to improvements in tyres, especially the increased use of 'dead' rubber for tyre treads (see also ref. 12).

Another analysis which studied the age and sex of drivers involved in skidding showed some differences between the sexes but, for both, the indications were that with increasing age and experience drivers learn to avoid skidding to some extent. Motor-cycles are about 10 times more likely to be involved in personal-injury skidding accidents than cars; this is due to the difficulties that arise when braking motor-cycles in an emergency and to the greater vulnerability of motor-cyclists to injury.

When the carriageway is wet there is a tendency for cars with greater engine capacities to have higher skidding rates. Similarly there is a slight trend towards higher skidding rates for motor-cycles with larger engine capacities—with the notable exception of scooters. Scooters have the highest skidding rate notwithstanding that they have engine capacities of less than 150 cc.

Most skidding accidents occur where drivers must brake or corner, thereby demanding high frictional properties of the road surface. Simply by improving the resistance to skidding at locations where these accidents tend to cluster, considerable reductions in accidents can be achieved at a very low cost.

Accidents at night. It appears that, other things being equal, the ratio of road accidents in darkness to those in daylight is about 1·5 to 2 in dry weather and about 3 to 4 in wet weather.[13]

Effects of good street lighting. The influence of good street lighting has been studied[14] by comparing the accident frequencies on stretches of roadway before and after the introduction of better lighting. The results of these studies can be summarized as follows:

1. Good lighting reduces accident rates on most roadways, particularly on those which have high night-to-day accident ratios, lower design standards, and intersections. It is difficult to substantiate claims of substantial safety improvements resulting from continuous motorway lighting.

2. There is substantial evidence to show that the reduction in pedestrian accidents is greater than the reduction of other types of injury accident.

3. There were no significant differences between the accident reductions for fluorescent, mercury and sodium lighting.

4. The total savings in accident costs can very often more than pay for the capital and running costs of the lighting installations.

Motor-cyclist accidents. Some one-fifth of the road fatalities in Britain are motor-cyclists or their passengers; most are young men under 30 years of age. When compared with their numbers on the road, motor-cycles are also the greatest source of injury to pedestrians of all ages, but especially to older people. The results of many studies into motor cyclist-associated accidents have resulted in the following conclusions:

1. The personal-injury accident rate of solo motor-cycles is about 4 times that for cars. The rate for motor-cycles with sidecars is about twice that for cars.

2. The risk of a motor-cyclist being killed per kilometre travelled is about 20 times that of the chance of a car driver being killed.

3. Ninety-seven per cent of the casualties which occur as a result of collisions between motor-cycles and motor vehicles are the motor-cyclists.

4. The risk of death or serious injury to a pillion passenger is about 5 per cent greater than the risk to the driver.

5. Motor-cyclists with less than six months' experience have about twice as many accidents per head and per kilometre as those with more experience.

6. Most of the motor-cyclists who are killed, and many of those who are injured, receive head injuries. The risk of injury to the parts of the head covered by a helmet is 30 to 40 per cent less for those wearing a safety helmet

than for those without. The risk of death or serious injury due to head injury alone is 20 to 30 per cent less for those wearing a helmet.

Driver characteristics. The most complex and least understood element of every road problem is the human one. Repeated studies indicate that errors of judgement are important factors in over 90 per cent of all highway accidents; it is therefore the intention of this discussion to present briefly some of the factors which directly or indirectly influence the manner and rate at which errors occur.

Vision. Good vision is a pre-requisite for safe driving since it accelerates the process of perception-reaction to traffic situations. The present driving test requires the motorist to be able to read a motor car number plate at a distance of 23 m. It is likely that this simple test is inadequate, since it is a test for visual acuity only at the time of the initial test, and reveals nothing about other important visual factors such as depth perception, field of vision, night blindness and colour blindness. It is possible to test easily and quickly for these visual skills and persons found lacking can be detected and encouraged to have a more thorough examination by an eye specialist.

Visual Acuity is the ability to focus quickly and to see clearly without a blur. The American Optometric Association recommend 20/40 vision as the minimum visual acuity for safe driving; this means that drivers should be able to read at 20 feet letters on a standard test chart that should normally be read at 40 feet. A traffic sign with 127 mm high letters can be read at about 85·5 m by a driver with 20/20 vision. Drivers with 20/40 or 20/50 vision must approach to within 34·5 and 27·5 m respectively in order to read the same sign.

As well as having good visual acuity, a driver should also have good *Depth Perception.* This visual skill requires good teamwork of both eyes to enable the driver to judge relative distances and to locate objects correctly in space. Very many overtaking accidents could be avoided if drivers realized that their depth perception capabilities were weak and that they needed to take extra care when overtaking other vehicles.

A third important visual faculty is the *Field of Vision.* Although most acute vision is subtended by a 3-deg. cone, it is still fairly satisfactory up to about 20 deg. Traffic signs should be located within a 10-deg. cone, which is the level beyond which the visual acuity for legibility rapidly decreases; this is roughly the area covered by the width of the hand held at arms length. Although acute vision is limited to about 20 deg, most drivers have sufficient peripheral vision to enable them to perceive objects contained in a cone of between 120 and 160 deg.—which is why the driver whose eyes are focussed on the road ahead is able to notice and avoid side hazards.

Some drivers have what might be termed 'tunnel vision'; it is most acute in the direction straight ahead, but little or nothing is noticed to the right and left or up and down. A driver who has tunnel vision, and neither knows it nor realises its importance, is a potential danger to all other road users. He is particularly dangerous on high-speed roads when overtaking manoeuvres are being carried out, and on busy city streets where vehicles are continuously parking and unparking and pedestrians are likely to step unexpectedly on to the carriageway. Early detection of tunnel vision need not necessarily disqualify a driver; instead he can be trained to move his head continually when driving in order to overcome this liability.

It is also necessary for a driver to have good *Night Vision.* This demands three important visual skills:

1. The ability to see efficiently under low illumination. The recognition of details is easiest when the level of illumination is that which is visually available in ordinary daylight; this unfortunately cannot normally be provided at night time. In addition many drivers also suffer from 'night blindness' in that dusk blots out for them as much as late darkness does for others.

2. The ability to see against headlight glare. There is considerable variation in the degree of 'glare-out' experienced by individuals subjected to the same amount of glaring light, and in the time required for the recovery of sensitivity. Normally the eye adapts itself quickly when going from darkness to light but can take up to 6 seconds to adjust from light to darkness—as, for instance, when a vehicle is entering an unlit tunnel or the motorist is dazzled by opposing headlights. Older persons are more susceptible to glare-out, as are persons with poor visual acuity.

Practically, the angle between the line of vision and the glare source is of the utmost importance. When the glaring light is only 1 deg to the side of the line of vision the effect is 3 times that when the angle is 5 deg.

3. The ability to distinguish between various colours. Although it has not been possible to establish a relationship between colour blindness and accidents, the heavy reliance placed on colour in traffic signs emphasizes the importance of this visual skill. Colour-blind people have particular difficulty in distinguishing between green and red. To overcome this one important measure has been to standardize the positions of the different coloured lamps in traffic signals.

Hearing. There appears to be no relationship between poor hearing and driving accidents. In fact American data suggest that most deaf mutes are very careful drivers. Hearing sensitivity is apparently more related to operating practices, such as judging when to change gears, than to highway safety.

Fatigue. Obviously a tired driver has longer perception-reaction times and is more liable to commit an error of judgement. What is not so clear, however, is the effect of fatigue brought on by prolonged driving as compared to the fatigue caused by such features as poor living conditions and extreme anxieties. It is interesting to note that two large insurance companies in the United States reported in 1949 that about 60 per cent of all long-distance lorry accidents happened in the first $3\frac{1}{2}$ hours of driving. The U.S. National Safety Council reported similarly that most driver-asleep accidents happened after the driver was at the wheel for only a few hours.

For least tiring driving the motorist should take a rest after, at the very most, one hour's driving. Drinking coffee helps to ward off drowsiness and fatigue; this is substantiated by laboratory studies which indicate that caffeine slightly speeds up reaction time.

Alcoholic drinks. The effect of alcohol is that of a general anaesthetic; it is a drug which depresses the central nervous system, affecting first the brain and then the spinal cord. It makes a person less sensitive to and aware of sensations of various kinds. As well as dulling the sensations themselves, even moderate amounts of alcohol prolong the reaction times to sensations of sight, touch and hearing. This is of considerable importance to a driver, since increasing the reaction time significantly increases the distance required to stop a vehicle. Alcohol also impairs night vision, as it increases the time required for the eyes to adjust from light to darkness.

Upon intake alcohol is first absorbed into the bloodstream and distributed through all the fluids and tissues of the body in proportion to their water contents. The level of intoxication attained by any individual then depends on the concentration of alcohol in the brain and central nervous system—and not just on the total amount of alcohol taken into the stomach—and this is closely related to the alcohol concentration in the blood. This latter concentration is further dependent on the rate at which alcohol is absorbed into the blood and the rate at which it is burnt up by the body or excreted in the urine, breath or perspiration.

Many factors affect the rate of absorption and elimination of alcohol in the body, and these result in variations not only between individuals but also in the same individual at different times. The rate of absorption into the bloodstream depends on the driver's constitution, the time since the last meal, the food consumed, and how quickly the alcoholic beverage is taken; beer and light wines are absorbed less quickly than spirits, and all drink is absorbed more slowly if taken on a full stomach or with food. Alcohol is eliminated from the blood at a more or less standard hourly rate of about 15 mg of alcohol per 100 ml blood; again, however, there are variations between individuals, and the elimination rate tends to be faster if the initial concentration is high. Tests have shown that one pint of typical British beer will take from two to three hours to disappear from the blood, and seven pints will take about nine hours. If the latter content of alcohol is taken in spirit form it may take fifteen hours to disappear.

TABLE 7.4. *Minimum intake of alcohol in the form of beer or whisky needed to raise the blood-alcohol level in a 70 kg man to a given concentration*

Blood-alcohol concentration mg/100 ml	*Beer or stout, pints*	*Single whiskys*
20	½	1
40	1	2
55	1½	3
75	2	4
100	3	6
150	4	8

It is not possible to state how much any person will need to drink before becoming 'incapable' of driving. It is possible, however, to give the *minimum* amounts that the *average* person must drink in order to attain a certain alcoholic concentration in the blood. These data, shown in Table 7.4, are based on tables published by the British Medical Association.[15]

The accident risks associated with certain concentrations of alcohol are indicated in Table 7.5, which lists some of the results of three comprehensive investigations into the problems associated with drinking and driving. While the values obtained in these studies are comparatively different, they all show quite clearly that it is considerably more dangerous to drive after drinking heavily. The Grand Rapids study also showed that the increase in risk resulting from high alcohol levels is greater for young and elderly drivers than for middle-aged ones, while at low alcohol levels it is greater for those who drink rarely than for regular drinkers.

TABLE 7.5. *Relative risks to drivers associated with drinking alcohol*[16]

Blood-alcohol concentration mg/100 ml	*Toronto, Canada*	*Grand Rapids, U.S.A.*	*Bratislava, Czechoslovakia*
30	1	1	1
30–99	1·5	1·4	7
100–149	3	6	31
>150	10	19	128

The Licensing Act of 1872, which made it an offence in Great Britain to be 'drunk while in charge on any highway or other public place of any carriage, horse, cattle or steam engine' (extended in 1925 to include 'any mechanically propelled vehicle'), and all of its successors prior to 1967, were relatively ineffective as traffic safety measures—primarily because of the lack of a legal definition for the word 'drunk'. A major step forward in accident prevention occurred with the passing of the Road Safety Act, 1967 which made it an offence for anybody to be in charge of a vehicle whose blood alcohol or urine alcohol content was above 80 mg/100 ml or 107 mg/100 ml, respectively. The effect of this Act was to bring about the biggest reduction in road casualties to have occurred in recent years. It is not possible to give a precise quantitative estimate of this effect because of the presence of other simultaneously occurring factors, but the sudden improvements coinciding with the new law, the greater reductions in casualties at drinking times and at week-ends than at other times (e.g. see Figure 7.4), the reduction in the proportion of those killed who had blood alcohol levels of 80 mg/100 ml or more, and the sudden drop in the casualty rate, all point to the reduction in driving after drinking brought about by the Act as being the main reason for the change. All this occurred with no diminution either in the amount of traffic or in the consumption of alcohol.

There is evidence now, however, that the effect of the new law (and its associated publicity and enforcement) declined after the first year of operation. Nevertheless, it must be pointed out that the residual benefits in succeeding years were still greater than any other recent safety measure.

Accident proneness. It is commonly held that some drivers are more prone to accidents than others and many papers have been published giving the results of studies into this assumption. While these have indicated that certain groups of people do certainly tend to have more accidents than others, it

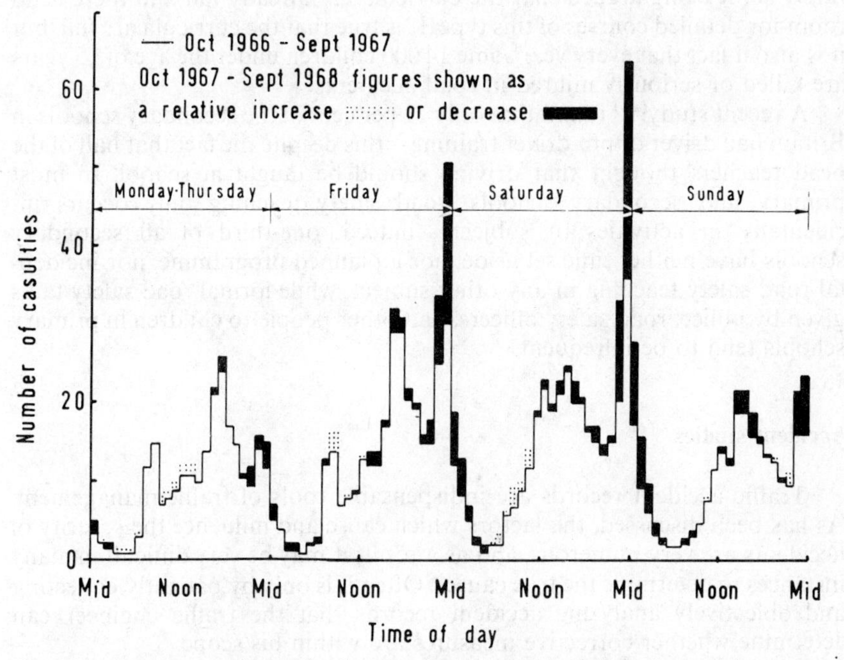

Fig. 7.4. Average number of fatal and serious casualties (all road users) by hour of day in Great Britain 1966–1968[17]

is very difficult to use the conclusions to evaluate individuals who are accident prone. In general, however, it would appear[18] that accident repeaters react against authority, have aggressive tendencies, and may be irresponsible and socially maladjusted.

Young people also appear to be more prone to accidents than their elders. Thus, one British study[19] found that the casualty rate (i.e. casualties per unit distance travelled) for male car drivers under 25 was at least $2\frac{1}{2}$ times that of their elders. This is in spite of the fact that youths are at their greatest potential with regard to operating skill. Apparently it is the lack of maturity rather than lack of skill that is the basis of this type of accident proneness.

Driver training. As with any other skill, a driver develops particular habits which result in certain reactions when confronted with specific traffic situations. It is important therefore that bad driving habits, which might lead to undesirable reactions, be anticipated and eliminated before they can result in accidents. This can only be done by means of proper driver training, good driving tests and continuous public propaganda regarding highway safety.

Logically, education in road safety should be given initially by parents to their children of pre-school age. When the children start to attend formal schools it is this writer's opinion that road-safety education should be continued in a determined fashion in the regular curricula. Unfortunately this is rarely so, it being argued that the curricula are already full and there is no room for detailed courses of this type. It is true that the curricula are full, but it is also a fact that every year some 14 000 children under the age of 15 years are killed or seriously injured in road accidents.

A recent study[20] found that only 11 per cent of the secondary schools in Britain had driver or pre-driver training—this despite the fact that half of the head teachers thought that driving should be taught at school. In most primary and secondary schools, road safety teaching only occurs incidentally in activities or subjects—indeed, one-third of all secondary schools have neither time set aside, nor a planned programme, nor incidental road safety teaching in any other subject, while formal road safety talks given by police, road safety officers, and other people to children in primary schools tend to be infrequent.

Accident studies

Traffic accident records are indispensable tools of traffic management. As has been discussed, the factors which cause and influence the severity of accidents are very numerous, and as a result it may be very difficult in many instances to determine the true causes. Often it is only by patiently collecting and objectively analyzing accident records that the traffic engineer can determine whether corrective measures are within his scope.

Six basic steps may be recommended[21] in any detailed study of accident experience at selected locations in a community. They are as follows: 1. Obtain adequate vehicle accident records. 2. Select high-accident-frequency locations in order of severity. 3. Prepare collision diagrams, and sometimes physical-condition diagrams, for each selected location. 4. Summarize facts. 5. Supplement accident data with field observations during hours when most collisions have been reported. 6. Analyze the summarized facts and field data, and prescribe remedial treatment.

The following discussion is concerned with the means by which these steps can be utilized to treat, for example, intersections with high incidence of accidents.

Collecting accident records. The first step in any detailed accident study requires the collection for examination of all road accident records within the study area. In Britain this information is readily available from the police authorities. In addition to their accident-location file, the local police usually maintain 'spot' maps on which accidents are shown by means of differently coloured pins. Where these are available they give very useful and quick information regarding accident trouble spots.

Selecting dangerous locations. On the basis of the collected evidence it is possible to isolate locations having large numbers of accidents. One simple method requires the drawing-up of an ordered list of intersections having, say, 10 or more accidents during a 12-month period. An alternative and more desirable method assigns a rating number to each accident and the weighted result is used instead of the total number of accidents in isolating the trouble spots. This latter method accepts accident severity as the most important consideration by assigning, say, 12 points to each fatal accident, 3 points to a personal-injury accident, and 1 point to a damage-only accident.

While there can never be any such thing as an 'acceptable' number of accidents, it is primarily a matter for local decision as to what constitutes a tolerable number of accidents at any given location. One rough guide, however, is that 5 to 10 accidents at any location is a cause for further investigation.

Preparing collision diagrams. The collision diagram is a most valuable tool, since it indicates graphically the nature of the accident record at any particular location. An example of a collision diagram is shown in Fig. 7.5. The diagram need not be drawn to scale but should show by arrow indications the movements which led to each accident. The date and hour of each accident are shown alongside one of the arrows. If weather and visibility collisions are important these are also indicated. Colour coding can be used if further information is required, e.g. whether the vehicle skidded, severe or minor injuries, etc.

In many instances it may be desirable to supplement the collision diagram with a scaled drawing (usually 1:2500) or photograph illustrating the location of road signs and markings, pedestrian crossings, traffic signals, bus stops, parking locations, sight obstructions and fronting land uses. This will be most useful if it is felt that physical features are influencing the accident experience.

Summarizing facts. It is normally helpful to obtain summaries of the following: the total number of each type of accident, the number of collisions on each arm of the intersection, the number of accidents during various periods of the day, and the number of accidents occurring under different weather conditions. It is very probable that at this stage certain trends will appear which may indicate physical features which need correction. For instance, it may be that a large number of accidents involve skidding, thereby suggesting that the carriageway should be immediately resurfaced.

Field observations. Visiting the intersection, preferably during the hours when the greatest number of accidents occur, will usually provide further valuable data for analysis. A study of the behaviour of traffic at the site and observation of the physical and operational features controlling traffic will very often suggest immediate remedial measures.

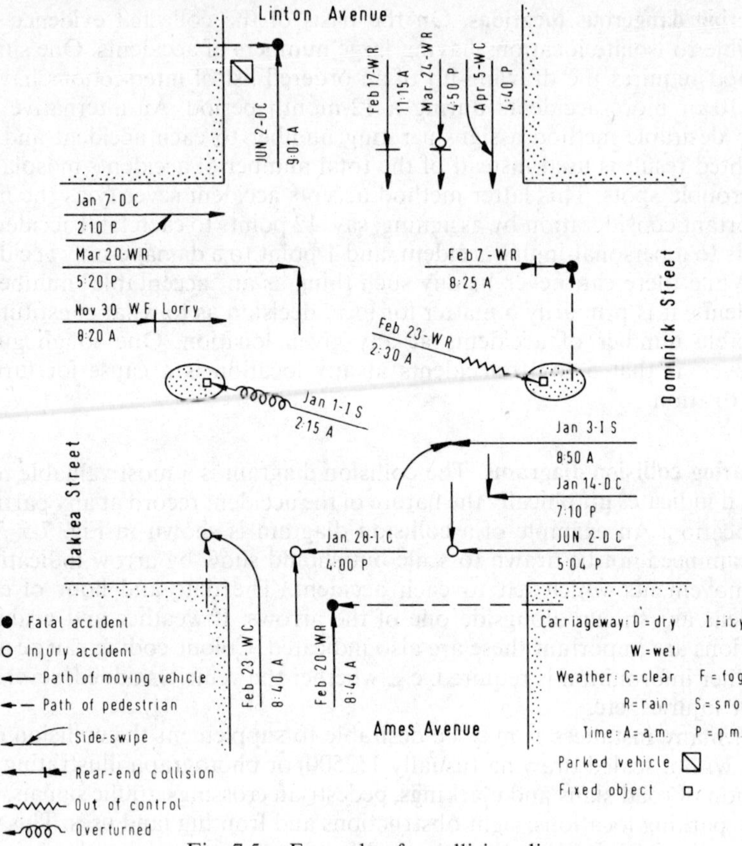

Fig. 7.5. Example of a collision diagram

Final analysis. On the basis of the information obtained from the office and field studies it should be possible for the accident investigator to propose positive recommendations leading to a substantial reduction in accidents. In the great majority of cases it will be found that these recommendations consist of requests for standard traffic management measures such as the installation of traffic signals or pedestrian crossings, the prohibition of certain turning movements, increased usage of signs and markings, an improvement in lighting facilities, limitation of parking or, possibly, the need for channelization of the intersection.

PEDESTRIAN CONSIDERATIONS

Traffic management is primarily concerned with two aspects of pedestrian control. The first, and by far the more important, of these is the control necessary to ensure pedestrian safety. The extent to which this is necessary is reflected in the data in Fig. 7.1, which show the extent to which road accident casualties are pedestrians. The high rate of pedestrian fatalities is a measure of the vulnerability of the pedestrian when involved in a conflict with a motor vehicle.

A second reason necessitating pedestrian control is the effect of pedestrian movement on the ease of movement of vehicles—or perhaps, as might be preferred, the effect of vehicle flow on the ease of movement of pedestrians. British law allows the pedestrian to have unrestricted access to any part of the highway system with the exception of special roads such as motorways. In the congested central areas of cities this right is often exercised to the extent that vehicular traffic may be brought to a standstill as continuous streams of pedestrians cross the road at unauthorized locations. In suburban areas, on the other hand, the pedestrian's desire to cross the road may often be foiled by a continuous stream of high-speed motor vehicles on their way into or out from town.

Pedestrian accidents

Nearly all pedestrian casualties happen on the carriageway; this indicates very strongly the desirability of segregating pedestrians from vehicles. Over 90 per cent of the casualties occur in built-up areas; this again might be expected in the light of the many more opportunities for pedestrian-vehicle conflicts in these areas.

Although most pedestrian accidents occur in built-up areas, the majority of pedestrian fatalities happen in rural areas—primarily because of the higher vehicle speeds. Obviously there is a great need for the pedestrian to be kept off the carriageway in rural areas. Unfortunately, however, most British rural roads do not have hard shoulders and so people are forced to walk on the carriageways. Many accidents could be avoided if pedestrians could be educated to walk facing the traffic and, at night-time, carry a flashlight or wear some reflective coating material. 'Hitch-hikers' should be discouraged from soliciting rides since they are a source of casualties themselves as well as being the cause of accidents through vehicles stopping on the carriageway to pick them up.

The data in Table 7.6 show the population groupings that are most susceptible to pedestrian casualties. It can be deduced from these data that the most vulnerable are the pre-15 and post-60 groups. Older people are more vulnerable because their sense of anticipation and agility is less than that of younger people; also their recuperative powers are less. Among the young, the high risk can perhaps be expected—but not accepted—because of the impetuousness of youth and the use made by children of streets as playgrounds. Certainly these statistics make it clear that special consideration should be given to the initiation of pedestrian management measures on roadways adjacent to such facilities as schools and old people's hostels.

The manner in which Table 7.6 is given takes account of changing population and changing traffic and hence it is possible to see that, although

TABLE 7.6. *Pedestrians killed and seriously injured per 100 000 living per 10^9 vehicle-kilometres in various years*[4]

Ages	1959	1964	1969
0–4	0·47	0·37	0·31
5–9	0·90	0·80	0·65
10–14	0·35	0·37	0·38
15–59	0·21	0·17	0·14
60+	0·78	0·54	0·37
All ages	0·39	0·32	0·26

the actual *number* of pedestrians injured has continued to rise, the *rate* for each group (except the 15–59's) has declined—which indicates that risks to these groups of pedestrians have generally been rising more slowly than the amount of traffic.

Management measures

Most pedestrian management measures are aimed at segregating the pedestrian from the vehicular traffic. When complete physical segregation is not possible, controls are aimed at restricting pedestrian movement on the carriageway to particular locations, and, if possible, during particular times. Measures in use are pedestrian channelization, signals, and crossings.

Pedestrian channelization. By pedestrian channelization is meant the use of footpaths in conjunction with guard-rails or barriers so that pedestrians are kept off the carriageway at certain locations. With the exception of the regulations affecting the special roads, there is no specific law which says that pedestrians must use the footpath and not the carriageway. Thus in congested locations, or where the pathway is cracked and uneven, or when the pedestrian simply wishes to cross the road, he is at liberty to step on to the carriageway at any time and at any place.

In busy shopping centres it is probable that many accidents could be avoided if adequate footpath widths were provided. One guide to pedestrian footpath requirements states that the capacity of an open footway may be taken as 33 to 49 persons per minute after deducting approximately 1 m 'dead width' in shopping streets and 0·5 m elsewhere. There is no published guide as to when the surface quality of a footpath is unsatisfactory, but it cannot be accepted that people will walk on a broken or uneven footpath when a smooth and flat carriageway is beside it. In order to induce people to remain on the footway, therefore, its surface must be at least equal in merit to that of the carriageway.

At dangerous locations and to avoid unwanted interruptions to traffic, guard-rails will need to be used both to keep pedestrians on the footpaths and to canalize the stream of pedestrian traffic wishing to cross the road. The guard-rails should be inset about 0·5 m from the kerb both to prevent contact between the rails and the passing vehicles and to provide places of refuge for the people who do stray on to the carriageway and as a result may be imperilled by finding themselves on the wrong side of the railings and debarred from access to the footpath.

Traffic signals. Traffic signals are used in a variety of ways to control pedestrian movement across the carriageway. By far the most widely used procedure is simply to allow the pedestrians to cross 'with the lights' when opposing vehicular traffic is normally brought to a standstill at a junction. Although this is quite efficient in the great majority of cases, problems may arise through conflicts between the pedestrian flow and turning vehicles. When this occurs a separate pedestrian phase may have to be included in the signal cycle. When pedestrian volumes are very high, vehicular traffic is moderate, and streets are so narrow that it is not possible to have separate traffic lanes for turning and straight-ahead traffic, consideration should be given to the use of an all-red 'scramble' period during which the pedestrians can take the shortest way across the intersection rather than the traditional rectangular route.

There can be however, considerable disadvantages to providing separate pedestrian phases at important intersections. The most critical of these is concerned with the signal times required to accommodate both the pedestrians and the moving vehicles. On the one hand, the pedestrian phase must be sufficiently long to ensure that it is completely safe for the pedestrian to cross; on the other hand, the consequent reduction of time available for the other traffic movements often necessitates a substantial lengthening of the signal cycle so that the vehicular traffic can be accommodated. The net result is that the signal cycle may be so long that pedestrians do not wait for the period allotted to them and cross during traffic phases. If the cycle length is reduced to satisfy the pedestrian requirements, then the free movement of vehicles may be impaired.

Pedestrian crossings. Besides the crossings controlled by traffic signals, there are two other types of special interest to highway engineers: the completely segregated crossing and the at-grade zebra crossing.

Segregated crossings. Ideally all crossings should be of this type since, as the name implies, there is no possibility of conflict between the pedestrian and the vehicle, since the traversing of the carriageway is carried out by means of a subway or bridge. Unfortunately this ideal type of crossing also happens to be the most expensive.

Studies of the extent to which existing segregated facilities were used when installed have given the very interesting results illustrated in Fig. 7.6. The main point to be noted is that, assuming the segregated facility is clean and illuminated and that usage does not require strenuous effort, pedestrians will only use the segregated crossing provided that the route via the new facility is quicker than the ground route. In the case of a bridge crossing, it is

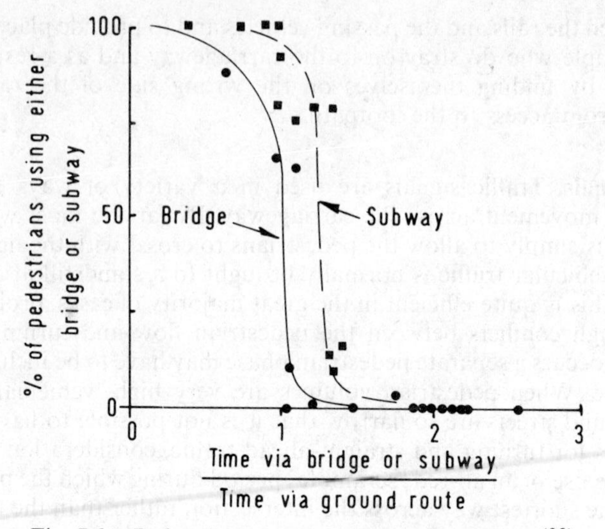

Fig. 7.6. Pedestrian usage of segregated crossings[22]

not possible to get almost complete usage until the crossing time is three-fourths that of the ground route. Subways (which are more expensive) require only a very small time-saving in order to ensure 100 per cent usage.

TABLE 7.7. *Walking speeds of pedestrians on ramps, stairs and on the level*[23]

Type of pedestrian movement	*Average speed of adults*
Level walk	1·52 m/sec
Slopes up and down, up to 1:10	1·22–1·37 m/sec
Stairs	0·15 m/sec vertically

Before constructing a segregated crossing proper studies should be initiated to ensure that it meets pedestrian desires. Ideally the crossing should be located at a point that automatically guarantees a swifter passage; alternatively the need for guard-rails to lengthen the pedestrian path via the ground route should be also examined. Desirably there should be no traffic signal, stop sign control, or existing pedestrian bridge or underpass within about 200 m of the proposed location.[28] If an underpass is being considered, thought should be given to any possible need for its supervision as this type of crossing can be subject to vandalism and other criminal acts. Preliminary estimates of the success of a particular type of facility can be made on the basis of the data in Table 7.7.

Zebra crossings. A zebra crossing is simply an uncontrolled portion of the carriageway that is reserved for the use of pedestrians crossing the road. This strip of the carriageway is marked with alternate black and white stripes parallel to the centre line of the road; the beginning and end of each crossing

TABLE 7.8. *Risks relative to 'elsewhere' on given stretches of road*[24]

Pedestrian crossed	*Crossing located within 18·3 m of a junction in*		*Crossing located >18·3 m from a junction in*	
	Provincial towns	*London*	*Provincial towns*	*London*
On zebra	1·40	0·59	0·97	1·21
Within 45·7 m of zebra	5·75	2·34	0·86	1·43
Elsewhere	1·94	1·32	1·00*	1·00*
At light-controlled crossing	0·89	0·36	—	—
Within 47·5 m of l.c. crossing	4·72	3·75	2·50	1·61

* Arbitrarily taken as unity in both studies.

is marked by flashing beacons. Unlike the mandatory traffic signals, the usefulness of the zebra crossing is dependent on the extent to which the motorist is willing to yield the right-of-way to a pedestrian stepping on the crossing.

There is doubt in some engineers' minds as to the exact value of zebra crossings in relation to pedestrian safety. The dilemma is to a certain extent illustrated in Table 7.8, which shows the results of two separate pedestrian accident studies, one in London and the other in four provincial towns. Both studies showed that near junctions the most dangerous place to cross the road is within 45·7 m of an 'official' crossing, and the safest is via the crossing. Furthermore, the risk to younger (<16 years) and older (>60 years) people was considerably greater than for the 16–60 age group, while the overall risk to pedestrians crossing the road increased as the vehicle flow increased. What have yet to be defined are the circumstances whereby the *combined* risk on and within 45·7 m of a zebra crossing will be always greater than 'elsewhere'. In other words, one queries the concept of installing a zebra crossing at any given location without deep consideration—for it may be that the accident situation could be made worse rather than better.

Criteria for establishing pedestrian controls

There are no firm rules for the establishment of pedestrian management measures. In urban areas it is customary to have footpaths on both sides of the street and management measures are aimed at keeping the pedestrian off the carriageway except at the designated crossings. In this respect guard-rails should be used at locations such as exits from schools, recreation grounds, footpaths or passages (in order to prevent children from running heedlessly on to the carriageway), along busy shopping streets, and adjacent to zebra crossings, signals and segregated crossings.

TABLE 7.9. *A.A.S.H.O. recommendations with respect to footway requirements in rural areas*

Vehicular traffic, vehicles/h	*Pedestrians per day suggested for construction of footways when the roadway design speed is:*	
	48–80 *km/h*	96–112 *km/h*
Footway, one side		
30–100	150	100
>100	100	50
Footway, both sides		
50–100	500	300
>100	300	200

The need for regulated crossings in urban areas is most often indicated by the accident statistics at particular locations, although criteria based on the economic savings have also been suggested.[23] Considerably more research has yet to be carried out, however, before these latter criteria can be used with full confidence. The disadvantage of this and similar types of analyses of the crossing problem is that they all depend on some necessarily arbitrary assumptions concerning the level of pedestrian delay that can be tolerated, the value of pedestrian time, and the extent to which pedestrians are prepared to suffer delays. In many instances investigation may show that it is not a pedestrian crossing that is required to reduce pedestrian accidents but simply better highway illumination; this is especially so if the traffic volumes are relatively low. Once, however, the vehicular volumes are in excess of about 300–400 vehicles per hour, it may be found that pedestrian crossings in conjunction with guard-rails are required. If both the vehicular and pedestrian volumes are very heavy, consideration should be given to using segregated crossings.

In rural areas the problem is more abstract and less easy to solve. All that can be said is that if there is any substantial pedestrian traffic, footpaths must be provided along the side of the road. Some American recommendations in this respect are illustrated in Table 7.9.

SPEED RESTRICTIONS

Excessive speeding is the single most widely *blamed* cause of accidents. As a result there has been a trend in recent years to place upper speed limits on highways in this and other countries.

The term 'excessive speeding' as used in this context refers to the situation where a motorist travells at a speed greater than that which the traffic and roadway conditions can safely allow. By this definition a speed of 40 km/h may be regarded as excessive on a crowded shopping street, whereas 112 km/h on a non-crowded motorway can be a safe cruising speed. If motorists could be relied on to adjust their speeds according to the prevailing circumstances, then there would be no need for speed restrictions.

Unfortunately, however, a significant portion of the motoring population cannot be relied upon and so speed limits are necessary.

Consequences of high speeds

An early arrival at a destination and the psychological thrill of driving at high speed are two considerations which result in a motorist travelling at high speed. What are not always so obvious, however, are the reasons why higher vehicle speeds involve greater risks.

Stopping distances. A most important consequence of increasing speed is that the total distance required to bring the vehicle to rest is considerably increased (see the chapter on Geometric Design). In urban areas in particular, therefore, vehicle speeds should be reasonably low so that the motorist will be able to bring his vehicle to a stop when necessary to avoid colliding with the many to-be-expected interferences on the carriageway.

As also discussed in detail previously, the distance required to safely overtake a vehicle is substantially increased the higher the overtaken vehicle's speed. This is a most important consideration on the rural roads of this country. So many of these roads were never really designed, and have many blind hills and curves. Many needless accidents happen at these locations because motorists attempt to overtake blindly.

Even where the highways were scientifically designed, the original design criteria may not meet modern requirements. A basic design criterion which illustrates how standards change is the one for eye-height. In this country at this time the distances required for safe overtaking are measured between two consecutive points 1·05 m above the carriageway; prior to the mid-1960's it used to be 1·14 m—which was the eye-height of an average driver in a small sports car prior to World War 2. Design and styling trends in the motor industry over the past 25 years now mean, therefore, that on certain otherwise well-designed roads many cars do not have available to them the sight distances required for safe overtaking at high speeds.

Vehicle separation. The minimum safe separation between successive vehicles—which depends on the type of vehicle, its braking efficiency, and the driver's perception-reaction time—also increases with vehicle speed. The separations which should be maintained by vehicles, in the state in which they are normally met on the road, in order to give an overall 95 per cent chance of avoiding a rear-end collision when the vehicle ahead slams on the brakes, are given in Table 7.10. As can be seen, not only do the required clearances greatly increase with speed, but commercial vehicle requirements are considerably greater than for cars.

Observations on the road have shown that vehicles travel much closer together than the values suggested in Table 7.10, and separations become increasingly insufficient at higher speeds. This is probably the main reason why so many accidents involve the following vehicle colliding with the rear of the preceding one.

TABLE 7.10. *Desirable clearance distances between moving vehicles in order to avoid rear-end collisions*

Speed, km/h	*Desired clearance distance, m*		
	Car behind car	*Commercial vehicle behind commercial vehicle*	*Commercial vehicle behind car*
48	24·4	39·6	48·8
64	39·6	64·0	79·2
80	54·9	94·5	118·9
90	73·2	131·5	167·7

Resistance to skidding. The skidding resistance offered by the carriageway surface when wet decreases as the vehicle speed increases. This is because the film of water trapped between the tyre and the carriageway has less time to be squeezed either to the side or into the tyre crevices of a swiftly moving vehicle. Results of tests made on a variety of surfaces throughout a range of speeds up to 161 km/h have shown that the skidding resistance at the highest speeds sometimes fell to less than half its value at 32 km/h, depending on the surface being tested. Indeed some roads have surfaces which are perfectly satisfactory at 32 km/h but become slippery when a vehicle is travelling at 125 km/h or more.

Sign legibility. As will be discussed in detail later, a given size of road-sign lettering can become inadequate at higher speeds. This means that at high speeds the message of a sign may be missed or not fully understood within the distance available for acting upon it.

Pedestrian risk. The pedestrian's estimate of the speed of a vehicle becomes less reliable when the speed is high. The data in Table 7.11 represent the

TABLE 7.11. *Onlooker's estimate of the speed of oncoming vehicles, based on 100 observations at each speed*

	Speed of oncoming vehicle, km/h				
	32	48	64	80	96
Would have been hit	2	7	9	11	22
Would have been a near miss	8	7	7	12	13
Total number of misjudgements	10	14	16	23	35

results of one study which attempted to determine the reliability of the onlooker's estimate of the speeds of vehicles. In this study pedestrians stationed at the side of the test road were asked to push buttons at the last instant at which they deemed it safe to cross the carriageway, in the face of an opposing vehicle. As can be seen from this table the accuracy of the pedestrians' judgement deteriorated as the speed was increased; the sharpest deterioration took place as the vehicles travelled at speeds of 80–96 km/h.

Actual observations on the road have confirmed the results shown in Table 7.11. These further showed that the speeds of smaller vehicles and those travelling on the far side of the road were underestimated more often.

Miscellaneous effects. The risk of mechanical failure in a vehicle due to metal fatigue, overheating, burst tyre, and such-like mechanical failures is more likely at sustained high speeds than low ones. Loss of control, when something unexpected does occur, is much more likely at the higher vehicle speeds.

Mandatory speed limits

From the previous discussion it is clear that there can be a substantial deterioration in the principal factors controlling safety when vehicle speeds are increased. It is logical therefore to conclude that accident numbers and severity should be decreased if speeds are restricted on highways where safety requirements make it desirable. It should be pointed out, however, that while this conclusion appears to be substantiated by the results of European investigations (see, for example, Table 7.12), this is not the case with regard to American studies.[25] Why there should be such a difference cannot be explained at this time.

TABLE 7.12. *Effects or urban speed limits on accidents in various countries*[26]

Country	Date imposed	Limit, km/h	% change	
			Fatal and serious	All injury
Great Britain	1935	48	−15	−3
Great Britain (various sites)	1945–53	48		−10
Sweden	1955	50	−11	
N. Ireland	1956	48	−23	−24
Netherlands	1957	50	−10	−6
Germany	1957	50	−30*	−18
London (various sites)	1958	64	−28	−19
Jersey	1959	64	−47	−8
Switzerland	1959	60	−21*	−6

* Fatalities only.

Types of speed limit. Two types of speed limit are in general use throughout the world. For want of better terminology they will be termed the 'reasonable' and 'absolute' speed limits.

Reasonable limit. With this type of restriction no numerical speed limit is specified but instead dependence is placed on the driver to adjust his speed to the roadway conditions. Under this restriction a motorist driving on a motorway in heavy fog at, say, 48 km/h might be summoned for exceeding the reasonable speed limit; on a clear day with little traffic on the motorway, this same motorist would be well within the law when travelling at 112 km/h.

In theory the reasonable speed limit is ideal. It is flexible and allows the enforcing officials to adjust the limit according to the conditions. In addition it is a concept which the motorist likes; there is nothing more aggravating than to be on an open road but forced to maintain an artificially low speed.

In practice, however, this type of restriction is inefficient. To be successful it requires the co-operation of the motoring public and this cannot always be achieved. In addition it relies heavily upon the judgement of the enforcing officials as to what constitute reasonable speeds for the conditions—thus in essence making them both policemen and judges. At worst this can lead to abuse, while at best it leads to understandable and degrading arguments in court.

Absolute limit. With this type of restriction a numerical speed limit is specified for a road or group of roads. If a vehicle exceeds this limit it breaks the law and there is little that the motorist can do to challenge it in court. While this is a decided advantage from the enforcement aspect, it lacks flexibility and often results in ruffled motorist feelings due to unreasonable speed restrictions being placed on obviously higher speed roads.

Throughout the world the trend is towards utilizing absolute speed limits on roads. In general, the following trends are discernible:

1. Speed limits are applied to both urban and rural roads all through the year.

2. Different upper speed limits are used for different types of highway in different environmental locations. In addition, it is now being suggested that lower speed limits should be imposed on high-speed roads of motorway calibre. A typical lower speed for a rural motorway would be 64 km/h. Its purpose would be to reduce accidents resulting from vehicles moving at such slow speeds that other vehicles are impeded and have to take risks in order to maintain reasonable speeds.

3. Different speed limits are used for different types of vehicle. Thus speed limits for commercial vehicles on rural highways are generally 8 to 16 km/h less than for passenger cars. Cars pulling caravans, oversize vehicles, etc., have even lower speed limits imposed on them.

4. Different speed limits may be in force during day and night.

Observance of absolute limits. A measure of the extent to which drivers respect speed regulations can be obtained by comparing the number of vehicles which travel at higher speeds before and after the initiation of a speed limit. Observations made on the highways of a number of countries are shown in Table 7.13 and they indicate that the proportion of vehicles travelling faster than the limit is always considerably less after the speed limit comes into operation. Complementarily, another study showed that, when the speed limit was raised from 48 to 64 km/h on some major roads in the London area, the proportion of vehicles exceeding 64 km/h rose from 9 to 14 per cent.

The single greatest hindrance to motorist observance of speed restrictions is the establishment of a general speed limit on specific roads or road sections where obviously it is inappropriate. This can be avoided, however, if a speed distribution study is carried out on the highway before imposition of the speed limit. Assuming that the great majority of drivers travel at reasonable speeds, then, say, the 85 percentile speed might be used as a guide to the most desirable speed limit. In no instance, however, should the posted speed limit ever exceed the design speed of the road.

Detailed recommendations regarding the choice of speed limit for particular locations are available in the literature.[27]

TABLE 7.13. *Motorist observance of speed limits*

Speed limit	*Percentage of vehicles exceeding limit*	
	Before	*After*
64 km/h (London area)	39	24
48 km/h (various sites in Britain)		
Private cars	52	46
Goods vehicles	33	28
80 km/h (week-ends in Britain)	28	11
48 km/h (urban roads, N. Ireland)	59	16
80 km/h (Belgium)	51	38
50 km/h (Netherlands)	77	40
80 km/h (week-ends in France)	40	24
90 km/h (week-ends in France)	22	13

Special considerations. Certain other highway sections will normally require special speed regulations that are different from the general limits imposed. An obvious example of this is the section having an excessively high accident record, where investigations into the causes of these accidents reveal that the existing speed restriction should be either raised or lowered. Transition sections with 'medium' speed limits should always be interposed when a high speed rural highway is about to enter a built-up area. Construction zones, as well as being carefully signed, should also be posted for suitably low speeds. Roadways alongside schools and such-like institutions should be given special attention during term times.

Enforcement. As indicated in Table 7.13, there is always an element of the motoring public who require supervision to ensure that they obey the speed regulations. It cannot be emphasized too strongly here that these regulations should be enforced, since the whole basis of law is dependent on the respect shown to its rules and regulations. Indeed it is much better to have no speed limits at all than have ones that are flouted and not enforced.

Two procedures are generally used to enforce the regulations. The first of these utilizes 'speed traps' to check vehicle speeds at particular locations, the actual measurements being made with a radar speedmeter. The second is by means of police patrol cars which pace the speed violators. This is the most effective method when long road distances have to be policed; it can however be dangerous on congested highways when the violator is travelling at a high speed.

Advisory speed signs

Very many road accidents occur because of excess speeds at highway bends. Although not the general custom in Britain, it is practice in some other countries for advisory speed limits to be posted prior to dangerous highway curves so as to give motorists some guidance as to what speeds may be used to safely negotiate these curves. This practice can reduce accidents, particularly at road locations used by high numbers of non-local drivers.

Difficulty arises, however, in defining exactly what is a safe speed. One suggestion[(29)] is that the selected speed should be consistent with the maximum comfortable radial acceleration for the bend in question. Another simpler proposal[(30)] is that the advisory speed should be the 85 percentile 'before' speed of free-moving private cars; for level rural roads this can be estimated from the following equation:

$$S_{85} = 99{\cdot}34 - 2{\cdot}48X$$

where S_{85} = 85 percentile speed, km/h
and X = curvature of bend, deg.

SAFETY FENCES

For many years safety fences have been erected at locations where vehicles may accidently leave the carriageway and be subjected to considerable danger. Where properly installed, safety fences (a) serve to deflect the erring vehicle away from the dangerous locations and/or (b) they have a desirable psychological effect in providing the nervous motorist with a feeling of security when traversing apparently dangerous highway sections.

Before discussing the most important type of safety fence, the crash barrier, mention will be made of the use of *Guide Posts* or *Post Delineators*, alongside highways. Post delineators are intended to *inform* the motorist regarding the shape and limitations of the highway. To do so, they are normally white in colour and have inserted reflector buttons to show the essential delineation. They are expected to break when struck by a vehicle. It should be noted, however, that, at the time of writing, no significant data are available regarding how effective they actually are in helping the driving task—this although they are widely used on all types of highway. Details regarding European practice with these delineators are readily available in the literature.[(31)]

Another type of 'safety fence' is the *Anti-glare Fence*. As its name suggests, this fencing is used at locations where motorists are subject to dangerous '*glare-out*' from opposing vehicle headlights. These fences, which are usually of light mesh construction, are most commonly placed on the central reservations of high speed roads. To overcome the criticism that these fences result in a significant loss of daylight amenity, hedges of Rosa Multiflora Japonica have been planted in (wide) central reserves to serve the same function.

If the highway location is potentially very dangerous, a *Crash Barrier* safety fence may be required which will physically prevent vehicles from continuing in an undesirable direction.

Principles of crash barrier design

Ideally a crash barrier fence should present a continuous smooth face to an impacting vehicle, so that the vehicle is redirected, without overturning, to a course that is nearly parallel to the barrier face and with a lateral deceleration which is tolerable to the motorist. To achieve these aims the vehicle must be redirected without rotation about either its horizontal or vertical axis (that is, without 'spinning out' or overturning), and the rate of lateral deceleration must be such as to cause the minimum risk of injury to the passengers.

In practice the happenings at a barrier fence are so complicated that it has not yet been possible to devise a theoretical treatment which represents what actually does occur. As a result safety barrier research is usually carried out in full-scale road tests. The following discussion must therefore be regarded as a theoretical description based on a greatly simplified model of what occurs during an actual collision.

Lateral rotation. In Fig. 7.7(a) a vehicle of mass m is colliding with the barrier at an angle θ. The horizontal forces acting on the vehicle are the normal reaction of the barrier R, which tends to turn the vehicle anti-clockwise and $F+\mu R$ which tends to turn it clockwise. In this diagram the sliding friction between the vehicle and the barrier is represented by μR, and F is the average value of the other forces produced by, for instance, striking posts or other obstructions.

If the barrier is to fulfil its function of redirecting the vehicle parallel to the safety fence, then the net effect of the moments of these forces must be clockwise. This condition is given by

$$\frac{F+\mu R}{R} < \frac{x}{y}$$

$$< \frac{c \cos\theta - b \sin\theta}{c \sin\theta + b \cos\theta}$$

Thus it can be seen that if F or μR is large, the condition may not be satisfied and the vehicle may spin out of control with a clockwise rotation. When $\tan\theta = c/b$, the vehicle will spin out from the barrier, regardless of the value of $F+\mu R$. For a typical car the value of this critical angle is about 70 deg. If, however, $F=0$ (i.e. the vehicle does not strike any posts), then for $\mu = 0{\cdot}01$ (steel on steel) and $\mu = 0{\cdot}3$ (steel on concrete), $\theta = 64$ deg. and 54 deg., respectively. In other words, as long as the impact angle is kept below 54 deg. for a concrete barrier, and 64 deg. for a steel barrier, then spin-out will only occur if appreciable retardation is imparted to the vehicle by posts or other barrier components—which is why, to ensure that retardation is small however severe the impact, barrier posts should be made weak in the direction of the line of the barrier.

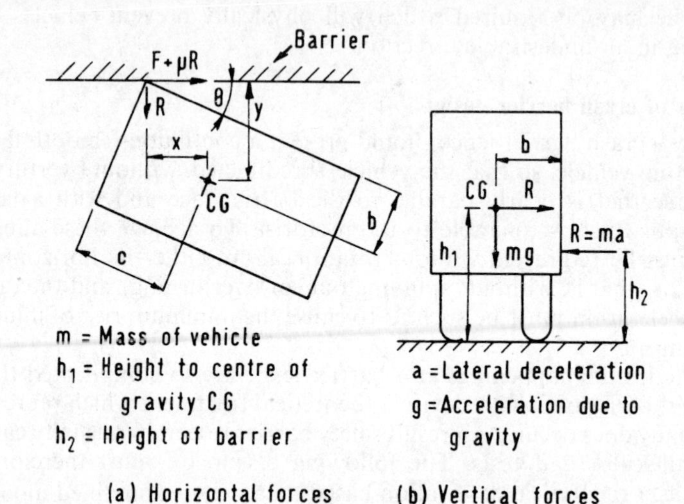

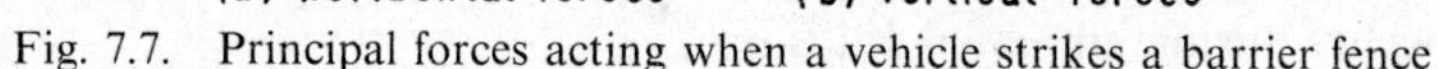

Fig. 7.7. Principal forces acting when a vehicle strikes a barrier fence

Lateral deceleration. While an impact is taking place the barrier will deflect and the vehicle will crumble to a certain extent. If it does not bounce off or climb but instead swings and scrapes along the barrier, then the centre of gravity of the vehicle will move through a total lateral distance of $(y-b+d)$, where d is the sum of the deflection of the barrier and the lateral crumpling of the vehicle. The most important quantity from the point of view of the motorist is the average lateral deceleration of the vehicle; this is given by

$$\frac{(v \sin\theta)^2}{2[c\sin\theta + b(\cos\theta - 1) + d]}$$

where v is the approach velocity.

This equation is actually a reflection of the risk of injury to the motorist. Death can result if the human body is subjected to excessive deceleration—and the amount which he can withstand is in turn influenced by whether or not he is wearing a safety harness.

If the barrier consists of a rail or cable with supporting posts, the tension in the barrier is given approximately by

$$T = \frac{ma}{2\sin\alpha}$$

where α is the angle of the rail or cable relative to its original position.

Overturning. If the vehicle is assumed to be a rigid body, the principal vertical forces acting on it are as shown in Fig. 7.7(b). From this it can be seen that the vehicle will tend to overturn if

$$ma(h_1 - h_2) > mgb$$

and

$$a > \frac{gb}{h_1 - h_2}$$

Hence if the effective height of the guard rail is equal to or greater than the centre of gravity of the vehicle, then the vehicle cannot overturn.

Types of crash barrier

The multitude of crash barriers available commercially can be divided into three main types: 1. Rigid concrete. 2. Flexible wire rope. 3. Steel beam. The following brief discussion on these types of barrier is based on an excellent publication available in the literature.[32]

Rigid concrete barriers. Concrete barriers have been used for a considerable length of time, although now their usage is generally being phased-out on high-speed roads, primarily because the rigidity of the concrete results in peak deceleration rates which can result in (otherwise avoidable) fatalities. A further disadvantage is that even minor scrapes can result in extensive damage to the vehicle bodywork. Limited success has been achieved in introducing greater flexibility into concrete barriers by the incorporation of reinforcing steel; this, however, has also created the new problem of how to repair the barrier after impact.

In general, concrete barriers are now being limited to low-speed highly trafficked roads where the high risk associated with a vehicle crossing the central reservation outweighs the probable rise in the cost of damage-only accidents. They should never be used on high-speed roads unless it is absolutely essential to prevent a vehicle encroaching, whatever the effect on the vehicle's occupants.

Flexible wire rope barriers. Cable barriers in various forms have been in use for nearly 35 years as a means of easily and efficiently stopping an out-of-control vehicle. The main advantage of this barrier is that, because of its great flexibility, a cable can 'slowly' decelerate a crashing vehicle and redirect it most easily along a path parallel to the barrier. In addition, it is comparatively simple to fix the height of the different cables so as to cater for the greatest number of the different vehicles in use today; care has to be taken, however, to ensure that the cable height is not so great that it rides up over the bonnet of a small car, e.g. a sports car.

Probably their greatest disadvantage is that, as yet, cable barriers cannot be used at locations with narrow safe clearances on high-speed roads because of their potential for deflection under impact. In full-scale tests, deflections of up to 4·25 m have been recorded and this, for example, would normally disqualify their usage on the central reservations of British motorways. A further point is that in order to allow for any movement, the cable barrier must use some form of automatic tensioning device, and this further increases the cost of this barrier. Because of the tension in the rope and the nature of the posts, the barrier cannot be used at road locations where the radius of curvature is less than 850 m.

Steel beam barriers. Barriers based on a corrugated steel beam are the most common types used in Britain, and possibly in the world. These beams, of 10, 11 or 12 gauge steel, are corrugated longitudinally to give greater lateral strength with a thinner section. As well as giving a strong section the corrugations, which occur most commonly as varients of a W-section, also increase the distance between the impacting vehicle and the supporting posts.

A main advantage of these barriers is that they present a broad face to traffic and are effective with a wide range of vehicles. Their impact duration times vary from type to type but generally lie between the times for concrete and cable barriers. A method of lengthening the duration time that is now widely used is the development of the strong-beam, weak-post system whereby the posts separate from the beam during the impact. Steel beam barriers are suitable for installation at most sites, and will cater for nearly all types of impact.

Particular attention must be paid to the end treatment of these barriers, as an end-impact will usually be of a serious nature. The safest course in this instance is to continue the barrier beyond the sphere of influence of the traffic.

Use of barrier fences

Barrier fences can be classified as either edge barriers or median barriers. Edge barriers are primarily used to prevent vehicles from hurtling off the roadway and down the sides of steep embankments. They should not be used where it is economically possible to provide embankment slopes of 4 horizontal to 1 vertical or flatter, since a driver forced on to such slopes has a much better chance of regaining control over his vehicle than he has of walking away uninjured from a collision with a barrier fence. Except at lay-bys, these fences should be always placed at a fixed distance from the edge of the carriageway so as to avoid confusion as to their location during bad weather. Edge barriers should be highly visible if they are to be fully effective.

The separation of carriageways by a central reservation serves to reduce, but normally does not eliminate, collisions between opposing vehicles. If head-on collisions are to be nearly eliminated, central reservations of about 15 m are required to enable erring drivers to regain control of their vehicles. In Britain it is the rare highway that has such wide medians; in fact most of the new motorway construction is provided with 4·5 m-wide central reservations. The inadequacy of this narrow width has given rise to a call for median barrier fences as an economical way of eliminating accidents which occur because vehicles traverse the median and collide with vehicles on the other carriageway.

Notwithstanding that it is now government policy to install crash barriers on all British motorways, it is doubtful whether they are justified on all roads at all locations. This can be explained by reference to the data illustrated in Fig. 7.8. As can be seen, a substantial number of these accidents involved vehicles intruding on to the central reservation; of the 220 involved,

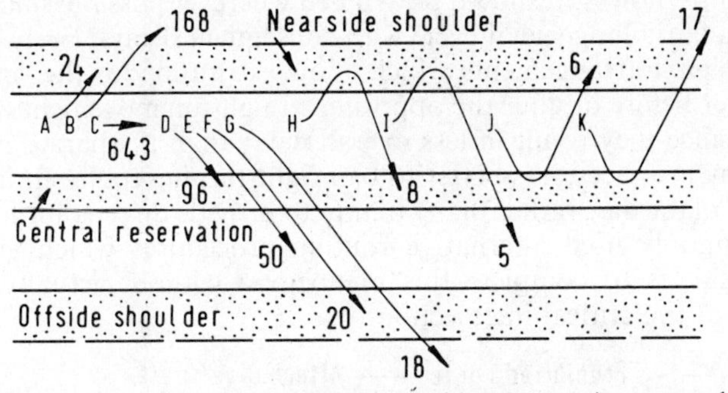

Fig. 7.8. Movements of vehicles involved in accidents on uninterrupted sections of the M1 motorway during the years 1960 and 1961[33]

93 traversed it entirely. Of these 93 movements, *only* 12 *resulted in collisions with vehicles on the opposing carriageway*. Now the important factor to be considered is what would have happened if a barrier fence had been installed within the median. Firstly, it is certain that all 93 crossover movements would have resulted in collisions with the safety fence, while some of the remaining 127 would almost certainly have hit the barrier also. Secondly, it can reasonably be expected that there were many unreported cases where vehicles entered the central reservation but their drivers regained control and were able to steer back on to the carriageway before an accident happened. If a barrier had been located on the median, then again it is very probable that some of these vehicles would have struck it.

Thus the problem is simply reduced to whether the saving in the number of collisions on the opposing carriageway justifies the likely increase in the overall accident rate. Research in the United States indicates that median barriers will certainly reduce the overall accident rate when the average daily traffic flow is above 130,000 vehicles. To have a significant effect on the cross-median accident rate, it appears that barriers can be used with advantage at places where the average daily flow is in excess of 60,000 veh/day.

RESTRICTION OF TURNING MOVEMENTS

Right-turns

Although a certain intersection may work most efficiently during off-peak periods, serious congestion can often be caused during rush periods by right-turning traffic (in Britain). This can be particularly serious when opposing right-turning vehicles ‘lock’ and introduce temporary stoppages of all movements through the intersection. Even though locking may not occur, a few right-turning vehicles can cause a disproportionate loss of capacity. Studies at signal-controlled intersections carried out in Britain have shown that there is usually a loss in approach road saturation flow of $\frac{3}{4}$ per cent for each 1 per cent of right-turning movements.

If there is a heavy right-turning movement, the vehicles may be accommodated at intersections with signal control by inserting an extra phase in the cycle; this, however, should be avoided wherever possible since it usually results in a fairly long signal cycle with consequent delays. Early cut-off and late-start signal arrangements which allow extra time for the right-turning traffic either before or after the opposing straight-on movements are usually preferred since they result in less overall delay than a separate phase.

In many instances, it is better to ban right-turning traffic entirely during all or part of the day, rather than attempt to provide directly for it. There are three commonly used alternative routing procedures which allow right-turning vehicles to complete this manoeuvre without actually making a right-turn at the critical intersection:

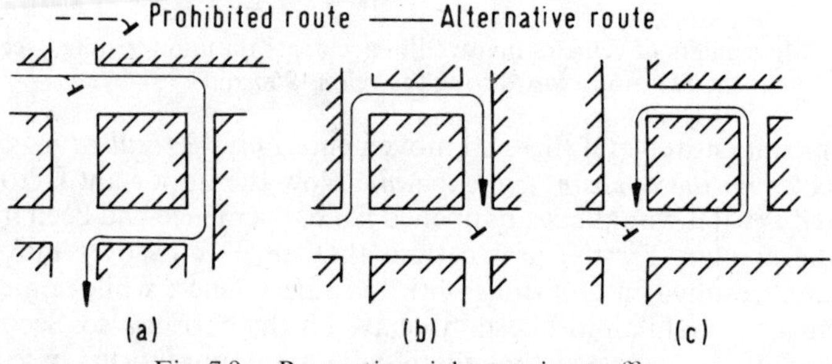

Fig. 7.9. Re-routing right-turning traffic

1. Figure 7.9(a) illustrates the diversion of the right to an intersection further along the road where there is more capacity. This routing is most useful for dealing with a difficult right-turn from a minor on to a major road; the right-turning movement then takes place at a minor-minor intersection.

2. Figure 7.9(b) illustrates the diversion to the left before the congested intersection in what is known as a G-turn. This is most suitable for a right turn off the major road, since it results in a left turn off the major road and a straight-over movement at the critical intersection. The diversion involves two right turns at minor intersections and care has to be taken to see that these do not create extra difficulties. A particularly important consideration is careful signing, so that non-local motorists do not overshoot the initial left turning.

3. Figure 7.9(c) illustrates what is known as a Q-turn. This is a diversion to the left beyond the intersection which requires three left turns. Although this is regarded as the least obstructive diversion, it does require the motorist to travel twice through the original intersection, thereby increasing the total volume of traffic handled there.

The main difficulty associated with the introduction of right-turn restrictions is that of finding alternative routes that are suitable not only with regard to width and structure, but also to amenity. Problems can also arise with buses which, because of their need to serve particular objectives, usually wish to return to the original route as quickly as possible; hence they are extra sensitive to turning restrictions. In any particular instance where this is critical it may be possible for the buses to be exempted from the right-turn ban. Although this may cause some minor confusion, a few vehicles turning right without opposition from other right-turners can usually be accommodated without loss of efficiency.

Of particular importance is the attention that must be paid to the signing of the diversions. Normally the turning restrictions are indicated to the motorist by conventional signs mounted on posts, but if the prohibitions are in force only during the peak periods of the day then they can be advantageously indicated by overhead neon signs which light up only during the critical times.

Left turns

Left turns are not usually considered to be obstructive to traffic movement and hence they are rarely banned. Left-turn bans may, however, be utilized occasionally to reduce the possibility of pedestrian-vehicle conflicts when the number of persons crossing the minor street is unusually heavy. In these cases of course, the ease with which vehicles can move through the intersection is also improved.

ONE-WAY STREETS

One-way street systems are those in which motor-vehicle movement on any carriageway within the system is limited to one direction; they are generally considered to be one of the simplest tools for relieving traffic congestion without expensive reconstruction or excessive policing. Their most effective usage is in the congested central areas of cities where the possibilities of utilizing more extensive aids to traffic movement are often very limited.

Advantages of one-way systems

The primary reason for making streets one-way is to improve traffic movement. Although one-way operation is normally also accompanied by a reduction in accidents, safety is rarely the main reason for its introduction. The improvements are brought about in a number of ways.

Increased capacity. The conversion from two-way to one-way operation can increase the capacity of a street by anything from zero to 100 per cent, depending on such local conditions as the distribution of traffic, turning movements, street widths, etc. Some capacity figures for one-way streets are given in Table 7.14. If the data in this table are compared with the two-way

TABLE 7.14. *Recommended capacities of one-way streets*[34]

Description of one-way street	*Capacity, in p.c.u./h, of carriageways of width, m*											
	6·1	6·7	7·3	7·9	8·5	9·1	9·75	10·4	11·0	11·6	12·2	14·6
All purpose road, no frontage access, no standing vehicles permitted and negligible cross-traffic	2000	2200	2400	2600	2800	3000	3200	3400	3600	3800	4000	4800
All purpose street with 'no waiting' restriction on parking of vehicles, and high capacity junctions	1300	1450	1600	1800	1950	2150	2300	2450	2650	2850	3000	3700
All purpose streets with capacity restricted by waiting vehicles and junctions	800	950	1100	1300	1450	1650	1800	1950	2150	2350	2500	3200

capacities given in Table 7.4, direct effects of making streets one-way can be deduced.

Motorists find it more convenient and less confusing to drive where all vehicles are moving in the same direction, thereby enabling more efficient usage to be made of the carriageway. In addition, odd lanes which could not be utilized under two-way working can now be fully used under one-way operation; this is of considerable importance in this country because so many towns have inherited streets which are not wide enough for four lanes of traffic but have widths well in excess of that required for two lanes.

Slow-moving or stationary vehicles are also more easily overtaken when one-way operation is in effect. More important, traffic flow at intersections is very considerably reduced because of the elimination of some right-turning movements and because any extra road width can now be used more efficiently to speed vehicle movement.

Increased speed. The more even flow of traffic allowed by the removal of opposing vehicles permits higher operating speeds. The higher speeds can be taken advantage of by linked signal systems which can be designed to benefit from them. For example, on the Baker Street/Gloucester Place one-way scheme in London the average speeds were more than doubled from 12·9 to 27·4 km/h, while the volume of traffic was only slightly increased.[35]

Not only are the vehicle speeds increased, but both the journey times and the variability of journey times also reduced. This latter factor is of considerable importance to public transport facilities since regularity of service helps to attract and keep passengers. For instance, on introduction of the Baker Street/Gloucester Place one-way scheme mentioned above the coefficient of variation of bus journey times—this is the ratio of the difference between the maximum and minimum journey times to the average journey time—dropped from 0·37 to 0·17, indicating more consistency in the journey times due to the initiation of the one-way operation.

Increased safety. The introduction of a one-way street scheme generally results in a reduction in the number of accidents, particularly between-intersection accidents. Accidents of the head-on variety are eliminated because of the removal of an opposing traffic stream. Accidents due to bad road lighting should also be reduced since there is now no headlight glare problem. Certain types of accident at intersections are greatly reduced because of the reduction in the number of possible points of conflict (see Fig. 7.10).

Pedestrian accidents at intersections can also be reduced, since pedestrians can be given a fully protected crossing while traffic emerges from a side road (see Fig. 7.11) without loss of time in the signal cycle. In addition signal-controlled pedestrian crossings can often be provided at other points without interfering with vehicle progression. On one-way streets controlled by linked signals, pedestrians are usually able to make safe crossings at intermediate intersections during gaps created by the signal timing.

Economic savings. One-way street operation is also generally held[36] to give

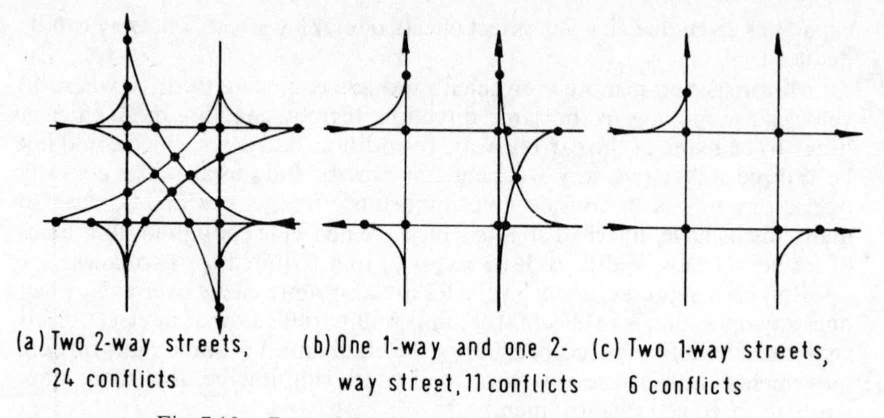

Fig. 7.10. Potential points of conflict at an intersection

economic savings which are well in excess of the initial costs of the schemes. These savings arise from a lowering of motorist journey times and a saving in delays. (It should also be pointed out, however, that one major economic analysis of the effect of one-way schemes in London[37] has suggested that this need not necessarily be the case.)

One-way systems should also result in economic savings due to a lessening need for police-control at congested intersections and streets. Definitive studies regarding savings in police manhours due to the introduction of one-way systems have yet to be published in this country. However, it is reported[38] that, about the year 1950, some 725 of the 3500 km of streets in Philadelphia were one-way and that these one-way streets required 75 per cent less police-officer control than two-way streets. A similar report came from the city of New York which at that time had 2400 km of one-way streets; these required 50 per cent less enforcement than two-way streets.

Increased parking facilities. Normally the parking problem is also the benefi-

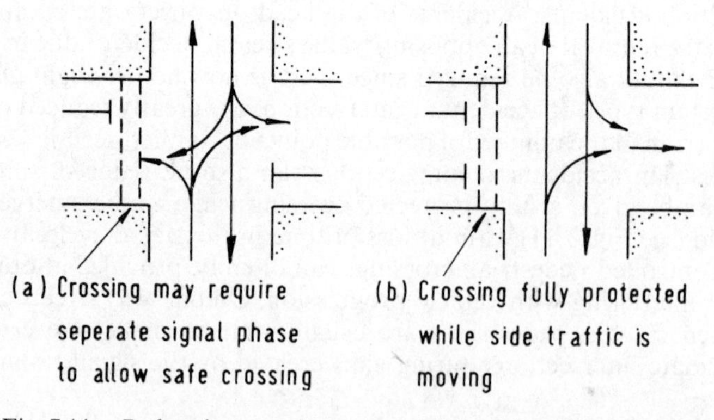

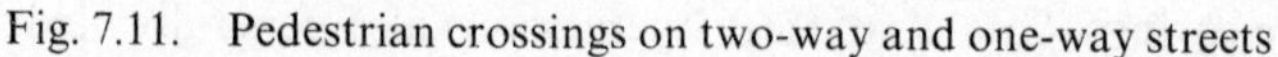

Fig. 7.11. Pedestrian crossings on two-way and one-way streets

ciary of significant relief when one-way operation is introduced into the central area of a city. At the very least, parking and unparking manoeuvres tend to be less dangerous and obstructive when all vehicles face in the same direction. Driving across opposing traffic in a two-way street in order to park on the 'wrong' side of the street is eliminated with one-way operation. In many instances it may be possible to allow parking on one side of streets hitherto considered too narrow for parking, just as on wider streets parking on both sides may become allowable once they become one-way.

Disadvantages of one-way systems

While the advantages associated with one-way operation normally by far outweigh the disadvantages, it sometimes happens that the disadvantages can be extremely significant. Some of the more obvious ones are as follows.

Increased travel distance. With the installation of a one-way scheme it is usually necessary for motorists to travel farther to reach their destinations. This is because a pair of parallel one-way streets becomes a dual-carriageway with buildings on the central reservation, and each block of buildings becomes a roundabout; thus the motorist cannot head directly to his destination but must take a longer and more indirect route. (Obviously, the effect of this increase in distance should be offset by the decrease in journey time which results from increasing the journey speed.)

Another important inconvenience is that public transport stops for opposing directions of travel will have to be relocated on different streets. Cases will arise where these stops may be so far apart that potential bus passengers will baulk at walking what they consider to be excessive distances.

Loss of amenity. One-way traffic operation sometimes involves vehicles using residential streets with consequent loss of amenity to the surroundings. This can be very detrimental when there are large numbers of heavy commercial vehicles present in the traffic stream. Hospitals and schools can also be affected adversely by the changeover.

Loss of business. The introduction of a one-way system is often opposed by local trading interests who fear that business will be detrimentally affected. While this is a possibility which must always be taken into account, it is rarely that there is evidence to support this contention. This factor can be expected to be more deeply felt when buses are taken away from parades of shops.

Increased severity of accidents. One-way operation tends to increase the numbers of vehicles weaving between lanes and to result in higher speeds. While there is usually a reduction in the total number of accidents, the higher speeds at the moments of impact will normally cause an increase in the severity of non-head-on accidents.

Possible confusion. The introduction of a one-way scheme often results in confusion to motorists, particularly just after the scheme has been initiated. With the use of intensive publicity and proper law enforcement, it will be found, however, that local motorists soon change their driving habits; strangers to the system may still be confused, however, and may violate the one-way rule so that head-on collisions can result.

Difficulties of introduction

The most obvious prerequisite to a one-way scheme is that the existing road system should be capable of being modified. For instance, if part of the traffic is displaced from one street there must be a complementary street available to take the removed vehicles. With the grid-iron pattern of streets this can usually be considered as automatically assured; with the more irregular patterns normally prevalent in the urban areas of this country, the availability of a complementary street cannot be taken for granted, e.g. in a linear town that is strung-out along one main roadway. In other instances the complementary street may be much narrower than the main street, and the result is an unbalanced inefficient flow of traffic.

Probably the greatest problems associated with establishing one-way operation occur at the ends of the one-way roads. Ideally each complementary pair should converge to form a Y intersection, but of course this is not always possible. If both of the parallel roads end on a cross-street, then this will be required to carry a considerable extra burden of traffic, and the two intersections will have to handle much more turning traffic; indeed, if the cross-street is an important street, it may be required to carry an unbearable burden, with the result that congestion ensues. In many instances the only way to avoid this is to extend the one-way system beyond the area originally intended.

Inter-connection between two complementary roads can also raise problems. The two streets should not be too far apart—usually not more than about 125 m—or be at such different levels that heavily-travelled connecting roads have steep gradients. Care must also be taken to ensure that the connecting roads do not become overloaded or unusable.

Signing. It is absolutely essential that the one-way scheme should be thoroughly signed at all points where the motorist may have a decision to make. 'No Entry' signs are required at all terminals of one-way streets, and 'One-way' and/or 'Two-way' traffic signs should be placed at the entrances and exits of all the intersections within the scheme. Where necessary supplementary 'No left turn' or 'No right turn' signs should also be displayed.

These signs should be located where they can easily be seen by the motorist. In most cases this will necessitate signs on both sides of the one-way carriageway.

TIDAL-FLOW OPERATION

With the ever-increasing use of the private car for travelling to and from work, all towns now experience the familiar morning and evening rush periods. In the morning the in-bound lanes are filled to capacity, while in the evening the same is true of the out-bound lanes. The term 'tidal flow' has been given to traffic exhibiting this unbalanced characteristic at peak periods. Tidal-flow operation refers to the traffic management process whereby the carriageway area is shared between the two directions of travel in near proportion to the flow in each direction; as a result the area assigned to a given direction varies with the time of day.

There are two obvious methods whereby surplus road space can be transferred. First and perhaps most apparent is on the route with more than two traffic lanes, so that a greater number of lanes can be allocated to the in-bound traffic in the morning and to the out-bound traffic in the evening. The second method requires that there be two or more routes or bridges capable of serving the same origins and destinations, so that one of them can be made completely one-way—in-bound in the morning and out-bound in the evening—with the displaced lighter flow being diverted to the other routes or bridges.

Introduction requirements

For maximum efficiency of operation the number of traffic lanes allocated to each direction of travel should correspond as closely as possible to the ratio of the flows at the peak periods. In addition adequate provision must be made for the lighter flow on the two-way road.

This latter factor means that streets with three or four lanes are best adapted to complete reversal so that they are completely in-bound in the morning and out-bound in the afternoon. A three-lane road with two lanes reserved for the heavier flow and one for the lighter will usually give about the right ratio of lanes for many traffic situations, but, if peak period clearway restrictions are not applied and enforced, the single lane may be frequently blocked and the light traffic halted. The four-lane road is also more suitable to complete reversal because of the difficulty of keeping a single lane continually open. In addition it is rarely that the ratio of three lanes to one represents the ratio of the traffic flow.

On a five-lane road two-way tidal-flow operation can be readily carried out, since three lanes can be applied to the heavier flow and two lanes to the minor flow. During off-peak hours the median lane should be converted into a temporary central reservation and no traffic allowed on it.

On a six-lane road, four lanes are usually given to the heavier flow during the peak hours. During the off-peak hours traffic can be allowed on three lanes in each direction. If flows are sufficiently low during off-peak hours, it may be possible to allow parking on the outside lanes.

Management techniques

Whenever there is a need for tidal-flow operation on a particular route, there is also a need for the imposition of clearway restrictions. At the very least, parking should be prohibited on the side of the major flow during the critical hours. Right-turns from the predominant flow should be forbidden at important intersections because of the restrictions which they place on the free movement of traffic. Depending on the volume of the minor flow, right-turning may however be allowed at non-critical intersections.

A consequence of tidal-flow operation is that central pedestrian refuges have to be removed. A positive feature of the operation is that it lends itself to the linking of the traffic signals in the predominant direction of travel. The linked system will, however, need to be adjusted according to whether it is peak-period or off-peak traffic flow that is being carried.

Controlling the reversible lanes. There are a number of ways of draining and then controlling the reversible lanes so that tidal-flow operation can be safely and efficiently carried out. The three most commonly used procedures require the use of signs, traffic light signals, and movable barriers.

Signing. This method, which is perhaps the one most widely used, requires permanent signs to be placed at intervals along the route so that drivers and pedestrians alike are informed regarding the number of lanes operating in each direction at stated times of the day. Immediately prior to the change in direction police patrol cars are used to clear the way of opposing traffic.

This procedure requires the least initial outlay of capital and so is viewed with favour in many cost-conscious communities. It is most successfully utilized on reversible streets with few visitors in the traffic stream, as regular users can be relied on to know the rules and keep the lanes operating correctly once they have been established. The main drawback of this method is that police vehicles are continuously tied down during the peak periods. Frequently traffic cones, prefaced with 'Keep left' signs, are placed along lane lines to differentiate between the directions of travel: again, however, while this can be very successful on a temporary basis, it is most uneconomical over a long period of time because of the manual labour required to lay and remove the cones.

The signs should be illuminated internally during the periods of tidal flow, the message being varied according to whether it is the morning or evening peak. These signs attract attention and serve better the needs of the irregular users of the tidal-flow route.

Overhead traffic signals. Traffic-light signals suspended over the centre of each reversible lane at the beginning and end of each reversible section are the most flexible means of controlling tidal-flow operation. Intermediate signal lights can be suspended as necessary at intersections along the route. The signals can be of the normal pattern with a face for each direction of travel or, as is more desirable, of a pattern comprising, say a green arrow or a red cross spelled out in energized retro-reflectors, e.g. a driver facing a downward-pointing green arrow is allowed to drive in the lane beneath the arrow, but cannot do so when the red cross is facing him. If the signal shows red in both directions—e.g. during off-peak hours—it means that the reversible lane is being used as a temporary central reservation.

Signals can very easily drain opposing traffic from a reversible lane. The easiest way is to bring on the red signals consecutively from the far end while keeping the running direction amber until all the opposing signals are red. The transitions should be carried out at times that are well clear of the peak periods. When compared with signing control, traffic signals are a relatively expensive method of traffic management. While the initial cost of installing the lights may be high, however, the long-term costs are low.

Movable barriers. The outstanding example of the use of movable barriers to control tidal-flow movement is found on the Lake Shore Drive in Chicago. This is an eight-lane urban motorway that can be operated as six lanes in-bound and two out-bound, or four lanes in each direction, or six lanes out-bound and two in-bound. Three lifting concrete barrier kerbs, each roughly 0·5 m wide and capable of being hydraulically raised a height of 200 mm above the carriageway, are used to separate the reversible lanes. When the barriers are retracted and flush with the carriageway they perform the same function as lane lines. When used as physical dividers the barriers are raised very slowly so that vehicles can easily steer clear as the lift commences.

Movable barriers are extremely expensive to install and maintain in good working order. These costs are only justified when a new motorway is to be constructed in an urban area and the initial and maintenance cost can be favourably balanced against savings in property acquisition due to less lanes being required. In the Chicago illustration, 12 lanes on a conventional dual-carriageway would be necessary to provide the same capacity and ease of movement as on the existing eight lanes with tidal-flow operation.

CLOSING SIDE STREETS

In instances where there are a number of lightly-travelled side streets along an important route, consideration may be given to closing a number of these to vehicular traffic. The conditions under which this management procedure might be utilized are best illustrated by comparing the advantages and disadvantages associated with closing side roads.

Advantages of closing side streets

1. *Improvements in journey time and running speed.* Vehicle running speeds and journey times are closely associated with the number of side street connections along a route. Normally, if some of the streets are closed off, there is a reduction in the factors restricting vehicle movement and vehicle speeds will be increased and journey times decreased.

2. *A reduction in accidents.* Many accidents happen at intersections, and a disproportionate number of those occurring in urban areas happen at lightly-travelled junctions. Many of these would be avoided if side streets were closed. Similarly, a substantial reduction in accidents can be expected when cross-road intersections are converted to staggered junctions.

3. *Usage of linked signals.* When utilizing linked signals it is desirable that the signalized intersections be about, say, 275 m apart. On heavily-travelled roads it may be worthwhile considering the closure of a number of intermediate side streets in order to make use of the facilities offered by linked traffic signals.

4. *Usage for parking.* In certain instances a closed side street can also be used as a car park, especially for long-term parking. Care has to be taken, however, that as a result large numbers of vehicles do not attempt to exit onto the main street at one time and without control. If this happens the resulting restriction to flow may outweigh the other advantages.

5. *Usage for pedestrian precincts.* The advantages associated with turning busy shopping streets into pedestrian precincts are numerous and obvious. In housing areas, where children use the streets as playgrounds, the closing of certain side streets to vehicular traffic will reduce the number of particular types of accidents.

Disadvantages of closing side streets

1. *Congestion at intersections.* When particular side streets are closed, then the displaced vehicles will move to other streets. The net result may be that the increase in volume at the remaining side road connections may necessitate signal-control measures that otherwise might not be necessary.

2. *Increased parking on main roads.* When side roads which provide rear accesses to buildings fronting on a major road are completely closed to vehicular traffic, there is an automatic tendency for these vehicles to park on the main road. This can result in decreases in traffic speed and flow on the major road and increases in accidents associated with parking and unparking of vehicles.

3. *Interference with other management measures.* In certain instances it may be more advantageous to use a side street as part of a right-turn diversion route rather than close it entirely.

4. *Non-availability as an alternative route.* When major road intersections are congested or when road works are being carried out, it is often found that certain side roads compose the only alternative route for traffic.

ROADSIDE TRAFFIC SIGNS

In this age of the motor car, more and more people travel to farther and farther places. One of the problems associated with these movements is the difficulty which motorists from different countries, speaking with different tongues, have in understanding traffic regulations on highways far away from home. As a result the tendency in recent years has been to move towards the international standardization of traffic management devices. This is reflected in the United Nations conferences convened in Geneva in 1949 and in Vienna in 1968 to conclude a new world-wide convention on road and motor transport. Although, because the divergencies between countries were too great, it was not possible to obtain a complete standardization of traffic signs at these conferences, one important principle on which all countries agreed was that whenever possible signs should communicate their message by means of *graphic symbols* and not by inscriptions incomprehensible to foreign motorists.

Definition of traffic signs

In the Road Traffic Act of 1960, traffic signs are stated to be 'any object or device (whether fixed or portable) for conveying to traffic on roads, or any specified description of traffic, warnings, information, requirements, restrictions or prohibitions of any description specified by regulations . . . and any line or mark on a road for so conveying such warnings, information, requirements, restrictions or prohibitions'. Thus, by definition, traffic signs are composed of roadside signs, traffic signals, carriageway markings, retroreflecting road studs, and other such indications on or adjacent to the roadway. Some of the more important of these items are discussed separately later in this chapter; the present discussion is concerned primarily with roadside traffic signs.

Principles of signing

In order to obtain the greatest efficiency of usage from traffic signs in general, and roadside signs in particular, the following principles of signing have been enunciated;[39]

1. The signs must be designed for the foreseeable traffic conditions and speeds on the roads on which they are to be used.
2. They should be conspicuous, so that they will attract the attention of drivers at a sufficient distance and be easily recognizable as traffic signs.
3. They should contain only essential information and their significance should be clear at a glance, so that the driver's attention is not distracted from the task of driving.
4. They should be legible from sufficiently far away to be read without diverting the gaze through too great an angle.
5. They should be placed so that they are obscured as little as possible by vehicles and other objects.
6. They should be designed and sited so that after reading the sign the driver is left with sufficient time to take any necessary action with safety.
7. The signs should be effective by night and day.

Types of roadside sign

The system of traffic signs used in Britain comprises three classes, namely:

A. *Informatory signs*
   1. Advance direction signs and direction signs.
   2. Place and route identification signs.
   3. Other informatory signs.

B. *Warning signs*

C. *Signs giving definite instructions*
   1. Mandatory signs.
   2. Prohibitory signs.

Informatory signs

Informatory signs guide the road-user along established routes, inform regarding intersecting highways, direct to towns, villages and other important destinations, identify rivers, parks and historical sites, and generally help the road user along his way. Unlike most other types, informatory signs do not lose effectiveness by over-use and should be erected wherever there is any doubt.

The following discussion relates to the factors which must be taken into account when designing and locating informatory signs. While this is primarily concerned with directional signs—which account for only about 20 per cent of all traffic signs—it should be kept in mind that many of the considerations apply also to the design of warning, mandatory and prohibitory signs.

Lettering. A driver trying to read a direction sign starts scanning it as soon as the words are legible. When the sought-for word is found, he then returns his gaze to the road ahead. It has been found that this operation takes about $\frac{1}{3}N+2$ sec., if there are N words on the sign. Of course, on average, the actual time may be about half of this because the required word is not always last; however, in the worst case, this relationship is applicable.

During the period of scanning and reading the vehicle is getting closer to the sign and the driver's eyes are diverging further and further away from the straight-ahead position, reaching a maximum with the finding of the required word. In Britain, the United States and Germany, the maximum divergence from the line of sight is used as a measure of the amount of distraction produced by the sign. Because of the sharp fall in visual acuity at points more than 5° from the line of sight, it was decided that the maximum divergence should be limited to 10°, so that the edge of the carriageway up to about 15·25 m ahead of the vehicle would be still within 5° of the line of sight.

Another factor of considerable importance is the visual acuity of drivers. Research has shown that the great majority of drivers can read lower-case letters with an x-height of 25·4 mm at a distance of about 15·25 m, so it is possible to use this relationship as the basis of letter-height design on signs. (The x-height of a lower case script is the height of the letters such as u, m or x).

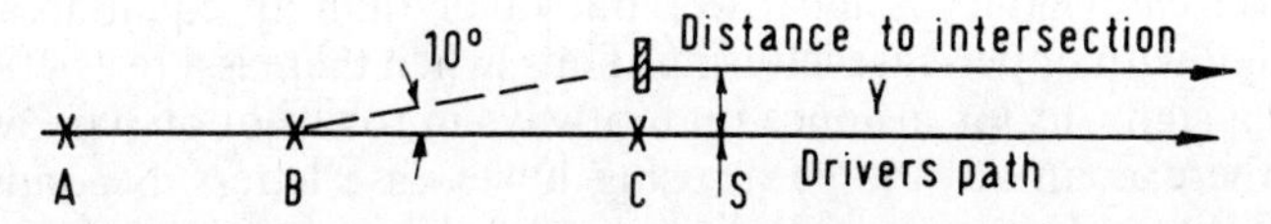

Fig. 7.12. Calculation of letter size for advance direction signs

Determination of letter height. Figure 7.12 illustrates a typical situation where a sign is positioned so that it makes a lateral displacement of S metres from the driver's path. In order to maintain the maximum divergence criterion of 10°, the driver must have completed reading the message by the time he reaches the point B which is a distance of $S \cot 10°$ m from the sign. If the vehicle speed is V km/h and it takes $\frac{1}{3}N+2$ sec to read the sign, then the driver must begin reading at point A, which is $0{\cdot}278V(N/3+2)$ metres from point B. Thus the driver must begin reading at a total distance from the sign of

$$AB+BC = S \cot 10° + 0{\cdot}278\,V\left(\frac{N}{3}+2\right)$$

If the driver can read a 2·54 mm high letter at a distance of 15·25 m then the required letter height is

$$x = \frac{S \cot 10° + 0{\cdot}278\,V\left(\frac{N}{3}+2\right)}{15{\cdot}25} \times 25{\cdot}4$$

where x = height of lower case letter, mm
S = offset distance to centre of sign, m. (In the case of a multi-lane highway the vehicle should be assumed to be in the fast lane)
V = speed of the vehicle, km/h
and N = number of words on the sign.

In these equations the speed value V should normally be the design speed of the road, and not, as is suggested by some, the averave speed; it is logical to assume that if a road has a design speed then the signs should also be tied to that speed. On ‘old’ roads the 85 percentile speed is probably the most suitable value to take.

Upper case versus lower case. There has been a good deal of argument about the relative merits of upper versus lower-case letters, and conflicting claims are often made about the merits of the two types. One investigation which compared the distances at which signs of equal area but with various types of lettering could be read,[(40)] showed that there was no significant difference between the results obtained with good lower-case, upper-case without serifs and upper-case with serifs lettering. As a result, the choice can simply be considered as a matter of preference or based on aesthetics.

Two practical points of interest emerged from this study. First, a sign with lower-case letters is narrower but taller than an equal-area equally-legible sign with upper-case lettering. Thus, when there is a restriction on the width of a sign—as for instance on footways in towns or on narrow country roads—there is an advantage in using lower-case letters. Secondly, words like Stop, Slow, Danger and Police, present themselves more forcibly when written as STOP, SLOW, DANGER and POLICE. This indicates the advisability of using upper-case letters on mandatory and warning signs with written messages.

Layout. Bigger signs cost more money, and so studies have been carried out to find the most economical layout for directional signs. Two of the layouts considered are shown in Fig. 7.13. By far the best results were obtained with the stack type of layout; its only disadvantage was at intersections with more than four arms where confusion occurred. It was concluded that the stack type is adequate for straight-forward intersections but where they are complicated the map layout should be used.

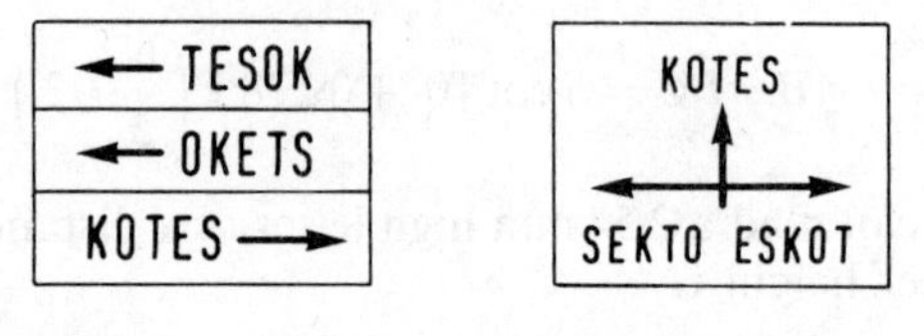

Fig. 7.13. Examples of sign layout

Note that both signs have exactly the same total area but the letters in the stack layout are larger.

Colour. The choice of sign colour is to a large degree limited. To be easily read letters must be of a light colour on a dark background or a dark colour on a light background. If the sign is a small one—e.g. 1·85 m^2 or less—the background colour affects the ease with which the sign can be picked out from its surroundings (see Table 7.15). For small signs there is a disadvantage in using a dark background, but larger signs—especially those on motorways—are sufficiently conspicuous.

From an economic aspect the dark background is to be preferred since it is cheaper to reflectorize letters rather than the whole area of a sign. In urban areas, however, notwithstanding the cost, smaller signs with light backgrounds should be used where large signs will be out of scale with their surroundings.

British practice is for directional signs on motorways to have white letters on a blue background. The network of other trunk and principal routes utilizes white lettering on a dark green background and route numbers in yellow.

TABLE 7.15. *Areas at which colours are equally conspicuous at a distance of 230 m*[41]

Background colour	*Sign area, m^2*
White	1·49
Red	1·67
Blue	1·86
Green	2·04
Black	3·34

Other main features of the signs are the use of lower-case letters and, where possible, map-type advance direction signs.

Reflectorization. The type of reflectorization used is very important. *Diffuse reflectors* such as flat paint reflect the incident light-rays in all directions, while *Specular reflectors* such as glossy paint act as mirrors reflecting the light so that the angle of incidence is equal to the angle of reflection. *Retro-directive reflectors*, which are the most suitable for traffic signs, show maximum brilliance when viewed from the light source, since they throw back the light towards the source. The efficiency of reflectorization is expressed as the

$$\text{Relative intensity} = \frac{\text{Reflected light intensity}}{\text{Incidence light intensity}}$$

at specified acceptance (incident) and divergence (reflective) angles. The acceptance angle is the angle between the incidence ray and the axis of the sign normal to its surface. The divergence angle is the angle between the incident ray and the direction in which the car is going. Table 7.16 shows that from the point of view of reflectorization, the closer the sign is to the edge of the carriageway, the greater the intensity of reflectorization.

TABLE 7.16. *Characteristic relative intensities of retrodirective reflectors*[42]

Acceptance angle, degree / *Angle of divergence, degree*	0	5	10	20
0·33	3·1	3·1	2·7	1·6
0·5	1·5	1·5	1·4	0·8
1	0·2	0·2	0·2	0·1

Location. Desirable though it may be from the reflectorization aspect, signs must not be located too close to the edge of the carriageway. A sign that is too close will not only become spattered with mud, thereby severely reducing its reflectivity, but it may constitute a hazard to traffic. Wherever possible therefore the edge of a directional sign should never be closer than

1·8 m to the carriageway edge. In addition the lower edge of the sign should be as close as possible to the eye-height of the driver, i.e. 1·05 m; of course if the sign is over a footpath there should be a clearance of at least 2 m.

Advance direction signs should be sited sufficiently far in advance of the intersection for the motorist to make the appropriate manoeuvre—which may necessitate stopping—without endangering himself or other vehicles on the roadway. If the sign is located as in Fig. 7.12 the driver will have finished reading the sign at a distance of at least S cot 10° or 5·7S metres from the sign. If the sign is assumed to be Y metres from the intersection, then the distance $Y + 5{\cdot}7S$ must be equal to or greater than the minimum safe stopping sight distance.

On multi-lane roads overhead signs can be used advantageously to direct traffic. On heavily-travelled roads and at complicated intersections these signs are mounted at least 5·5 m above the traffic lanes to which the information applies. Advantages associated with the use of overhead signs are a clear view of the message—most important in urban areas—greater mounting height giving greater visibility at curves and hills, the ability to provide larger lettering, and the elimination of the need for the driver to move his eyes horizontally away from the line of travel. Overhead signs have the disadvantage of being very costly as compared to side signs; this is because of the larger structure needed to support the overhead sign. To avoid being lost against the surrounding sky, they usually have white letters imposed on a black background.

The placing of advance direction signs before an intersection does not eliminate the need for additional directions at the intersection. Not only does the driver's memory need reinforcing but on emerging from the junction the driver should be reassured by means of confirmatory signs that he is on the right road to his destination.

Amenity. It is difficult to reconcile the signing needs of the road-user with the aesthetic requirements of his environment. From the aesthetic point of view the signs should be as unobtrusive as possible; from the motorist's viewpoint they should be as conspicuous as possible. The artist designing a sign wishes it to be a thing of beauty and is likely to want a larger sign than is necessary for legibility because of the need for generous margins and bigger spaces between words; the highway authority is concerned with the cost of the signs and wants them to be as small as possible.

The approach taken towards solving this problem is for the highway engineer to state the minimum requirements that will meet the road-users' needs and for the designer to produce the most economical design within these limitations. It is fortunate that the interests of amenity and of the traffic management engineer generally coincide. Because it is important that signs command the respect of the road-user they are only used where necessary; this is of considerable importance with warning and mandatory signs since overuse and mis-application destroy respect for these signs. In addition they should be maintained in good condition and removed when no longer needed.

Warning signs and signs giving definite instructions

The most numerous and probably most important of the signs are the warning and regulative ones. The major signing systems in use throughout the world utilize combinations of shapes and colours to distinguish various classes of these signs. British signing practice conforms generally to the European system as agreed at an international meeting held in Geneva in May 1971. This European Agreement on Road Signs is stricter than the Vienna Convention and determines precisely the shapes, colours and minimal dimensions of all signs.

Warning signs. In the European system warning signs are distinguished by an equilateral triangle with a red border, and a white or yellow centre with a black or dark blue symbol. Typically, warning signs are used at such locations as approaches to intersections not previously indicated by advance direction signs, dangerous bends or hills, concealed or unguarded level crossings, near schools, pedestrian crossings, converging lanes and other such locations where the motorist requires warning of hazardous conditions on or adjacent to the carriageway.

Mandatory signs. Most mandatory signs are circular with white or light-coloured symbols on a blue background. Important exceptions are the 'STOP' and 'YIELD' signs which have distinct shapes and colours as well as capital letters in order to produce a more immediate impact on the observer. For example, the European stop sign is octagonal in shape with a red ground and a white or yellow border and 'STOP' symbol; the latter must be at least one-third the height of the sign.

Mandatory signs give definite positive instructions when it is necessary for the motorist to take some positive action. The two most important are the stop and give-way signs. The stop sign is used only at intersections where the visibility is so bad that it is imperative for the motorist to stop on every occasion. It should not be used indiscriminately; if the sign is used where stopping is never or rarely necessary, its impact on drivers will be depreciated and it will tend to be ignored, with serious consequences to road safety. The give-way sign is used at intersections where control is not exercised by traffic signals, police or stop signs, but where there is need for drivers on minor roads to proceed so as to not cause inconvenience or danger to traffic on major roads. It is most commonly used on minor roads at junctions in heavily-travelled rural areas.

Prohibitory signs. As the name implies, these signs generally give definite negative orders which prohibit the motorist from making particular manoeuvres. Typical examples of these are the 'no right turn', and 'no entry' signs. With the exception of the 'no entry' and the waiting restriction signs, all prohibitory signs are circular with a red or white centre; the symbols or inscriptions are black or dark blue.

CARRIAGEWAY DELINEATION

Carriageway delineators (including markings) have definite and important functions to perform in a proper scheme of traffic management. In many instances they are used to supplement the regulations or warnings of traffic signs or signals. In other situations, they are used to obtain results, entirely on their own merits, that cannot be obtained with other devices.

Carriageway delineators have, however, several definite limitations to their effectiveness. They may be obliterated by snow or dirt, obscured when the volume of traffic is heavy, and some are not readily visible when the carriageway is wet. Markings are not very durable when subject to heavy traffic wear and must be replaced at frequent intervals; in addition, they cannot be utilized at all on unsurfaced carriageways.

The following discussion will concentrate on the longitudinal delineators which feed information continuously to the driver. Details of the regulatory nature of delineators (e.g. as used at stop signs, to indicate parking/non-parking at kerbs, etc.) or of the sizes and spacings of particular markings, lettering, etc., as used in any given country will *not* be discussed. For these the reader is referred to the appropriate manuals.[e.g. 43, 44]

Reasons for longitudinal delineation

Generally it can be said that the various forms of longitudinal delineation have three main functions: 1. To characterize the road. 2. To provide route guidance. 3. To act as a tracking reference.

By *characterization* is meant the situation where particular forms of delineation are used to provide the driver with information about the nature of the road which leads him to expectations regarding the ease of the driving task. For example, if the centre-line on a carriageway is marked in yellow, then it could indicate to the driver that he is on a two-way carriageway, whereas a white centre-line (or lane-line) would indicate a one-way carriageway.

The term *route guidance* is used to define the situation where carriageway delineation might be used to direct the motorist into one of, say, two alternative routes. Consider, for example, a simple Y-junction at which all three legs appear to be of the same route hierarchy so that the motorist entering the junction along its stem is unable to discern which of the two remaining legs is the major route, and which is the minor. If, however, a particular form (or colour) of marking is used on the stem and continued through the appropriate (left or right) leg, then it would be clear to the driver which route should be followed.

In relation to the *tracking reference* function it should be remembered that a motorist has two basic types of motion continually under his control to ensure that his path remains with the road at night or day; these are speed and direction. While at first one might think that a driver judges his speed by looking at his vehicle's speedometer, the fact is that in practice he prefers to use subjective judgement by noting the rate at which he passes objects by the side of the road—and at night, particularly on rural roads, these reference objects are not visible. Of greater real importance is directional control since[45] '*in driving the task is not exactly that of following a line, but rather of remaining at a constant lateral displacement from it, the line being either the kerb or a white guide line.*' It has been shown by experiment that the centre and edge of the carriageway are two of the principal locations used by drivers in directional guidance, and hence strong emphasis at these locations, especially at night, can greatly simplify the driving task. Furthermore, the higher the vehicle speed, the more useful is this longitudinal contrast delineation.

Also at night, the uncertainty caused by the glare of oncoming vehicles can be heightened by a lack of delineation. For example, a driver meeting an oncoming vehicle is inclined to look at the carriageway edge so as to avoid looking directly at the headlights. If, however, the road edge is indistinct, the driver must glance back periodically at the centreline to check his lateral position—but in so doing, his eyes become more 'light-adapted' so that when he looks back again, the carriageway edge appears even more indistinct so he must look again at the centreline to regain his lateral bearings. The situation can, in fact, develop to the stage (in heavy traffic flow) where the driver cannot discern the centre-line because of the glare, and has to orient himself by looking directly at the oncoming vehicles' headlights—which explains why the deliberate delineation of the edge of the carriageway must help this situation.

Types of delineator

Basically there are two types of delineators used on the carriageway: markings and road studs.

Marking materials. The two most widely used materials are conventional paints and hot-applied thermoplastics (including spray-plastics). Between 80–90 per cent of the roadlines laid in Britain are thermoplastic, whereas on the Continent the reverse is generally true.

Although more expensive than paint, thermoplastic has the *advantages* that (*a*) it has a longer life, (*b*) the ability to fill the interstices of rough-textured roads, whereas paint soon wears from the surface-dressing peaks and the interstices fill with dirt, (*c*) a high temperature of application which enables it to fuse with a bituminous road surface, sometimes even when the road is cold or slightly damp, (*d*) it is proud of the road surface, and this assists visibility on a wet night by facilitating drainage of the water film, (*e*) it contains 60 per cent sand and a binder which ensures good skid resistance as it erodes, (*f*) the material has a rough surface when laid which aids immediate diffusion, and (*g*) it sets almost immediately after being laid.

The *disadvantages* of thermoplastic are (*a*) it has a greater initial cost as compared with paint, (*b*) rapid application on a large scale is more difficult because of the large bulk of material which has to be melted down, (*c*) care is needed to avoid an undue build-up of thickness by successive applications, as this can be hazardous to motorcyclists, (*d*) adhesion is usually poor on concrete road surfaces, and (*e*) on dirty roads carrying light traffic, thermoplastic discolours more readily than paint which is relatively smooth and glossy.

Reflectorization. The reflectorization of a paint or thermoplastic road marking is achieved by the addition of tiny glass spheres ('ballotini') which are premixed and dispensed ('dusted') on to the surface of the line material as it is being laid. Premixing, very often followed by dusting, is the usual practice with thermoplastic; dusting is the more common practice with paint, although reflectivity only exists as long as the beads remain in place on the surface. An advantage of premixing is that as the binding material becomes worn by traffic, further beads are exposed to reflect the light from the vehicle's headlights back to the driver.

While the excellent reflectivity properties of lines containing ballotini are well recognized in dry weather, there is some doubt about their absolute effectiveness in geographical areas subject to regular rain and fog. Experiments have shown that the reflectivity of the lines decreases as rain falls, the water film tending to reflect the light from the headlights away from the driver. When the rain is heavy enough to completely submerge the beads, as easily happens with paint, the line can become practically invisible.

Road studs. The 'ideal' answer to the wet reflection problem is without doubt to supplement the reflectorized road lines with reflectorized road studs. Road studs used on their own are of relatively little value to the driver under normal daytime conditions, unless they have coloured shells and are spaced very close together.

The reflective road stud most commonly used in Britain is the *cat's-eye*. Each cat's-eye consists of two parts: a metal base embedded in the road, and a separate rubber pad insert into each side of which (for two-way roads), or in one side (for one-way carriageways), two longitudinal biconvex reflectors are fixed. As vehicle tyres pass over the rubber pad, its centre part is depressed so that the faces of the reflectors are automatically wiped by the front part of the pad, thus giving the stud its well-known self-cleaning property. The length of time before the insert rubber pad must be replaced depends very much on the speed and density of the traffic, as well as on the lateral location of the road stud on the carriageway; however, measurements have shown that the reflectivity of a cat's-eye on a centre-line of a high-speed road can fall to 50 per cent of its original value after twelve months.

Another type of road stud which is now rapidly coming into favour is the corner-cube type of reflector. The reason for this name is that the individual reflectors in each face of the stud consist of three sides of a cube, and the headlight ray is reflected from all three sides before returning to the eye of the motorist.

When both are new, objective measurements have shown that a corner-cube road stud can return as much as 20 times more light than a cat's-eye at a distance of a 100 m or more; however, when viewed more obliquely, e.g. at distances of 30 m or less, its superiority is not as great. Even though abrasion from tyres soon causes the face of a corner-cube reflector to become etched with a network of fine scratches which cause diffusion of some of the light (typically the reflecting power can be reduced to about 10 per cent of its original value in a year), yet at long distances it can be still considerably brighter than a cat's-eye after the same period of time.

Fig. 7.14 shows the delineation used on British motorways, including at junctions. Note the different types of lines and studs used at the different locations, each intended to convey a message to the motorist regarding that position.

Practical justification for use

Because of their expense, which is recurrent, carriageway delineators must be justified. Although there is little objective research evidence available re centre-lines and lane-lines, there is no doubt but that they are most effective traffic management tools. Edge-lines are more controversial, although in recent years a certain amount of evidence has been collected regarding these.

Lane and centre-lines.

Centre-lines indicate the division of the travelled way carrying traffic in opposing directions. They are usually denoted by broken single lines, their function being to act as guide lines which may be crossed at the discretion of the driver. At locations such as a carriageway-width transition or where an extra uphill traffic lane is provided, the 'centre' line may not be actually located at the geometrical centre of the road although it is still known under the same name. While they are desirable on all surfaced highways, centre-lines should most certainly be used on all major rural highways having an even number of lanes and on as many urban streets as possible—particularly narrow ones—and on all four- and six-lane roads.

Lane-lines are particularly useful in the organization of traffic into its proper channels, and in increasing the efficiency of carriageway usage at congested locations. They should be used on all major rural highways with three or more lanes, on one-way roadways, at the approaches to all important intersections and pedestrian crossings and on congested urban streets where the carriageway will accommodate more lanes of traffic than would be the case without the use of lane-lines. Lane-lines are usually indicated by broken lines, the marks being considerably shorter than the gaps. In certain situations warning markings may be used in place of lane lines in order to perform the same function.

Two parallel lines are used instead of centre-lines at horizontal and vertical curves on two- and three-lane highways where overtaking is prohibited because of restricted sight distances or other hazardous conditions. The parallel lines may be continuous or one may be broken and the other continuous. If the near line is broken, it indicates that the driver may cross the double white lines at his own discretion; if the nearside line is continuous, he should not cross at all.

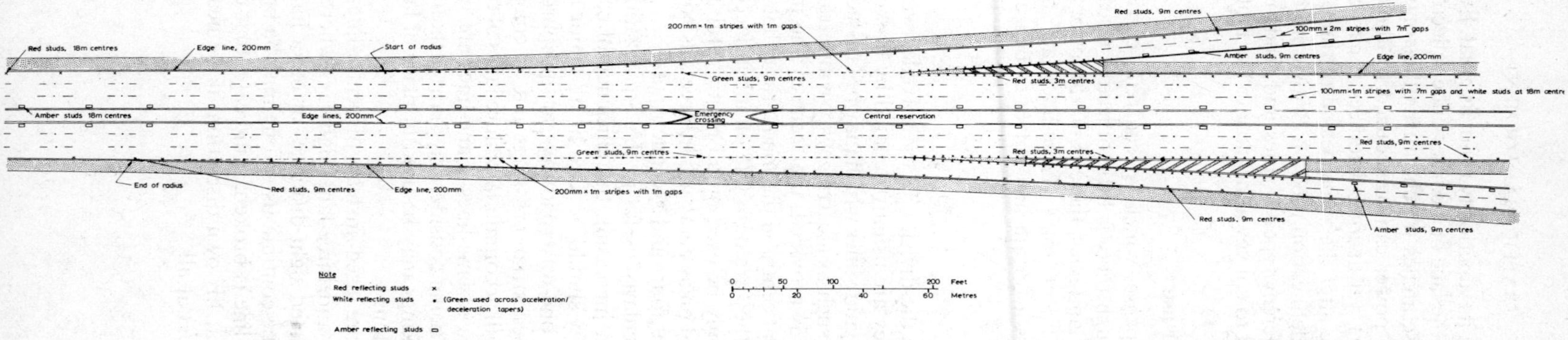

Fig. 7.14. Delineation practices on British motorways

Where the sight distances at vertical and horizontal curves are considered adequate, but only just so, a warning line, consisting of broken single lines with the mark approximately twice as long as the gap, may be used instead of the normal centre line. Warning lines are also used as lane markings at approaches to intersections and central refuges, except when the latter are within a double line section. The purpose of these lines is to warn the motorist to drive with particular care at these locations. Unlike the double lines, however, they do not prohibit the driver from carrying out any movement.

Edge-lines. As used in this context edge delineators are broken or solid longitudinal lines which indicate to vehicular traffic the location of the edge of the carriageway on unlighted rural roads, lay-bys and intersections. What simple logic, and the studies which have been carried out, would appear to suggest with regard to delineating the edge of the carriageway is as follows[31]:

1. Drivers like edge-lines. More important, they feel that they make the driving task easier and more secure and hence, they can enjoy the drive much more.

2. Edge-lines do not cause increases in the number or severity of accidents. It is likely that they reduce accidents, particularly at junctions—*provided* the junctions are at least partially unmarked.

3. They do little for the motorist during the day, when the contrast between the carriageway and the side of the road is such as to make it readily distinguishable.

4. When the shoulder or verge is not in good contrast with the carriageway surface, a properly located edge-line can be very effective in ensuring that the motorist stays on the designated travelway both by day and night.

5. On a normal width, single carriageway road at night, an edge-line may result in vehicles moving small distances in a lateral direction towards the centre-line. The amount of movement will likely be influenced by the location of the edge-line relative to the edge of the carriageway. There is no evidence to suggest that this small amount of lateral movement is a cause for concern in relation to accidents.

6. Under normal dry weather circumstances by day or night, edge-lining has no practical effect on speed. No data are available regarding either speed or placement under wet weather conditions.

7. Whether the edge-line is dashed or continuous has little or no effect on either vehicle placement or speed. This assumes that the dashed line gives the general appearance of a near-continuous line when the motorist is travelling 'with the traffic' on the roadway.

8. Edge-lining cannot but result in less roadside maintenance being required.

9. Edge-lining should help reduce pedestrian accidents in rural areas, by more clearly designating the part of the roadway where the person on foot may not walk.

TRAFFIC SIGNALS

Road-traffic signalling devices have been utilized on the highways of the world for over 100 years. The first traffic signals, which were manually operated, were installed at the intersection of Bridge Street and New Palace Yard outside the Houses of Parliament in London in 1868. This installation, which incorporated semaphore arms and red and green lights illuminated by gas, was short-lived, however, due to an explosion which injured the policeman on duty and brought the experiment to a hasty conclusion. After this unfortunate episode interest in traffic-signal development languished until about 1914 when an electrically-controlled traffic signal device was used in Cleveland, Ohio; it is this device that can perhaps be truly regarded as the basis of modern traffic signals. In 1926 the first modern type of signal was introduced into this country with an installation at a busy road intersection in Wolverhampton.

Since these early days the traffic signal has become one of the principal tools of the highway engineer engaged in traffic management. Traffic signals, when properly located and operated, usually have one or more of the following *advantages*:

1. They provide for orderly movement of traffic. Where proper physical layouts and management measures are used, they can increase considerably the traffic-handling capacity of an intersection.
2. They reduce the frequency of right-angle and pedestrian accidents.
3. Under conditions of favourable spacing they can be co-ordinated to provide for continuous or nearly continuous movement of traffic at a given speed along a particular route.
4. They can be used to interrupt heavy traffic at given intervals in order to permit other vehicles or pedestrians to cross speedily and in safety.
5. When compared to the cost of police control they represent a considerable economy at intersections where there is a need for some definite means of assigning right-of-way.
6. Signals have standard indications which drivers can very easily follow. At night or in foggy weather they are more easily understood than the hand signals of a policeman.
7. Unlike manual controllers, who frequently delay main road traffic while straggling vehicles negotiate an intersection, automatic signal control is impartial in assigning right-of-way.

Although automatic traffic signals are usually preferable to police control, they also have their *disadvantages*:

1. They can increase *total* vehicle delay at intersections. This is especially noticeable during off-peak hours.
2. When improperly located and/or timed they may cause unnecessary delay, thereby decreasing motorist respect of this and other management tools.
3. They usually cause an increase in the frequency of rear-end collisions.
4. They are not normally capable of granting right-of-way to emergency vehicles such as ambulances and fire engines.

5. Failure of the installation, although infrequent, may lead to serious and widespread traffic difficulties, especially during peak traffic periods.

Terminology

Before discussing traffic signals as such it is convenient to define some of the terms associated with the use of this form of intersection control.

Traffic signal. Generally this can be defined as a device which is manually, electrically or mechanically operated so that traffic is alternately directed to stop and permitted to proceed. This device includes a signal head containing different coloured lanterns, all wiring and the control mechanisms.

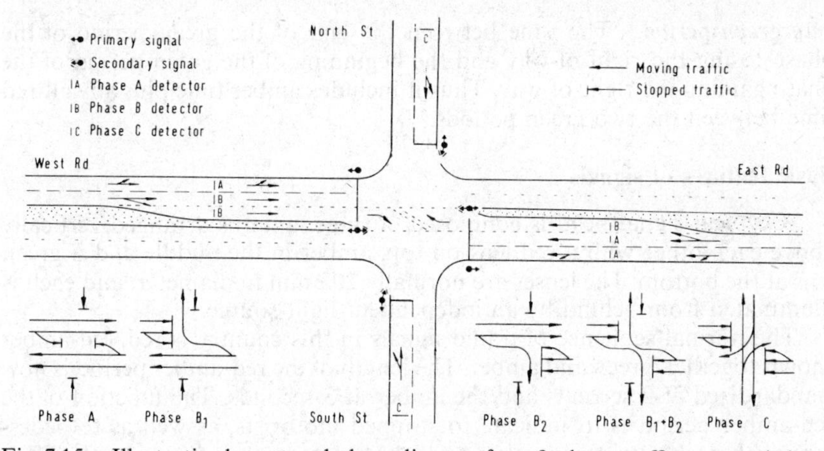

Fig. 7.15. Illustrative layout and phase diagram for a 3-phase traffic-actuated signal installation on a dual carriageway

Traffic signal controller. This is the complete electrical mechanism, usually installed in a cabinet at the side of the road, which controls the operation of the traffic signals. It includes the timing equipment operating the mechanism which lights the lanterns in the signal heads. Some of the different types of controllers are as follows:

Automatic controller. A self-operating mechanism which operates the traffic signals automatically. Automatic controllers are usually fitted with a facility switch which enables the signals to be changed manually by depressing and releasing a push button.

Pretimed controller. An automatic controller for supervising the operation of traffic signals in accordance with a predetermined fixed-time cycle.

Traffic-actuated controller. An automatic mechanism for supervising the operation of traffic signals in accordance with the varying demands of traffic.

Master controller. An automatic controller for supervising a system of secondary controllers, maintaining definite interrelationships, or accomplishing other supervisory functions. The secondary controllers are the automatic controllers at the individual intersections.

Local controller. A mechanism for operating traffic signals at an intersection, or two or three adjacent intersections, which may be isolated or included in a signal system.

Traffic detector. A device by which vehicles or pedestrians are enabled to inform a traffic-actuated controller of their presence.

Time cycle. The time period required for one complete sequence of signal indications.

Traffic phase. A part of the time cycle allocated to any traffic movement or any combination of traffic movements receiving the right-of-way. An example of a time cycle composed of three phases is shown in Fig. 7.15.

Intergreen period. The time between the end of the green period of the phase losing the right-of-way and the beginning of the green period of the phase gaining the right-of-way. Thus it includes amber time plus any all-red time between the two green periods.

Basic features of signals

A typical signal head is composed of three lanterns arranged vertically above each other with a red lens on top, amber in the middle and a green lens at the bottom. The lenses are normally 203 mm in diameter, and each is illuminated from behind by an independent light source.

The normal sequence of traffic signals in this country is red, red-amber shown together, green and amber. The length of the red-amber period is now standardized at 2 seconds and the amber at 3 seconds. The function of the red-amber period is to indicate to stopped motorists, as well as to pedestrians, that the lights are about to change to green; they are therefore prepared to enter the intersection as soon as the lights change and so wasted or 'lost' time is kept to a minimum. The amber period similarly warns approaching traffic of a coming change in the signal indication; at the same time it acts as clearance interval for vehicles or pedestrians within the intersection as well as for those moving vehicles that are so close that to stop would be dangerous.

Roadway traffic signals used at intersections can be classified into the following groups:

1. Pretimed traffic signals.
2. Traffic-actuated signals.
   (*a*) Full traffic-actuated signals.
   (*b*) Semi-traffic-actuated signals.

The individual signals can be either operated in isolation—which is the more common—or as co-ordinated parts of a linked signal system.

Pretimed traffic signals. With basic pretimed traffic signal operation a consistent and regularly repeated sequence of signal indications is given to the traffic. Since the right-of-way is assigned on the basis of a predetermined time schedule or a series of such schedules, this type of signal is at its most efficient at intersections where traffic patterns are relatively stable over long periods of time. Among the *advantages* of pretimed signal operation are:

1. Consistent starting time and duration of cycle length facilitates linking of adjacent traffic signals. This linkage can permit progressive vehicle movement and speed control through many intersections.
2. Pretimed controllers are not dependent for operation on the movement of approaching vehicles past detectors. Thus a stopped vehicle or construction work cannot interfere with their proper usage.
3. Usually the cost of a pretimed installation is considerably less than that of a traffic-actuated installation.
4. The simplicity of the equipment results in lower maintenance costs.

Pretimed traffic signals have the following *disadvantages*:

1. Normally pretimed traffic-signals are geared to peak-time traffic requirements with the result that excessive and frustrating delays to vehicles will occur during off-peak times.
2. Pretimed signals cannot recognize or deal with short-period traffic demands.

Normally the disadvantages associated with pretimed signals outweigh the advantages and so they are rarely if ever used today in Britain.

Traffic-actuated signals. With vehicle-actuated signals, a detector pad is placed in the carriageway at some distance back from each stop-line and every vehicle approaching the intersection registers its approach by actuating the appropriate detector. The traffic signals are thus automatically adjusted to meet the needs of the traffic, however much they may vary from time to time throughout the day.

Traffic-actuated signals have the following *advantages*:

1. They provide maximum efficiency of movement at intersections where one or more of the traffic movements are subject to variation in volume.
2. They minimize vehicle delay, especially during off-peak hours.
3. Intersection capacity is usually increased since the more heavily travelled approaches are given favourable treatment.
4. They are especially useful at intersections requiring the use of more than two signal phases.
5. When compared with pretimed signals they tend to reduce the number of rear-end collisions which are associated with the arbitrary stopping of vehicles in the middle of a traffic stream.

Full traffic-actuated signals. This type of installation, which utilizes detectors placed at all approaches to the intersection, is that most widely used in Britain. The detectors are used to assign the right-of-way to a particular street as a result of the actuation on that street. In the event of continued actuation of a given detector, the right-of-way is—in the case of a 2-phase, 4-way intersection—transferred to the cross street on the expiration of a preselected maximum length of time. Right-of-way then remains with the cross street for at least a (variable) minimum predetermined period of time so that vehicles waiting between its detectors and the stop lines can enter the intersection.

After the minimum green period is ended, each new vehicle which crosses the detector extends the green period by an amount called the *Vehicle-extension Period.* The length of this extension is related to the speed of the vehicle and is sufficient to enable the vehicle to reach 3 to 6 m beyond the stop line. These green-time extensions are not directly additive; instead the time in the signal controller is only reset to a new value if the following extension is for a period that is longer than the unexpired time of the previous extension. When the interval between the vehicles crossing the detector becomes greater than the vehicle-extension period, or when a predetermined maximum green extension time has been expended, the right-of-way is returned to the other street where similar sequences can occur as required.

An example of a fully traffic-actuated 3-phase control at an intersection between a 2-lane street and a dual-carriageway is illustrated in Fig. 7.15. This method of control can be very useful when there are heavy right-hand turning movements from the major highway. Its operation is as follows:

1. Vehicles travelling straight ahead and turning to the left on both East and West Roads are given the right-of-way during phase A.
2. After the completion of a minimum length of green time, phase A can be interrupted by demands from vehicles on East Road who wish to turn to the north. This manoeuvre is carried out during the split-phase B_1.
3. Similarly phase A can be interrupted to allow vehicles to travel to the south during a separate split-phase B_2.
4. If there are vehicles waiting in the right-turning lanes on both East and West Roads, then phase A can be interrupted to allow a separate phase B_1+B_2 which permits only the right-turning movements.
5. At the end of phase B_1 or B_2 or B_1+B_2, phase C is initiated and the right-of-way is transferred to the vehicles on the cross streets, North and South.

Semi-traffic-actuated signals. This is a type of signal installation where detectors are provided for traffic-actuation on one or more, but not all, of the approaches to an intersection. Its usage is applicable at an intersection carrying a heavy volume of high-speed traffic on the major through route and with relatively light traffic on the minor approaches. Preference is given to traffic movement on the main road and the green aspect is transferred to the minor road only upon actuation of the detectors placed in the minor approaches. The length of the minor street green time varies with the traffic demand but cannot be extended beyond a maximum limit. When the required or maximum minor street phase has expired, the green aspect is automatically returned to the major street where it remains for at least a predetermined interval.

Linked signal systems

In dealing with the movement of heavy traffic volumes it is obviously desirable that the actions of the individual signals should be co-ordinated where possible. Proper co-ordination of signals can lead to more efficient traffic movement, less delay to vehicles at intersections, and an increase in the capacity of the linked route due to the ease with which vehicles can move through the linked intersections.

There are a great many ways of co-ordinating traffic signal operation. These can be classified into the four following main systems:

1. Simultaneous systems. 2. Alternate systems. 3. Progressive systems. 4. Area control systems.

Simultaneous system. In a simultaneous co-ordinated system all of the signals show the same indication to the same street at essentially the same time. At all intersections the signal timing is essentially the same and when orders are received from a master controller all indications change simultaneously, so that all signals show green to the major road traffic and red to all cross streets.

This is one of the early types of co-ordinated signal system, it being very easily adopted to usage with pretimed signals. It has limited applications in modern traffic signal practice because of the following disadvantages:

1. The simultaneous stopping of all traffic along the roadway prevents continuous movement of vehicles.

2. It promotes high speeds, especially during off-peak hours, as drivers attempt to pass as many intersections as possible while the lights are green.

3. The cycle length and green time allocations are usually governed by the requirements at one or two major intersections within the system; this often results in serious inefficiencies at the remaining intersections.

4. The net effect is low journey speeds and possibly a reduction in route capacity.

The simultaneous system can be very useful if only two adjacent intersections are to be linked. In this instance the green time given to the same major traffic flow during each cycle has to be relatively long so that there is ample time for a major portion of the main road traffic to clear through both intersections.

Alternate system. Under this type of operation all the signals are again operated by a master controller so that all the indications are changed simultaneously. It is different from the simultaneous system, however, in that adjacent signals or groups of signals show opposite aspects alternately along the major route. Typically the cycle lengths are the same at all signals, and the green and red periods within a cycle are also the same length. Thus, if the intersections are equi-distant apart, each block can be travelled in half the cycle time so that the driver meets a continuous green aspect along the route provided that he travels at the design speed.

With the irregular street systems prevalent in most British cities, the alternate signal system has limited possibilities for usage. Even in American cities, which tend to have gridiron road patterns, its use is limited to routes where the traffic on the cross streets is approximately equal in volume to the traffic on the through road. If the flows are unbalanced the result may be that the major route has too little green time and the minor roads have too much.

The alternate system does, however, limit speeding since motorists exceeding the design speed of the system have to stop at most of the intersections.

Flexible progressive system. This is the most satisfactory of the simple linking systems. The signals are so timed that the driver of a vehicle released by the green aspect at the first signal and proceeding along the major route at a predetermined speed will find that every signal changes from red to green on his approach. Thus, if he maintains the design speed, he can travel the entire length of the system without having to stop in front of one red aspect at any intersection.

Although the operation of the flexible progressive system requires that the cycle time for each intersection should be fixed, vehicle-actuated signals can also be used in the system. If vehicle-actuated signals are used, the local controllers at the individual intersections operate according to the overriding progressive plan while there are continuous demands from all detectors. When however the volume of traffic falls below a predetermined figure, each installation reverts to its original independent state and operates according to the individual requirements of each intersection.

Time-distance diagram. The phasing of the signals in a flexible progressive system can be determined with the aid of a time-distance diagram, a basic example of which is shown in Fig. 7.16. On this diagram the distances between intersections along the route to be controlled are plotted along the abscissa and the times along the ordinate axis. Ordinarily the problem is one of determining by trial and error, or mathematically, the best possible through-band speed and width for a given cycle length; in this simple example, however, it will be assumed that a design speed of 36·2 km/h has been decided on as being suitable.

First a sloping line is drawn from zero time starting at a point on the distance scale opposite the first intersection; this indicates the movement of the first vehicle released at the beginning of the green period and travelling north at the chosen speed. Since the selected speed is 36·2 km/h (10·06 m/sec), this line intersects a horizontal line drawn through the last intersection at 80 seconds from zero. The point where this line cuts the horizontal is taken as the beginning of a green period there also; the progress of the first vehicle travelling south at the same speed is then indicated by a second sloping line starting at this point and striking the ordinate axis at 160 seconds from zero. Under the progressive system the point where this north-to-south line strikes the axis is again the beginning of a main road green period, and so on. It therefore follows that this 160-second period should contain exactly one or more complete signal cycles.

In practice cycle lengths are rarely less than 25 seconds or more than 120 seconds. Short cycle lengths are undesirable because of possible insufficient running times as compared to stopping times. Often also, in urban areas, the minimum duration of the green period is governed by the length of time required by pedestrians to cross the road safely. A cycle time of more than 2

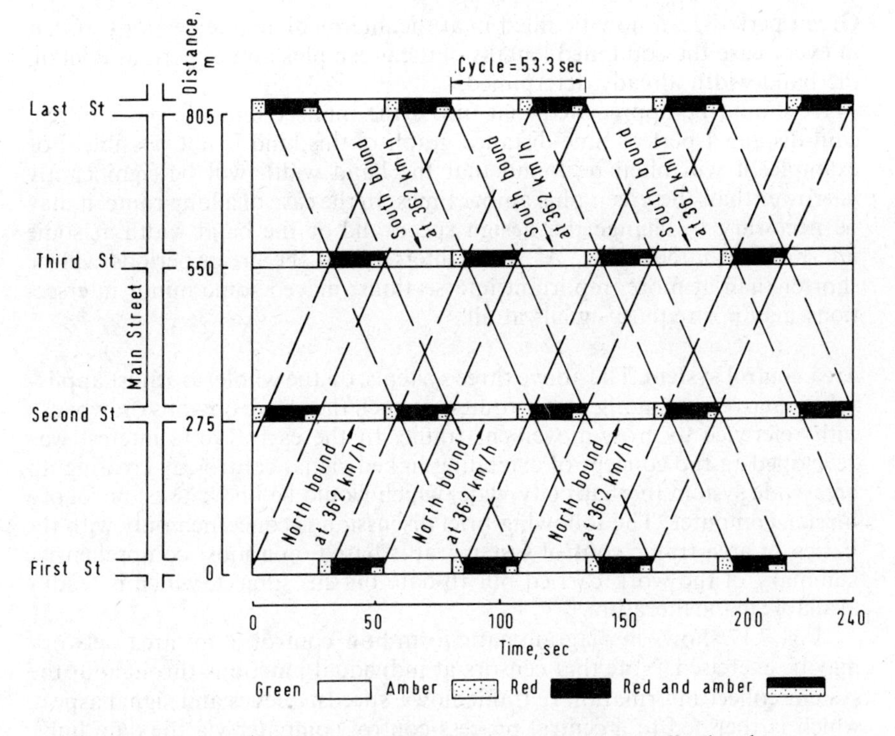

Fig. 7.16. Time-distance diagram for a linked system of traffic signals

minutes is rarely used as it may cause an undue accumulation of vehicles in the intersection approaches with resulting undesirable delays. In the example in Fig. 7.16, a cycle length of 53·3 seconds is arbitrarily chosen for illustrative purposes. Further northbound and southbound sloping lines can now be drawn from the ordinate axis at intervals of 53·3 seconds. Each of these represents the progress in each direction of the first vehicles to be released during all the green periods at the first and last intersections.

The conditions at each intersection must now be studied to determine the lengths of the green periods. In this instance let it be assumed that the conditions at the first and last intersections control the design, and these require that two-thirds of the cycle length should be in favour of the main road traffic and one-third in favour of the cross road traffic. Green and amber periods in accordance with this division are drawn on the horizontals from the two intersections, beginning at the progress line for the first vehicle in each direction. (If this is not exactly possible, the green periods should, anyway, include the moment at which the vehicles pass). Having filled in the green periods, the progress lines for the last vehicles in each direction can now be drawn through the end of the green periods. These lines are parallel to those already drawn, and the band between each pair represents the progress of a platoon of vehicles in each direction along the main road.

Green periods can now be filled in at the intermediate intersections so that in every case the combined lengths of the green plus amber periods contain the band-width already determined.

It should be appreciated that unless the intersections are more or less equi-distant a perfect time-distance graph of this kind is not possible. For example, it will often be found that the band width will be significantly narrower than the green plus amber times. In the case of a long route, it may be necessary to change the design speed and/or the band width at some intermediate intersection. At some intersections the green periods will be shorter than at more important intersections; indeed some minor intersections may not require signals at all.

Area control system. The above three systems, on the whole, are most applicable to movement along single routes; as such they have obvious drawbacks with reference to cross-movement traffic. In the early 1960's interest was developed in the concept of extending linked-signal control by creating an area-wide system (perhaps city-wide) which would be under the control of a master computer. The following brief discussion is concerned only with the basics of area traffic control systems; it is based on a most comprehensive summary of the work carried out to-date on this subject, which is readily available in the literature.(49)

Fig. 7.17 shows in diagrammatic form how control of an area network may be exercised. Note that censors at individual junctions throughout the system collect information *re* traffic flows, speeds, queues and signal aspect, which is then fed to a central process-control computer via the data links. This computer analyses the data and, with the aid of information from its store, provides commands which are routed back to the individual intersections to switch the traffic signals.

Control can be through the implementation of one or other of a number of overall plans that are prepared off-line in advance, and adopted without change for considerable numbers of signal cycle switches before the situation is re-assessed; only at the end of the predetermined cycles are the average values of current input data compared with the set of control plans, and the most appropriate one selected (see Section (a) of the Store in Fig. 7.17). Alternatively, the control may occur as a result of calculation carried out by the computer on receipt of each batch of fresh data at successive update intervals; typically these calculations would determine whether an immediate signal switch should take place, or whether a decision should be deferred until the calculation following the next up-date interval. With this latter type of control, cycle-length varies from one cycle to another, and greater responsiveness to traffic demands and flexibility of operation is achieved. The degree of flexibility is, of course, dependent on the type of algorithm employed (see Section (b) of the store in Fig. 7.17).

So far, area traffic control schemes have only been brought into use in a limited number of cities and all of these have been primarily regarded as experiments. In general, it is apparent that sufficient advantages have been derived from the area control schemes for them to be extended and for new

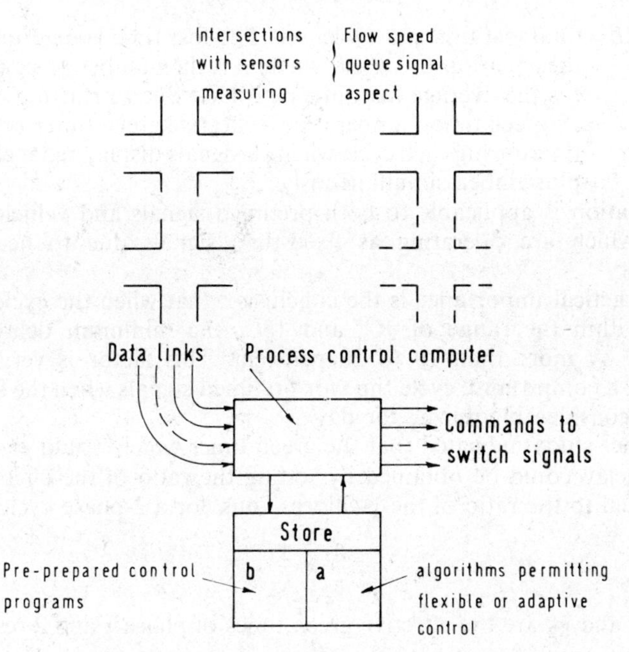

Fig. 7.17. Schematic of an area control sytem

ones to be implemented. In North America, however, the advantages accruing from computer control have been more spectacular than those in urban areas in Europe, i.e. fixed-time operation had been the practice there before, whereas traffic-actuated signals have been in use for very many years in Europe.

Consideration of present schemes also indicates that both the sampling frequency and the amount of data collected at each intersection will likely be reduced in the future. Furthermore, British results also suggest that increased sophistication in control policies may yield very little benefit over less complex ones.

Traffic signal settings

Studies[46] have shown that an individual intersection the cycle time for which the total delay at all approaches is a minimum is represented by

$$C_o = \frac{1{\cdot}5L+5}{1-y_1-y_2\ldots y_n}$$

where C_o = optimum cycle time, sec

$y_1, y_2, \ldots y_n$ = maximum ratios of actual flow to saturation flow for phases $1, 2, \ldots n$. (If the lost time and saturation flows are both different for the different approaches of the same phase, then each arm of the phase must be considered in turn as the 'predominant' one and the longest cycle deduced is then the optimum one.)

and L = total lost time per cycle, sec. The lost time is determined from the equation $L = nl + R$ where n is the number of signal phases, l is the average lost time per phase due to starting delays (i.e. l = green time + amber time − effective green time), and R is the time during each cycle when all signals display red including red plus amber, simultaneously.

This equation is applicable to both pretimed signals and vehicle-actuated signals which are operating as fixed-time signals due to heavy traffic demands.

Of practical importance is the conclusion that when the cycle length is varied within the range of $\frac{3}{4}C_o$ and $1\frac{1}{2}C_o$ the minimum delay is never exceeded by more than 10 to 20 per cent. This factor is very useful in deducing a compromise cycle time for pretimed signals when the traffic flow changes considerably during the day.

Further study indicated that the green times which would give the least overall delay could be obtained by setting the ratio of the effective green times equal to the ratio of the y-values. Thus, for a 2-phase cycle,

$$\frac{g_1}{g_2} = \frac{y_1}{y_2}$$

where g_1 and g_2 are the effective green times of phase 1 and 2 respectively.

Example of traffic signal setting determination. Measurements of actual flows and saturation flows at a particular 2-phase, 4-way intersection gave the following results:

	North arm	*South arm*	*East arm*	*West arm*
Flow, vehicles/h	600	450	900	750
Saturation flow, vehicles/h	2400	2000	3000	3000
Ratio, y	0·250	0·225	0·300	0·250
Design y-values	0·250		0·300	

Lost time: Starting delays = 2 seconds per phase (typical value).
All-red periods = 3 seconds at each change of right-of-way.
Red-amber periods = 3 seconds at each change of right-of-way.

PROBLEM. Determine the controller settings which will give the minimum overall delay to vehicles.

Solution. The total time per cycle when red or red-amber aspects are shown to all phases is 12 seconds. The total lost time is therefore

$$L = nl + R$$
$$= 2(2) + 12$$
$$= 16 \text{ seconds}$$

Then the optimum cycle time is obtained from

$$C_o = \frac{1{\cdot}5L+5}{1-y_1-y_2}$$

$$= \frac{1{\cdot}5(16)+5}{1-0{\cdot}250-0{\cdot}300}$$

$$= 64 \text{ seconds}$$

In one cycle, therefore, there will be $64-16=48$ seconds of total effective green time. This has to be divided between the North-South and East-West phases in the ratio

$$g_1 : g_2 :: 0{\cdot}250 : 0{\cdot}300$$

The effective green times of the N-S and E-W phases are therefore 22 and 26 seconds respectively. The controller settings then obtained by including the 2 seconds lost time in each phase are thus:

N-S phase = 21 seconds green + 3 seconds amber
E-W phase = 25 seconds green + 3 seconds amber

Queue lengths

In designing a signal control at an intersection it is important to know what extensions of the queue are likely to occur, especially if there are other intersections nearby. The queue at the beginning of the green period is usually the maximum in each cycle and its length has been given approximately as:

$$N = (\tfrac{1}{2}qr + qd) \text{ or } qr, \text{ whichever is the larger,}$$

where N = average number of vehicles in the queue,
r = red time, sec,
q = flow, vehicles/sec,
and d = average delay per vehicle, sec.

The average delay per vehicle on any particular arm of the intersection can be expressed to a close approximation by the equation:

$$d = \frac{c(1-\lambda)^2}{2(1-\lambda x)} + \frac{x^2}{2q(1-x)} - 0{\cdot}65\left(\frac{c}{q^2}\right)^{1/3} x^{(2+5\lambda)}$$

where d = average delay per vehicle, sec,
c = cycle time, sec,
λ = proportion of the cycle time that is effectively green for the arm of the intersection under consideration
$= g/c$,
q = flow, vehicles/sec,

and x = degree of saturation. This is the ratio of the actual flow to the maximum possible flow under the given signal settings. Thus $x = qc/gs$, where s is the saturation flow in vehicles per second, and g = effective green time.

The last term in the above equation gives a value which is between 5 and 15 per cent of the average delay in cases. As a result the following equation is often used in practice to obtain an approximation to the delay:

$$d = \frac{9}{10}\left[\frac{c(1-\lambda)^2}{2(1-\lambda x)} + \frac{x^2}{2q(1-x)}\right]$$

CLEARWAY SURVEILLANCE SYSTEMS

A special traffic management problem arises on certain heavily-travelled roads in urban areas. These roads may have narrow central reservations and no hard shoulders at all but yet are designed for or expected to carry very heavy volumes of traffic. Any blockage of one or more lanes, whether as a result of an accident, mechanical breakdown, or simply from an excess of traffic, can quickly result in a long queue of vehicles. For example, with a flow of 1500 p.c.u./h per lane a queue of 1 km in length could build up in about 7·5 min. Hence it is essential that a early notification be obtained of any abnormal traffic conditions so that prompt action can be taken to deal with the cause of the trouble and to prevent additional traffic from pouring on to the blocked road.

A basic example of a surveillance system which automatically provides this information is that used on a 3·2 km section of the M4 motorway westwards from the Chiswick Flyover; this was the first example of an urban motorway in this country.[47] This dual two-lane length is on a viaduct above the existing A4 highway and it has no shoulders for emergency use.

The aim of this surveillance system is to give early warning to a central traffic management station of any detrimental change in the traffic flow and to illuminate signs—either automatically or under police control—which will prevent additional traffic from entering the affected section. The system is based on 'presence' detectors placed approximately 365 m apart on the motorway. The information from these detectors is transmitted to the central control centre over channels in a frequency multiplex system. Diversionary signs are sited at various positions on the urban motorway and on the all-purpose road network in the vicinity of the motorway at convenient points for traffic diversion.

PRICING CONTROL OF TRAFFIC MOVEMENT

Britain is predominantly an urban-based nation and as a result it is in the urban areas that the impact of the motor vehicle is most felt. Much of the present congestion in these areas is due to the inability of the inherited street system to handle the ever-increasing traffic demands—especially during the peak periods of the day. Because of this the traffic management features described in this chapter have been and are being utilized in urban areas in order to improve traffic conditions on the existing road network. It is being increasingly recognized, however, that notwithstanding the advantages gained by using management techniques, some other and more drastic approach to the traffic problem in built-up areas may have to be taken within the near future. One of the more imaginative but, nevertheless, realistic suggestions in this respect is that a pricing mechanism be created which would influence the usage of motor vehicles at particular locations. This suggested pricing mechanism is in the nature of a congestion tax, in that vehicle owners would pay higher charges or taxes when they used the more congested roads and lower charges when they used uncongested ones. Thus, in theory, only vehicles which must travel on congested roads would be willing to pay for the privilege of doing so, while other vehicles would be influenced not to make unnecessary trips on these facilities—and consequently public transport would be very considerably helped.

At the time of writing a pricing control mechanism of this type and for this purpose is not in use anywhere throughout the world. The concept has been outlined(48), however, and appears to be quite feasible.

Operational requirements

In order to be workable and to come close to attaining its objectives, any road-pricing mechanism must satisfy the following basic requirements:

1. The charges must be closely related to the use made of the roads.

This is probably the most important consideration since it ensures that people using the congested roads a great deal will pay more than people who use them only rarely. It can be achieved by making the charges proportional either to the time spent on the congested roads or to the distance travelled on them.

2. It must be possible to vary prices for different roads (or areas), at different times of the day, week or year, and for different classes of vehicles.

This criterion is also a most important one. Thus the heavy commercial vehicles which contribute most significantly to congestion might be charged very highly for peak-hour use of the road. Again the price might be varied so that there would be higher charges for all vehicles, say, during the weeks immediately preceding Christmas, which might encourage people to shop earlier and thereby spread the congestion load.

3. Prices must be stable and readily ascertainable by road users before they embark upon a journey.

The object of road pricing is to influence the decision of the motorist *before* entering the congested area. Hence automatic systems which vary the price according to the existing congestion would be of little value.

4. The method must be simple for road users to understand and the police (or traffic wardens) to enforce.

5. The system must be accepted by the public as fair to all.

6. Payment in advance must be possible, although credit facilities might also be permissible under certain conditions.

In view of the large amounts of money which would be involved, payment in advance would be the only practical way of ensuring the applicability of the pricing mechanism.

7. Any equipment used must possess a high degree of reliability.

Any kind of equipment used to operate the pricing mechanism must be designed to last many years under conditions of tough usage. Any meters attached to vehicles must be robust and reasonably secure against fraud.

8. The method must be amenable to gradual introduction commencing with an experimental phase.

9. It must be capable of being applied to the whole country and to a future vehicle population of 30 to 40 million. It must also allow for temporary road users such as those entering from abroad.

10. The pricing-control mechanism should, if possible, indicate the strength of the demand for road space in different places so as to give guidance to the planning of new road improvements.

Indirect methods of charging

There are four possible 'indirect' ways of charging for the use of roads in congested locations. These are, by means of a differential fuel tax, a parking tax, a system of differential licensing, and a poll tax.

Differential fuel tax. It is technically possible to levy fuel tax at different rates in different areas, relating it to the amount of congestion in each area. The effectiveness of the differential would depend on the opportunities and incentives for avoidance, and on its ability to discriminate in detail against journeys on congested roads. Avoidance by the motorist could occur in two ways: (*a*) Taking-on fuel while travelling on ordinary journeys through low-tax areas. (*b*) Making special fuel-fetching journeys to low-tax areas. Avoidance of type (*a*) could not be stopped and must be accepted as an inherent weakness of the method. Avoidance of type (*b*) could be minimized by the careful planning of area boundaries and differentials.

In this urban-based country the differential fuel tax scheme commends itself most obviously to conurbations, where a high tax could be levied throughout most of the built-up areas and then reduced gradually as one moved away from these areas, starting in the outer suburbs. However, unless the high tax areas are so large that it is possible to both penalise regular users and make special fuel-fetching journeys uneconomical, this method of control must be inefficient, as well as discriminatory to such businesses as garages.

Parking tax. Parking restrictions and regulations can also be a means of reducing traffic congestion on the roads, since they can increase the difficulty, by putting up the price, of making a journey. Using parking charges to control congestion implies that the parking space charges would be raised to a level which would *considerably* lessen the number of journeys made in the congested areas. The aim of the parking tax would not be simply to force a certain amount of traffic away from the road, but rather to eliminate traffic that it is not prepared to pay the real cost of its being there.

A parking tax can give some beneficial relief to traffic congestion on the roads, but it can never be regarded as an entirely efficient solution. Broadly there are three main deficiencies associated with it. First of all, traffic which does *not park* within the congested area is not only unaffected by the parking tax but is actually encouraged by whatever reduction the scheme extracts from other traffic. In other words, traffic wishing to stop in the area would be kept out in order to make way for traffic passing through the area. Secondly, a vehicle coming from outside the town and contributing to congestion all the way into the central area pays the same parking tax as a vehicle coming only a short distance and hence, because the tax forms a smaller proportion of its total journey cost, it is less likely to be deterred from making the trip. In other words, the traffic that causes least roadway congestion is kept out in order to make way for the traffic that causes most congestion. Thirdly, a parking tax that is likely to be efficient in the peak hours is likely to be too severe in the off-peak hours.

Differential licences. Under the differential licensing scheme the road user would require a specially priced licence in order to drive within a particular congested area. The licences could be varied in two ways: (*a*) By area, with differentially priced licences for different areas. (*b*) By time, with different licences for different times of the day.

Area variations would require congested areas to be divided into zones of different classes such that, for example, areas of heavy congestion could be designated as 'red' zones, areas of moderate congestion 'blue', and less congested areas 'yellow'. Thus a relatively cheap yellow licence would provide access to yellow zones, a more expensive blue licence to blue and yellow zones, and a still more expensive red licence to all three classes of zone.

Time variations would distinguish between day and night usage, and possibly between peak and off-peak usage. A zone might be declared 'red' at the peaks, 'blue' during daytime off-peak hours and 'yellow' at night. Only vehicles holding the required licence colour would then be able to enter these zones during these times.

To be effective a system of *annual* differential licensing for a congested area would have to be expensive—£100 a year (1963 prices) might be the level required—and transferable so that people could obtain licences for a short time. The high value of the licence could give rise to problems of theft, fraud and finance. A much greater difficulty is the technical problem of enabling the general public to hire the licences for limited periods of time.

There are no purely technical difficulties associated with the introduction of a system of *daily* differential licensing. The licences could be numbered in large figures or letters to denote the day and sold in two kinds of books. The first of these books might be termed a daily book since it would consist of a number of licences for a given day; this would be for retailers of licences such as garages. The second type of book would contain one licence for each day of the month or year; this book would be for the use of individual vehicle owners. In both cases a rebate would be given for licences not detached from the book.

The main advantage of the daily licence is that it is simple and should not be difficult to implement. In addition it has the advantage over the parking tax of being applicable to all traffic entering the congested area, and not just those parking. It has the disadvantages that not only does the short journey pay the same as the long journey, but through traffic with an expensive licence would be encouraged to use zones where congestion has been decreased, thereby diminishing the apparent fairness and reasonableness of the system.

Poll tax. It has been suggested that congestion would be reduced if a poll tax was levied on all employees in congested areas. The concept envisages that organizations with large numbers of employees would either move out of these areas and be replaced by organizations with fewer employees, or that the tax would encourage further mechanization of labour such as lift-operating, book-keeping, dictation and retail service.

There are so many complex variables associated with the introduction of a poll tax that it is difficult to examine it objectively. On balance however, the effect of a poll tax on the volume of road traffic would only be of minor value. This conclusion is based on the premise that the people most affected by the tax would be in the lower income groups and these groups have the lowest level of car ownership.

Direct methods of charging

Whereas the indirect methods of charging rely on a tax on some product or service that is more or less closely correlated with road usage, the methods now to be discussed all make a direct charge to the motorist for usage of the congested roadway. The simplest example of direct charge control is the toll-gate which is used on bridges and tunnels and on some (foreign) motorways with few points of access. Although simple in concept, toll-gates cannot be considered for usage on ordinary roads in urban areas because of their cost and impedance to the flow of traffic. Direct methods that might be possible are off-vehicle recording systems and vehicle metering systems.

Off-vehicle recording systems. These systems are analogous to telephone charging methods in that every vehicle would be fitted with a piece of equipment which can be automatically identified and recorded by detectors placed in, over or beside the roadway at suitable pricing locations. The recording that a vehicle has passed a certain pricing location then, in theory, could be used in one of two ways. The first, termed 'point pricing', would involve setting up pricing points within congested areas so that vehicles can be debited with the appropriate charge when passing any pricing point. The alternative, 'continuous pricing', would involve setting up pricing points on the borders of congested areas and charging according to the time spent in the zone, as deduced from recordings at the zone entry and exit locations.

The utilization of any continuous pricing system cannot be considered practical for two reasons. First, there is the considerable technical problem of monitoring private entrances and parking places within pricing zones so that charging will either cease or be reduced when vehicles enter parking places. Secondly, there is the intolerable error associated with a vehicle being recorded as entering a zone and its corresponding record of exit being missed or mislaid; this could result in a vehicle being charged for an indefinite period.

It appears therefore that any feasible off-vehicle system could work only on a point-pricing basis. Such a system might comprise the following stages:

1. *Identification of vehicles at the pricing points.* Vehicles would have to carry identification units which would enable their presence to be detected and recorded by roadside apparatus. In addition the system would have to be capable of distinguishing between perhaps up to 40 million different identities. On technical grounds, these requirements could probably be met, but only at a relatively high cost.

2. *Transmission of information to a central computing station.* The simplest method would be to accumulate the records at each pricing point on magnetic tape and then transport these to the central station at regular intervals.

3. *Processing of data.* All charges would have to be sorted by vehicle identification number, and charges multiplied by an appropriate charging factor such as 1 for a private car, 2 for a commercial vehicle, etc. Bills could be issued at monthly or quarterly intervals.

4. *Collection of payment.* The difficulties and costs of bill collection would need careful study. If there were 10 million accounts and each year one account in a thousand failed to pay after the usual reminders, there would be some 10 000 cases a year to follow up, many of them through the courts. While some protection could be obtained by making renewal of vehicle licences conditional on all road charges being paid many difficult cases would still remain; in addition, it is perhaps sociologically undesirable to allow large numbers of people to become debtors for the considerable amounts which could be involved.

The advantages of a fully automatic system of road pricing with ample price flexibility, in which the road user has no more to do than periodically send a payment cheque, are obvious. However, the disadvantages raised in the above discussion suggest its impracticality. In addition, the utilization of any such system is a definite threat to the privacy of road users insofar as it would be possible for public authorities to automatically trace the vehicle movement of perhaps the entire populace. Notwithstanding the advantage of such a system in aiding for instance, crime detection, this would appear to be such an invasion of privacy as to render its usage most undesirable in any democratic community.

Vehicle metering systems. The on-vehicle metering systems can be divided into those that are driver-operated and those that are automatically operated.

Driver-operated meters. Systems using these meters could only work on a continuous charging basis. Zones would have to be defined in congested areas and then allocated colours which would indicate the prices charged. The colours, which would be displayed electrically at all points of entry and exit, would have to be capable of being switched so that, e.g., during off-peak hours or Sundays or public holidays, an expensive zone could be derated or perhaps dezoned entirely.

Every vehicle entering a congested zone would carry a road meter displayed on the windshield and within comfortable reach of the driver's hand. When the meter was switched on and the timing mechanism working it would show a coloured light, indicating the rate being charged. Thus a driver would switch to purple when entering a purple zone, and if he later drove into a pink zone he would switch his meter accordingly. One or two lower rates for parking could be added, thereby providing each vehicle with what is in effect a personal parking meter and removing the need for footpath meters. Different types of vehicle would carry meters with different charging rates.

The method by which the motorist would pay for his use of the roadway could vary with the type of timing mechanism in the meter. If a clockwork mechanism were used, the motorist would purchase a complete meter which, as it was used, would run down at the predetermined rates. When the meter expired, it would be returned to an authorized station and a complete new one purchased. If an electrical timing mechanism were used, it would not be necessary to obtain a complete new meter whenever the meter time ran out. Instead a cylindrical 'throw-away' electrolytic timer could be inserted into the meter by the driver, to form part of a circuit connecting the meter lamp with the vehicle battery. Also included would be a network of resistors so that the timer could run down at the correct rates. When the timer reached its designed end-point, the circuit would be broken and the meter lamp cut out. The timer could then be redrawn and a new timer easily inserted.

The meter with the disposable electrolytic timer is to be preferred to the mechanical one. Not only is it cheaper but it considerably reduces difficulties of payment. Timers could be bought over the counter from garages or post offices in units of, say, £1, £2 or £5, thereby eliminating the need for the meter to be read or removed for resetting. The revenue due to the government could be collected as an excise duty from the timer manufacturers.

Automatic meters. The principal drawback associated with the use of the driver-operated meter is that it adds to the responsibilities of both the motorist and the traffic authority. The automatic meter attempts to eliminate this by the placement of a control apparatus in the highway pavement which would activate and switch meters installed in passing vehicles. These meters could be used to operate within the point pricing and continuous pricing systems.

Point-pricing systems. In this type of system the meter carried by the vehicle would be basically an instrument capable of counting electrical impulses generated by electrical cables carrying very low currents and laid in the roadway at fixed pricing points. The vehicle meter could be a small unit, probably of the size and shape of a small book. Since it would have to be near the ground in order to pick up the signals, the meter could form a part of the vehicle number plate. In its simplest form, this meter would probably be in the form of a 'solid-state' counter, of the form used in computers. Although it could not be read in the usual way, the meter could be made to change colour to show how its economic life was expiring.

Two methods of payment suggest themselves for use under point-pricing operation. Either the meter could contain a given economic capacity so that a new one would have to be bought when this was exhausted, or else a credit meter could be issued that would be capable of being fixed permanently to the vehicle and then the vehicle would have to be taken at intervals to registered stations so that the meter could be read and paid for.

The automatic meter point-pricing system is essentially a sophisticated form of toll-gate control. Its principal advantage is that highly flexible pricing—and hence traffic control—is possible due to the feasibility of varying the price from point to point while the points themselves can be deployed as densely as required within the control area. A disadvantage is that unless the pricing points are spread in very large numbers they can to some extent be avoided by the irresponsible motorist. In addition, the fewer there are the less accurately do they relate to costs and the more arbitrary are their effects upon traffic.

Continuous pricing systems. Under automatic meter-operated systems of continuous pricing, vehicles would be charged continuously while within pricing zones, the charges commencing when the vehicles enter a zone and ending when they leave. The main difference between this system and the driver-operated one is that, whereas in the latter the motorist sets his meter to the appropriate rate, under the automatic system this function would be performed by a switching circuit which would operate in response to signals received from road-sited transmitters installed at the entry and exit points and, if required, at intermediate points within the zone. A manual over-ride control would also be provided to enable the motorist to select the parking rate when the vehicle is stopped.

SELECTED BIBLIOGRAPHY

1. BRITISH ROAD FEDERATION. *Basic Road Statistics.* London, The Federation, 1972.
2. DAWSON, R. F. F. The Cost of Road Accidents in 1970. *RRL Report* LR 396, Crowthorne, Berks., The Road Research Laboratory, 1971.
3. SMEED, R. J. and JEFFCOATE, G. O. Effects of changes in motorisation in various countries on the number of road fatalities, *Traffic Engineering and Control*, 1970, **12,** No. 3, 150–151.
4. JOHNSON, H. D. and GARWOOD, F. Notes on Road Accident Statistics. *RRL Report* LR 394, Crowthorne, Berks., The Road Research Laboratory, 1971.

5. CHARLESWORTH, G. Design for safety, *Proc. Conference on Engineering for Traffic*. London, Printerhall, 1963.
6. TANNER, J. C. Accidents at rural three-way intersections, *J. Inst. Highway Engrs*, 1953, **2,** No. 11, 56–57.
7. COLGATE, M. G. and TANNER, J. C. Accidents at Rural Three-Way Junctions. *RRL Report* LR 87, Crowthorne, Berks., The Road Research Laboratory, 1967.
8. WEBSTER, F. V. and R. F. NEWBY. Research into relative merits of roundabouts and traffic signal controlled intersections, *Proc. Inst. Civ. Engrs*, 1964, **27,** 47–75.
9. BOX, P. C. and ASSOCIATES, Ch. 4: Intersections. In MAYER, P. A. (Editor), *Traffic Control and Roadway Elements—Their Relationship to Highway Safety*, Washington, D.C., The Highway Users Federation for Safety and Mobility, 1970.
10. CODLING, P. J. Thick fog and Its Effect on Traffic Flow and Accidents. *RRL Report* LR 397, Crowthorne, Berks., The Road Research Laboratory, 1971.
11. SABEY, B. E. and STORIE, V. J. Skidding in Personal Injury Accidents in Great Britain in 1965 and 1966. *RRL Report* LR 173, Crowthorne, Berks., The Road Research Laboratory, 1968.
12. SABEY, B. E. The Road Surface in Relation to Friction and Wear of Tyres. *Road Tar*, 1969, **23,** No. 1, 15–21.
13. STARKS, H. J., F. GARWOOD, G. O. JEFFCOATE and R. J. SMEED. Research into highway traffic accidents, *Traffic Engineering and Control*, 1961, **3,** No. 8, 493–498.
14. CLEVELAND, D. E. Ch. 3: Illumination. In MAYER, P. A. (Editor), *Traffic Control and Roadway Elements—Their Relationship to Highway Safety*, Washington, D.C., Automotive Safety Foundation, 1969.
15. BRITISH MEDICAL ASSOCIATION. *Relation of Alcohol to Road Accidents*. London, The Association, 1960.
16. SMEED, R. J. Methods available to reduce the numbers of road casualties, *Intern. Road Safety and Traffic Review*, 1964, **12,** No. 4.
17. BEAUMONT, K. and NEWBY, R. F. Traffic Law and Road Safety Research in the United Kingdom—British Counter Measures. Paper presented at the National Road Safety Symposium held at Canberra, Australia on 14–16 March, 1972. To be published.
18. TILLMAN, W. A. The psychiatric and social approach to the detection of accident-prone drivers. M.S. Thesis, University of Western Ontario. London, Ontario, 1948.
19. JOHNSON, H. D. Ages of Car Driver Casualties in 1970. *RRL Report* LR 431, Crowthorne, Berks., The Road Research Laboratory, 1972.
20. COLBORNE, H. V. and SARGENT, K. J. A Survey of Road Safety in Schools: Education and Other Factors. *RRL Report* LR 388, Crowthorne, Berks., The Road Research Laboratory, 1971.
21. CLEVELAND, D. E. *Manual of Traffic Engineering Studies*. Washington, D.C., Institute of Traffic Engineers, 1964.
22. GARWOOD, F. and R. L. MOORE. Pedestrian accidents, *Traffic Engineering and Control*, 1962, **4,** No. 5, 274–276, 279.
23. ROAD RESEARCH LABORATORY. *Research on Road Traffic*. London, H.M.S.O., 1963.
24. JACOBS, G. D. and WILSON, D. G. A Study of Pedestrian Risk in Crossing Busy Roads in Four Towns. *RRL Report* LR 106, Crowthorne, Berks., The Road Research Laboratory, 1967.
25. CLEVELAND, D. E. Speed and speed control. In MAYER, P. A. (Editor), *Traffic Control and Roadway Elements—Their Relationship to Highway Safety*, Washington, D.C., Highway Users Federation for Safety and Mobility, 1970.

26. DUFF, J. T. Road accidents in urban areas, *Journal of the Institution of Highway Engineers*, 1968, **15,** No. 5, 61–69.
27. NEWBY, R. F. Effectiveness of speed limits on rural roads and motorways, *Traffic Engineering and Control*, 1970, **12,** No. 8, 424–427, 431.
28. INSTITUTE OF TRAFFIC ENGINEERS. Pedestrian overcrossings—Criteria and priorities, *Traffic Engineering*, pp. 34–39, 68, Oct. 1972.
29. RUTLEY, K. S. Advisory speed signs for bends, *TRRL Report* LR 461, Crowthorne, Berks., The Transport and Road Research Laboratory, 1972.
30. O'FLAHERTY, C. A. and COOMBE, R. D. Speeds on level rural roads—A multivariate approach, Pt. 3, *Traffic Engineering and Control*, 1971, 13, No. 3, 108–111.
31. O'FLAHERTY, C. A. Delineating the Edge of the Carriageway in Rural Areas. *Printerhall Ltd.*, Morley House, 26/30 Holborn Viaduct, London EC1A 2BP, 1972. 42pp. Also (in abridged form) in *Research and Development of Roads and Road Transport 1971*, Washington, D.C., The International Road Federation, 1971. pp. 403–437.
32. MORSE, G. and MORGAN, E. J. Highway Crash Barriers. *IPC Building and Contract Journals Ltd*, 32 Southwark Bridge Road, London, S.E.1., 1971. 22pp.
33. NEWBY, R. F. and H. D. JOHNSON. London–Birmingham motorway accidents, *Traffic Engineering and Control*, 1963, **4,** No. 10, 550–555.
34. MINISTRY OF TRANSPORT. *Urban Traffic Techniques*. London, H.M.S.O., 1965.
35. DUFF, J. T. Traffic management, *Proc. Conference on Engineering for Traffic*, London, Printerhall, 1963.
36. DUFF, J. T. One-way streets, *Traffic Engineering and Control*, 1963, **4,** No. 9, 518–520.
37. THOMSON, J. M. The value of traffic management, Journal of Transport Economics and Policy, 1968, **2,** No. 1, 3–32.
38. EVANS, H. K. *et al. Traffic Engineering Handbook*. New Haven, Conn., Institute of Traffic Engineers, 1950.
39. WORBOYS, Sir W. *et al. Traffic Signs: Report of the Committee on Traffic Signs for All-Purpose Roads*. London, H.M.S.O., 1963.
40. CHRISTIE, A. W. and K. S. RUTLEY. Relative effectiveness of some letter types designed for use on road traffic signs, *Roads and Road Construction*, 1961, **39,** No. 464, 239–244.
41. ODESCALCHI, P. Conspicuity of signs in rural surroundings, *Traffic Engineering and Control*, 1960, **2,** No. 7, 390–393, 397.
42. MATSON, T. M., W. S. SMITH and F. W. HURD. *Traffic Engineering*. New York and Maidenhead, McGraw-Hill, 1955.
43. *Traffic Signs Manual* (and *Supplements*). London, H.M.S.O.
44. *Manual of Uniform Traffic Control Devices for Streets and Highways*. Washington, D.C., U.S. Government Printing Office, 1971.
45. BIGGS, N. L. Directional guidance of motor vehicles—A preliminary survey and analysis, *Ergonomics*, 1966, **9,** No. 3.
46. WEBSTER, F. V. Traffic signal settings, *Road Research Technical Paper* No. 39. London, H.M.S.O., 1958.
47. DUFF, J. T. Traffic engineering techniques, *Proc. Conference on Transportation Engineering*. London, Institution of Civil Engineers, 1965.
48. SMEED, R. J. *et al. Road Pricing: The Economic and Technical Possibilities*. London, H.M.S.O., 1964.
49. HARTLEY, M. G. A systems approach to area traffic control, *Transportation Planning and Technology*, 1972, **1,** No. 1, 11–26.

8 Highway lighting

The single greatest justification for installing lighting facilities on roadways is the consequent reduction in night accidents. Numerous investigations which have been carried out both in this country and abroad have shown clearly that there are significant reductions in fatalities, personal injuries and property damages wherever and whenever non-motorways are properly lighted. In fact, analysis of accident studies suggests that up to half of the nightly road deaths and accidents might be avoided if adequate lighting was provided at the following critical areas of driver decision:[1] 1. Entrances and exits. 2. Interchanges and intersections. 3. Bridges, overpasses and viaducts. 4. Tunnels and underpasses. 5. Guide sign locations. 6. Dangerous hills and curves. 7. Heavily travelled sections in urban and rural areas, or where entrances, exits or interchanges occur within 1·6 km or less of each other. 8. Rest areas and connecting roads. 9. Railroad grade crossings. 10. Elevated and depressed roadways.

The installation of a proper road lighting system provides other fringe benefits which are not always appreciated. For instance, traffic flow during evening peak periods and at night is considerably improved, since drivers are made more confident in their movements and can more easily observe traffic management measures. The recent growth in late evening shopping is partly due to the commercial centres being attractively and well lighted. In addition, there is no doubt but that the provision of adequate street lighting is a powerful weapon in the fight against crime in urban areas.

In the limited space available here it is not possible to discuss all the possible lighting designs at the critical roadway areas listed above, so the approach taken in this chapter is to concentrate on the basic features which underlie the rational design of lighting installations on major traffic routes, i.e. the emphasis is laid on the principles rather than the practice of road lighting.

TERMINOLOGY

Before examining the basic concepts and considerations underlying the design of road lighting systems, it is necessary to explain some of the terms in common usage. These may be divided into two major groups: A. The *photometric* terms, which relate to light and its measurement units. B. The lighting *installation* terms, which relate to the physical equipment and its layout on the roadway.

Photometric terms

The following definitions have been made as simple as possible rather than absolutely precise and the units and abbreviations are those used in Great Britain.[2]

Luminous flux. This is the light given by a light source or received by a surface irrespective of the directions in which it is distributed. The unit of luminous flux is the lumen (lm); this is the flux emitted through a unit solid angle (a steradian) from a uniform point source of one candela. A steradian is equal to the solid angle subtended at the centre of a sphere by a unit area of its surface.

Luminous intensity. This describes the light-giving power (candlepower) of a lantern in any given direction. The unit of luminous intensity is the candela (cd); this is an international standard and is related to the luminous intensity of a 'black body' (one which absorbs all radiation incident on it) at the temperature of freezing of platinum.

Lower hemispherical flux. By this is meant the luminous flux emitted by a light source in all directions below the horizontal.

Illumination. This is the luminous flux *incident* on a surface per unit area. The unit of illumination is the lumen per square metre (lm/m^2) or lux.

Luminance. The term luminance is now used instead of the technical term 'brightness' to describe the rate at which light is reflected from a unit projected area of an illuminated surface in a given direction; it is the luminous intensity per unit projected area of the surface. (Whereas illumination is a measure of the amount of light flux falling on a surface, luminance is a measure of the amount of light which the area reflects towards the eye of the observer. In other words, the luminance varies not only with the amount of light reaching the surface but also in the manner in which it is reflected.) The usual unit is the candela per square metre (cd/m^2). Thus, if a very small portion of an illuminated surface has an intensity I candelas in a particular direction, and if the projection of the surface on a plane perpendicular to the given direction has an area D m^2, then the luminance in this direction is I/D cd/m^2.

Luminosity. Sometimes loosely called 'brightness', this is the visual *sensation* indicating that an area appears to emit more or less light; it correlates approximately with the term luminance. It is not measurable.

Mean hemispherical intensity. This is the downward luminous flux divided by 2π; it is the average intensity in the lower hemisphere.

Peak intensity ratio. The ratio of the maximum luminous intensity to the mean hemispherical intensity is called the peak intensity ratio.

Beam. This is the portion of the luminous flux emitted by a lantern which is contained by the solid angle subtended at the effective light centre of the lantern containing the maximum intensity but no intensity which is less than 90 per cent of the maximum intensity.

Installation terms

Lighting installation. This term refers to the entire equipment provided for lighting a highway section. It comprises the lamps, lanterns, means of support and the electrical and other auxiliaries.

Lighting system. This is an array of lanterns having a characteristic light distribution. Systems are commonly designated by the name of the light distribution, i.e. cut-off, semi-cut-off, non-cut-off.

Lamp. The bulb or light source is called the lamp.

Lantern. The lantern consists of the lamp together with its housing and such features as refractors, reflectors and diffusers which are integral with the lamp and housing.

Outreach. This is the horizontal distance measured between the centre of a lantern mounted on a bracket and the centre of the supporting column or wall face.

Overhang. The overhang is the horizontal distance between the centre of a lantern and the adjacent edge of the carriageway.

Mounting height. This is the vertical distance between the surface of the carriageway and the centre of the lantern.

Spacing. By this is meant the distance measured parallel to the centre line of the carriageway between successive lanterns in an installation. The successive lanterns may or may not be arranged on the same side of the carriageway.

Arrangement. The pattern according to which lanterns are sited on plan is termed the arrangement. In a *staggered* arrangement the lanterns are located alternately on either side of the carriageway. When the lanterns are placed on either side but opposite each other, the pattern is called an *opposite* arrangement. With a *central* arrangement the lanterns are sited in an axial line close to the centre of the carriageway, while a *single-side* arrangement is one in which the lanterns are placed on one side only of a carriageway.

Width of carriageway. This is the distance between kerb lines as measured in a direction at right-angles to the length of the carriageway.

Effective width of carriageway. This is the width of carriageway which it is intended to make bright. For a single carriageway the relation between effective width and carriageway width is as follows:

Arrangement	*Effective width*
Central	Carriageway width
Opposite plus central	Carriageway width
Staggered	Carriageway width minus overhang
Single-side	Carriageway width minus overhang
Opposite	Carriageway width minus twice overhang

Span. The part of the highway which lies between successive lanterns is called the span.

Geometry. The geometry of a lighting system refers to the interrelated linear dimensions and characteristics of the system, i.e. the spacing, mounting height, effective width, overhang and arrangement.

BASIC CONCEPTS AND CONSIDERATIONS

Objective

The first step in the planning and design of any road-lighting system is the understanding that the primary purpose of the system is to promote the safe and efficient movement of pedestrian and vehicular traffic. Ideally this means that the visibility provided should be equivalent to that in daylight so that the motorist can drive without the aid of headlights and still not impair the safety of other road users. Practically, this is achieved, not by imitating daylight, but by making the maximum use of available light to provide the revealing power or 'visibility' by which the road users can proceed in safety.

Visibility

From the above it can be seen that the basic objective of road lighting is different from that of interior lighting. With interior lighting, the aim is to reproduce daylight as closely as possible so that the forms and textures of objects are clearly seen. Thus interior lighting requires not only high but also even illumination. In road lighting these detailed qualities are unimportant; what is important is that the motorist should be able to discern clearly the presence and movements of any object on or adjacent to the roadway which may be a potential hazard. This is achieved not by having an even illumination on the road and its surroundings but by appearing to have an even luminance on the road surface *as it is seen by the motorist*.

Means of discernment. An object is visible to a driver if there is sufficient contrast of luminosity or colour between the object and its background or between different parts of the object. This implies that discernment is either by (a) silhouette, (b) reverse silhouette or (c) surface detail.

When an unlit object on the carriageway is discerned by a driver because its luminosity is less than its background, then it is said to be seen by silhouette. It might perhaps be expected that road lanterns would make dark objects bright—as do vehicle headlights—but in fact the usual effect is to make the road surface more luminous and the object is seen in dark silhouette. It is only occasionally that the luminosity of the object is greater than that of the surface and then visibility is by means of reverse silhouette. This occurs, for example, when a pedestrian in white clothing stands just behind a lantern and receives a very high amount of illumination on the side of him which faces the oncoming driver. When a high amount of illumination is directed on the side of an object facing the motorist but discernment is by means of variations in luminosity within the object itself rather than in contrast with its background, then discernment is said to be by surface detail. This is the principle underlying the design of illuminated roadside traffic signs.[3]

Light distribution on the carriageway

When designing a lighting system on a major traffic route the visibility criterion is that of discernment by silhouette. In Britain the general aim in urban areas is to provide a sufficient contrast between the object and the carriageway so that the results in most situations are at least adequate for safe driving at about 48 km/h without headlights. On residential and other 'non-traffic' routes the motorist is expected to use his headlights to help achieve the desirable level of visibility; on these roads the lighting is intended to suit the needs of the pedestrian rather than the needs of the motorist.

If the desired contrast is to be obtained on the major traffic routes, then the power and geometry of a lighting system must be such that it appears to the driver that the carriageway has high luminosity.

The bright patch. Each lantern in a lighting installation contributes a single 'bright patch' on the carriageway and in the ideal case the lanterns are sited so that the patches link up to cover the entire road surface. In general the shape and luminance of this bright area depends on the following six main factors:

1. The reflection properties of the surface.
2. The distribution of light from the lantern.
3. The power of the lantern.
4. The height of the light source.
5. The distance of the light source from the observer.
6. The height of the observer.

It is easiest to compare the reflection characteristics of carriageways by considering two extreme types of 'surfacings'. Let these be a near perfect diffuser and a mirror. Assuming that the light source has a uniform luminous intensity in all directions (which in practice it does not have), then the contours of equal illumination on the road are concentric circles. If the carriageway is very diffuse, then the ratio of luminance to illumination is a constant, the contours of equal luminance coincide with those of equal illumination, and the bright patch produced on the road is nearly a perfect circle. To the motorist, this circle appears as the narrow 'ellipse' on the road surface just beneath the lantern. On the mirror surface, however, the bright patch would appear simply as the image of the light source itself just in front of the vehicle.

In practice, no dry road surface gives either a mirror image or a perfect ellipse. Instead, as is indicated in Fig. 8.1, a bright area or 'patch' is obtained which is roughly T-shaped with the head of the T stretching across

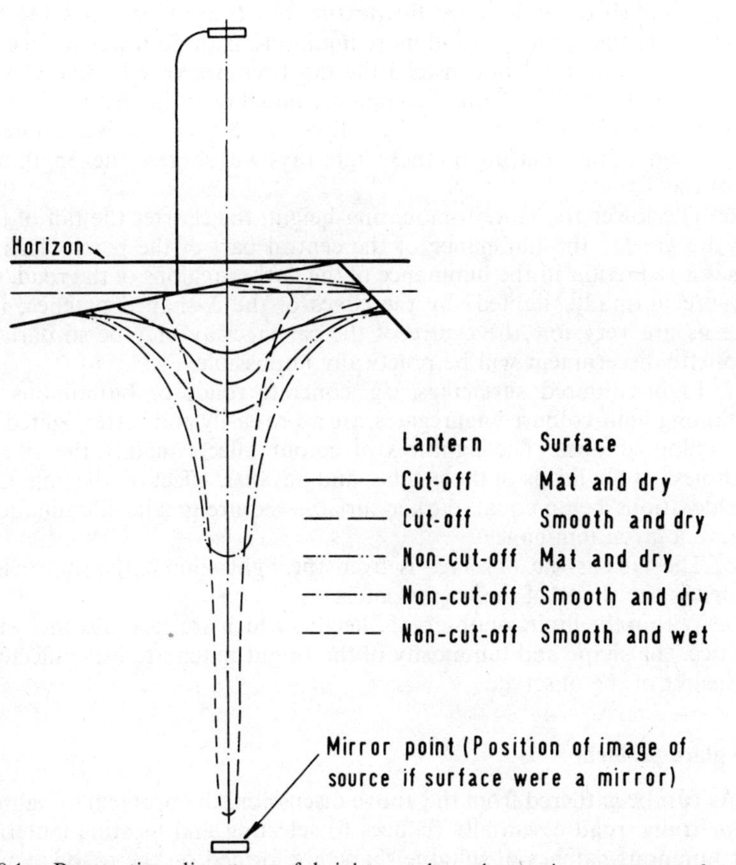

Fig. 8.1. Perspective diagrams of bright patches shown in contours of equal luminance formed on the carriageway by a single lantern, according to the type of its light distribution curve and the nature of the road surface (After ref. 4)

the road but reaching only a short distance behind the lantern, and the tail extending towards the observer. This T-shape has a number of very interesting characteristics:

1. The tail of the T always extends towards the observer, wherever he may be. This is due to the preferential reflection properties of the carriageway which only enable it to reflect the light back to the observer which reaches the surface in the direction *toward* him.

2. If the carriageway adjacent to a light is viewed from two different locations, then two different light patches will be seen.

3. The point of maximum luminous intensity occurs neither in the head of the T nor at the end of the tail, but somewhere in between.

4. The shape of the luminous area depends very much on the reflection characteristics of the road surface. A rough-textured surface gives a patch in which the head predominates and the tail is very small; as the road surface

becomes polished, or is more fine-textured, the head becomes less pronounced and the tail longer and more luminous. A smooth wet surface produces hardly any head but instead the tail becomes very luminous and is long and thin, e.g. the 'streaks' commonly noted on a flooded road surface.

5. Since the tail is formed by light rays which leave the lantern near to the horizontal, then cutting off these light rays will shorten the length of the tail of the T.

6. The lower the lantern-mounting height, the shorter the tail of the T and the greater the luminance of the central part of the patch. This also causes a reduction in the luminance of the darker regions of the road, since they are normally 'lighted' by the edges of the T-shaped patches; if the lanterns are very low the centre of the carriageway may be so dark that silhouette discernment will be practically impossible.

7. Light-coloured surfacings, e.g. concrete roads or bituminous ones containing light-coloured aggregates, are more easily and better lighted than dark-coloured ones. The lightness of colour affects mainly the size and brightness of the heads of the patches and has little effect on the tails. Other considerations being equal, darker surfaces require greater illumination to achieve a given luminance.

8. The farther the observer is from the light source, the more clearly defined is the shape of the bright patch.

9. Within the limits of observers' heights which are generally met with in practice, the shape and luminosity of the bright patch are little affected by the height of the observer.

The glare problem

As can be gathered from the above discussion, the problem of lighting a major traffic road essentially reduces to selecting and locating lanterns so that luminous patches of suitable shape are formed on the roadway which link up in such a way that the complete surface appears well-lighted to the driver. A major factor influencing the design therefore is the manner in which the light is distributed from the lantern. This brings into prominence the problem of glare.

If the total output of light from a point source is allowed to radiate with uniform luminous intensity in all directions, then not only will much of the flux be wasted, e.g. that which goes upwards, but the amount which falls on the carriageway at any point will vary inversely with the square of the distance of that point from the light source. To overcome these disadvantages the designers of road lanterns limit the directions in which the flux may be distributed and regulate the directional intensities in the vertical plane so as to obtain the desired luminosity on the roadway. To avoid flux wastage, lanterns are furnished with redirectional apparatus which receives the luminous flux emitted by the lamp and redistributes it in a desirable fashion. A bright patch having a particular luminosity is then obtained by varying the luminous intensity distribution so that the higher intensities are emitted at higher angles from the vertical.

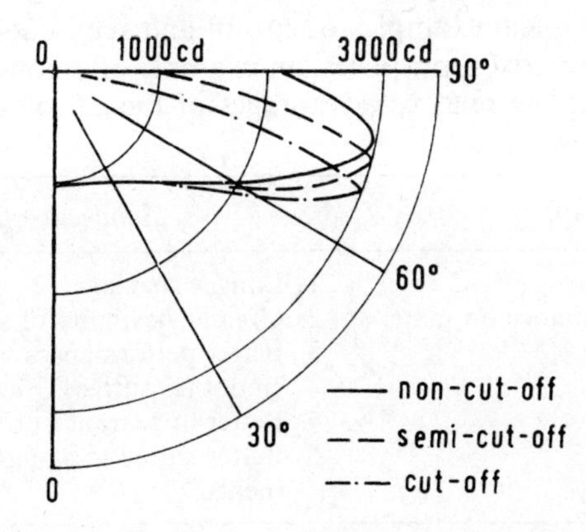

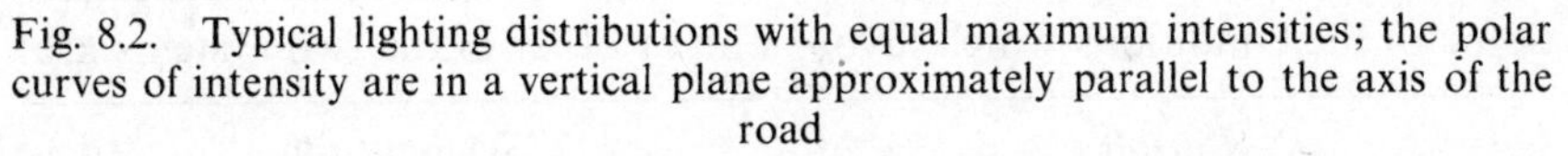
Fig. 8.2. Typical lighting distributions with equal maximum intensities; the polar curves of intensity are in a vertical plane approximately parallel to the axis of the road

The vertical angle at which the maximum luminous intensity is emitted may affect the safety of road users because of the glare effects to which the drivers of vehicles may be subjected. It is customary to describe two main types or effects of glare; these are *disability* glare and *discomfort* glare. Disability glare has been likened to the production of a veiling luminance or luminous fog over the whole visual field which effectively reduces the contrast between object and background, and therefore reduces the visibility. Discomfort glare is the term used to describe the sensations of distraction and annoyance which are experienced when glare sources are present in the field of view.

The effects of glare sources decrease the further they are removed from the line of sight and for this reason lanterns are normally placed above the driver's line of vision. Since a driver cannot usually see more than about 20 degrees above the horizontal because of the cut-off effect of the roof of his vehicle, it means that when the vehicle is close to a lighting column the luminous intensities emitted at angles of less than about 70 degrees from the downward vertical cannot reach his eyes and cause glare; there is therefore no necessity to limit the intensities below this angle. Glare is, however, caused by light leaving the lanterns within about 20 degrees of the horizontal and, in practice, the more lanterns that are visible to the driver, the greater the glare effect. To minimize this problem lighting engineers have designed lanterns that emit relatively low luminous intensities within this 20-degree glare-zone.

When the lanterns are designed so that there is a rapid reduction of luminous intensity between about 20 degrees below the horizontal and the horizontal, then they are said to belong to the *cut-off* system of lighting. If, however, there is a less severe reduction in intensity in the same region, the lanterns are characterized as belonging to the *semi-cut-off* system. If there is no reduction in intensity then the lanterns are said to be *non-cut-off*. Fig.

8.2 illustrates two typical examples of cut-off and semi-cut-off lighting distributions. For comparison purposes, an example of a non-cut-off distribution is also shown. The relative advantages of the cut-off and semi-cut-off systems are as follows:[2]

Cut-off	*Semi-cut-off*
1. Less glare 2. Better performance on matt surfaces.	1. Longer spacing. 2. Greater flexibility of siting. 3. Better performances with smoother surfaces. 4. Better appearance of buildings. 5. Better suited to staggered arrangements.

Some of the conditions which favour one or other of the two systems are:

Cut-off	*Semi-cut-off*
1. Matt carriageway surfaces. 2. Absence of buildings. 3. Presence of large trees. 4. Long straight sections. 5. Slight humps; bridges. 6. Few intersections and obstructions. 7. Adjacent to railways, docks or airports.	1. Smooth surfaces. 2. Buildings close to carriageway, especially those of architectural interest. 3. Many intersections and obstructions.

Arrangements of lanterns

On straight roads, the more commonly used arrangements of lanterns are single-side, staggered, central and opposite. While the arrangement chosen in any particular situation is primarily influenced by the width of road, a staggered side-mounted arrangement will generally give the 'best' light distribution.

The case for and against a central mounting arrangement involves considerations of appearance, visibility and safety, as well as of erection and maintenance. In favour of a central arrangement as compared to the side-mounted one is the fact that for a given number of lanterns per kilometre the central arrangement has half the spacing per row of lanterns. This means that the light patches overlap very well to produce a uniformly lighted roadway centre, the luminosity of which falls off smoothly to the sides. This results in an attractive appearance which makes the roadway very inviting to the motorist. It is, however, these very features which tend to cause the lighting engineer to hesitate about installing a central lighting arrangement. For instance, in wet weather, the driver of a vehicle is faced with a brilliant array of lamps in near-direct line with him, which distracts his attention from more important regions of the roadway and makes them appear less bright. The luminous centre invites the driver to increase his speed and to keep to the crown of the road. Most important of all, the sides of the carriageway are in comparative darkness, with the result that pedestrians or vehicles entering from side roads may be well on to the carriageway before the major road driver is aware of their presence. From the point of view of maintenance, central mounting has the additional disadvantage that the suspended lanterns can be difficult to service. On busy roads the maintenance crew may be exposed to some danger, especially if they have to work at night; in daylight hours, they are a source of obstruction to traffic while servicing the lanterns.

On balance, therefore, central arrangements are to be discouraged and it is only when special conditions justify them that they should be used. Typical instances of this are a carriageway flanked with trees on either side or a narrow road. When a central arrangement is utilized, it should be in conjunction with a semi-cut-off or cut-off lighting system, since both of these cause a reduction in the completeness of overlap of the light patches and hence the difference in luminance between the centre and the sides of the roadway is reduced.

Single-side mounting is also to be discouraged since it results in one side of the roadway being well-lighted, while the other side is in comparative darkness. This causes drivers to veer towards the lighted side, and the problem of safety again arises. When an opposite arrangement is utilized, a very uniform carriageway luminosity can be obtained, but this entails extra expenditure for additional lanterns.

Spacings

Most lanterns emit two main beams, one up and one down the roadway; thus the maximum intensities are emitted in vertical planes which are nearly parallel to the direction of the road. The spacing of the lanterns then depends on the length of the bright patches produced by these beams, and the extent to which it is desired that they should overlap. A typical cut-off lantern mounted 7·62 m above the road surface produces a patch which is only about 6 m long; therefore use of this mounting requires a lantern to be erected at least every 27·5–30 m along the road. At the other end of the scale, a non-cut-off lantern with a peak intensity at about 80 degrees, an intensity exceeding half the peak value at 86 degrees, and having an appreciable intensity right up to the horizontal, may produce light patches of nearly 46 m in length. For good lighting installations on important traffic routes a good general rule is that the spacing between lanterns on straight lengths of road should lie between 3 and 5 times the mounting height, depending on whether a cut-off or semi-cut-off light distribution is used. On sharp bends and at intersections the spacings will have to be closer than (and the illumination greater) than on straight roads.

Outreach and overhang

Lanterns are very often located so as to overhang the traffic lanes. This presents a rather pleasing roadway appearance to the motorist which is somewhat similar to that associated with a central arrangement of lanterns. On wide roadways with side lighting, overhanging the lanterns may be necessary in order to light the centre of the carriageway, which would otherwise appear unduly dark. It is not normally necessary to overhang carriageways less than about 9 m wide. Desirably the overhang distance should not exceed about 1·8 m on columns up to 9 m high, especially on heavily-travelled roads in built-up areas, as otherwise the footpaths and kerbs may be in undesirable shadow. Another practical consideration limiting the amount of overhang is that the lantern should be easily accessible for maintenance purposes. If the carriageway is very wide, it may therefore be necessary to use an additional central arrangement of lanterns to light the road.

The amount of outreach is governed by the extent of the overhang and by safety considerations affecting the location of the lighting columns. A lighting column close to the edge of the carriageway is a potential cause of accidents to vehicles which leave the roadway. Where there is a footpath close to the carriageway, the lighting columns should be located behind the footway. In some instances, it may be possible to support the lanterns on wall brackets.

Lamps

The lamps considered suitable for road lighting are of five main types. These are tungsten filament, sodium vapour, tubular fluorescent, high-pressure mercury fluorescent, and the uncorrected high-pressure mercury lamps.

The *tungsten lamp* is one in which the illumination is provided by passing an electric current through a filament of tungsten in order to raise it to incandescence. While this was at one time the most commonly used lamp in highway lighting practice, its use is now generally confined to residential streets. The use of tungsten lamps in residential streets is due to their low initial cost; this is important when one considers the great number of residential streets and the fact that there is a high level of malicious damage to lights in these streets. The fact that the light-producing efficiency and rated life of a tungsten lamp is low in comparison with the other types of lamp limits its use on heavily-trafficked roads with high standards of lighting.

A *sodium vapour lamp* is, in principle, one in which an electrical discharge takes place in a vapour of this metal. It consists of a discharge tube filled with a mixture of neon and argon gas in which the discharge takes place, and containing small drops of sodium distributed along the length of the tube. When heated by the discharge the sodium forms a vapour and the end-result is an electromagnetic radiation which gives a very characteristic yellow light. It is this colour that causes most debate regarding the sodium lamp. The monochromatic character of the light makes all objects which it illuminates appear more or less yellow, so that exact colour recognition is impossible. Its use therefore in central shopping areas is to be avoided. It has, however, a number of advantages which render its usage on main traffic routes most desirable. Most important of these is the fact that, with the exception of the tubular fluorescent lamps, the sodium vapour lamps are the least dazzling of all the types used for road lighting.[5] Visual acuity increases in monochromatic light owing to the elimination of the phenomenon of chromatic aberration, while the advantages of the yellow light as regards vision in foggy weather are very obvious to the motorist.

The *fluorescent lamp* consists of a long tube the inside of which is coated with a fluorescent powder. An electrical discharge is caused to take place in the tube and this causes the excitation of the fluorescent powder so that it emits a very pleasing white light. As well as providing good visibility, the fluorescent lamp is the least dazzling to drivers. In addition it is particularly suited to lighting shopping centres and other locations where colour appearance and colour rendering are important. Apart from its high cost, its principal disadvantages are the bulk and weight of the luminaire—because of the relatively low output per lamp at least two lamps are normally used in each luminaire—and the fact that the light output is noticeably affected by changes in the air temperature.

The principle of the *high-pressure mercury lamp* is similar to that of the sodium vapour lamp, except that mercury takes the place of sodium. The light emitted has a characteristic bluish-white colour which makes the surroundings more pleasing to the eye than the yellow of the sodium lamp. An improvement on the straightforward mercury lamp is the *high-pressure mercury fluorescent lamp.* The outer bulb of this lamp is coated internally with phosphorus which fluoresces when excited by the mercury discharge. The net result is an off-white light which gives a much more pleasing effect than the bluish-white of the regular lamp. Its usage is recommended at locations where colour appearance and colour rendering are important and where high illumination powers may be needed.

BRITISH PRACTICE

As with many other aspects of traffic engineering, the design of a modern road-lighting installation depends on the careful amalgamation of theoretical and practical knowledge. British practice in this respect is summed up in a Code of Practice[2,6,7,8] which lays down standardized methods of design which assure that lighting installations will be reasonably satisfactory.

The code divides the lighting requirements for main roads and streets into the following nine groups:

Group A1. Lighting for principal traffic routes. In this group is included lighting for very important roads which have high lighting requirements, e.g. through trunk roads.

Group A2. Lighting for normal traffic routes. This includes lighting for the generality of main roads having considerable vehicular and pedestrian traffic, e.g. urban radial roads not included in Group A1.

Group A3. Lighting for minor traffic routes. Lighting for roads such as main rural roads or minor urban roads carrying traffic not confined to the immediate locality—and which do not require lighting of the Group A2 standard—is included in this group.

Group B. Lighting for roads carrying only local traffic. This group covers the lighting of residential and unclassified roads which are not included in the previous groups.

Group C. Lighting for roundabouts and complex junctions.

Group D. Lighting for bridges and flyovers.

Group E. Lighting for tunnels and underpasses.

Group F. Lighting for roads with special requirements. In this group is included the lighting for roads such as motorways and roads near airports, docks and railways.

Group G. Lighting for town and city centres. These locations require high standards of lighting and usually demand a special design based on considerations of prestige and amenity.

As can be seen, the definitions of a number of these groups are rather vague and this is purposely so. The placing of a roadway into any group is left to the discretion of the local authority concerned, as it is generally considered that an outside body could not make the correct decision without having an intimate knowledge of particular and local circumstances.

The code does not consider it practicable to lay down levels of luminance, uniformity, visibility or the like because the lighting designer usually has neither the data on which to compute them nor the means of ensuring their permanence. Instead, he has to select a lamp and lantern and then decide on the geometry of the installation. The recommendations in the code are therefore in terms of the desired amount and distribution of light and of the geometry of the installation. An instance of a typical recommendation (metricated) from the code is given in Table 8.1. An illustration of how the data in the table can be used in practice is given in the following example, which is based on the code:

EXAMPLE PROBLEM. A main shopping street in a large suburb is 3·2 km long, has varied commercial buildings on both sides, and carries fairly heavy through-traffic including buses. The speed limit is 48 km/h. The road surfacing is an asphalt having medium texture and it is a little smoothed by traffic. The width between the kerbs is 13·72 m; the footways are 6·71 m, with flagstones and no verges. There are no trees or service roads. The junctions, some of which are important, are irregularly sited and there are slight curves. A mounting height of 7·62 m and an overhang of 1·82 m is standard in the district for traffic routes.

Step 1. Decide on the classification of the road. It is not sufficiently important for group A1, but is more important than group A3.

Decision—Use group A2 and design a little above the minimum standard if possible.

Step 2. Decide on type of system. Use of a cut-off system may involve

TABLE 8.1. *Recommendations for a semi-cut-off lighting system in accordance with a staggered arrangement on a straight section of single or dual carriageway*[2]

Group	Mounting height H m	Effective width, W m										Minimum light flux in lower hemisphere lm	Maximum overhang A m
		7·62	9·14	10·69	12·19	13·72	15·24	16·76	18·29	19·81	21·34		
		Maximum spacing, S m											
A1	7·62	30·5	25·9	21·3	18·3	16·8						6 250	1·82
	9·14	36·6	36·6	30·5	27·4	24·4	21·3	19·8				9 000	2·29
	10·69	42·7	42·7	42·7	38·1	33·5	30·5	27·4	24·4	22·9		12 250	2·59
	12·19	48·8	48·8	48·8	48·8	42·7	39·6	35·1	32·0	30·5	27·4	16 000	2·90
A2	7·62	33·5	30·5	25·9	22·9	19·8						5 600	1·82
	9·14	39·6	39·6	38·1	33·5	29·0	25·9	24·4				8 100	2·29
	10·69	47·2	47·2	47·2	45·7	39·6	36·6	33·5	30·5	27·4		11 000	2·59
	12·19	53·3	53·3	53·3	53·3	51·8	47·2	42·7	39·6	36·6	33·5	14 400	2·90
A3	7·62	36·6	36·6	32·0	27·4	24·4						5 000	1·82
	9·14	44·2	44·2	44·2	39·6	35·1	32·0	29·0				7 200	2·29
	10·69	51·8	51·8	51·8	51·8	47·2	42·7	39·6	36·6	33·5		9 800	2·59
	12·19	57·9	57·9	57·9	57·9	57·9	56·4	51·8	47·2	42·7	39·6	12 800	2·90

shorter and more rigid spacing. In this situation, with light-coloured buildings and shops and curves, a semi-cut-off system will not be objectionably glaring and has certain advantages. The road surface is suitable for a semi-cut-off-system.

Decision—Use a semi-cut-off system.

Step 3. Decide on lamp type. In this location, colour rendering is quite important.

Decision—Use either high-pressure mercury fluorescent lamps or tubular fluorescent lamps.

Step 4. Decide on probable lantern arrangement, column clearance, overhang and mounting height. It is important to emphasize the footways and kerbs on account of pedestrian traffic and stationary vehicles, and so the lanterns should be located at the sides of the road. On account of the existing footway arrangement it is only possible to set back the lighting columns to give 0·45 m clearance from the kerb; a greater clearance would involve unacceptable obstruction of the footpath. While a mounting height of 7·62 m is usual in the district, 9·14 m is also acceptable. A greater height would not only be aesthetically incongruous in this location, but it would be difficult to maintain with the local authority's existing equipment. An overhang of 1·82 m is standard in the district.

Decision—Use a staggered arrangement, an overhang of 1·82 m, an outreach of 2·29 m, and a mounting height of either 7·62 or 9·14 m.

Step 5. Determine spacing and light output.

(*i*) For high-pressure mercury fluorescent lamps (MBF/U):

Effective width = 13·72 − 1·82 = 11·9 m. Say 12·19 m.

From Table 8.1 (staggered, semi-cut-off), group A2, $W = 12{\cdot}19$ m:

For $H = 7{\cdot}62$ m, $S_{max.} = 22{\cdot}9$ m (44 lanterns/km) and the required light flux in the lower hemisphere = 5600 lm;

For $H = 9{\cdot}14$ m, $S_{max.} = 33{\cdot}5$ m (30 lanterns/km) and the required light flux in the lower hemisphere = 8100 lm.

The spacing at 7·62 m is inconveniently short and would give an expensive and over-crowded installation.

Possible decision—Use a 9·14 m mounting height and 33·5 m spacing.

The lamps will have to be 400 watt, which with a normal lantern give about 65 per cent of the lamp flux in the lower hemisphere, i.e. 12 500 lm. This is well above the minimum for the 9·14 m mounting height.

(*ii*) For tubular fluorescent lamps (MCF/U):

These lanterns have no overhang, hence $W = 13{\cdot}72$ m. Therefore, from Table 8.1:

For $H = 7{\cdot}62$ m, $S_{max.} = 19{\cdot}8$ m (51 lanterns/km) and the required light flux in the lower hemisphere = 5600 lm;

For $H = 9{\cdot}14$ m, $S_{max.} = 29$ m (35 lanterns/km) and the required light flux in the lower hemisphere = 8100 lm.

The spacing at 7·62 m is inconveniently short and will give an expensive and over-crowded installation.

Possible decision—Use a 9·14 m mounting height and 29 m spacing.

A 3-lamp 80 watt lantern will give about 8000 lm. This is just about acceptable for the mounting height. However, 4-lamp lanterns would be preferable as giving more than the minimum flux requirement.

FINAL DECISION—The choice must be made on the basis of local considerations and costs.

SELECTED BIBLIOGRAPHY

1. FRIEDE, H. A. Night accidents could be halved, *Traffic Engineering and Control*, 1962, **3,** No. 10, 629, 637.
2. C.P. 1004, Pts. 1 and 2: 1963. *Street Lighting*. London, British Standards Institution, 1963.
3. REID, J. A. The lighting of traffic signs and associated traffic control devices, *Public Lighting*, 1964, **29,** No. 127, 252–264.
4. WALDRAM, J. M. International recommendations for public thoroughfares, *Traffic Engineering and Control*, 1966, **7,** No. 12, 753, 755.
5. COHU, M. Public lighting by sodium vapour lamps, *Traffic Engineering and Control*, 1962, **4,** No. 3, 169–177.
6. C.P. 1004, Pts. 4, 6 and 8: 1967. *Street Lighting*. London, British Standards Institution, 1967.
7. C.P. 1004, Pts. 3 and 9: 1969. *Street Lighting*. London, British Standards Institution, 1969.
8. C.P. 1004, Pt. 7: 1971. *Street Lighting*. London, British Standards Institution, 1971.

Index

Accessibility, effect of parking on, 130
Accident studies, 300–2
Accidents
  and alcohol, 297–300
  and edgelines, 343
  and fatigue, 296–7
  and hearing, 296
  and highway type, 290–3
  and motor cyclists, 294
  and one-way streets, 323, 325
  and parking, 131–3
  and pedestrians, 288, 303–4
  and skidding, 293–4
  and street lighting, 294, 366
  and vision, 295–6
  at horizontal curves, 245
  at intersections, 233, 275, 290–2
  at night, 294
  costs, 197, 198
  data required for economic studies, 208–10
  effect on, of closing side streets, 330
    of season and day of week, 293–4
    of speed limits, 311–14
    of vehicle separation, 309
  general statistics for Britain, 287–9
  international comparisons, 289–90
  role of central reservation in preventing, 271, 318–19
  See also Pedestrian crossings
Administration of highways
  authorities, 42–9
  programming of schemes, 49–51
  See also Finance *and* Legislation
Advisory speed signs, 314
Alcohol and accidents, *see* Accidents
All-directional interchanges, 283
Alternate system of linked traffic signals, 349–50
Ancient roads, 1–3
Anti-glare fence, 314–15
  See also Lighting
Area traffic control, 352–3

Barriers, movable, for tidal flow operations, 329
Benefit-cost method of economic analysis, 213–14
Blind Jack of Knaresborough, 13
Braided interchange, 283
Brake reaction time, 238–9
Braking distance requirements, 239–41
Bright patch on the carriageway, due to lighting, 370–2
British highway system
  classification, 24, 30–3
  national highway network, 68–78
  nomenclature (identification), 32–3
  traffic characteristics, 74–8
  urban road patterns, 69–73
Business, effect of parking on, 130
By-passes about towns, 68–9

Camber of a carriageway, 273–4
Capacity of highways
  British studies, 222–3
  Department of Environment recommendations, 223–6
  factors causing reduction in, 218–21
  one-way streets, 321–3
Capacity of intersections
  priority, 226–9
  roundabouts, 229–33
  signalized, 233–7
  See also Traffic signal settings
Carriageway delineation
  reasons for, 338–9
  types of, 314, 339–41
Category analysis of trips generated, 120–1
Central reservation, 271–2, 318–19
  See also Crash barriers

Centre-lines, 341–3
Centrifugal force, 246–7
Channelization of pedestrians, 304–5
Channelized intersections, *see* Intersections at-grade
Class intervals, speed, 82–3
Classification of highways, 24, 30–3
Clearway surveillance systems, 356
Cloverleaf interchange, 283
Closing side streets
advantages, 329–30
disadvantages, 330
Collision diagrams, 301
Colour blindness, 296
Congestion, effects of parking on, 130
Contact strip detectors, 79–80, 93–4
Control-point comparisons in O–D surveys, 107–8
Conurbations, descriptions of, 60–1
Converted traffic, 115
Cordon-line comparisons in O–D surveys, 109
Cost of highway improvements
initial-cost items, 189–91
maintenance, 191, 193–4
Cost-benefit methods of economic analysis
benefit-cost ratio, 213–14
internal rate-of-return, 212–13
net present value, 211–12
short-term rate-of-return, 212
Crash barriers
design of, 315–17
types of, 317–18
usage, 318–19
Creeper lanes, 259–60
Cross-section elements of a highway
camber, 273–4
central reservations, 271–2
lay-bys, 273
shoulders, 272–3
side-slopes, 274
traffic lanes, 269–71
See also Accidents *and* Capacity of highways
Current traffic, 112–14
Cut-off lighting systems, 373–4

Dark Ages, roads in the, 8
Deceleration lanes, 276–7
Delay
at roundabouts, 207–8
at traffic signals, 355–6
data required for highway economic studies, 86, 207–8
studies, on highways, 86, 89–90
types of, fixed and operational, 89
Department of the Environment, organisation and functions, 42–7
Depth perception, 295
Design speed
for rural roads, 217–18
for urban roads, 218
Diamond interchange, 282
Design volume, constituents of the highway, 111–16
Desire-line graphs, 99
Development and Road Improvement Funds Act of 1909, 24
Development traffic, 116
Directional distribution of traffic flow. 77–8
Directional interchanges, 283
Driver characteristics, 295–300
See also Perception-reaction time
Driver training, 299–300
Driving test, introduction of, 26

Economic savings, due to one-way streets, 323–4
Economic studies for highways, *see* Cost of highway improvements *and* Cost-benefit methods of economic analysis
Edgelines on rural roads, 343
Elevated observer method of measuring speed, 87
Enoscope, 79
Environment
and one-way streets, 325
and parking, 129–30
and the motor car, 54–7
and traffic signs, 336

Fatigue and accidents, 296–7
Federal Aid Highway Act of 1944, American, 20
Field of vision, 295
Finance Acts of 1920, 1927 and 1936, 25, 27
Finance, highway, in Britain
grant system, 33–4
sources of revenue, 34–7

Flexible progressive system of linked traffic signals, 350–2
Fluorescent lamps, 377
Footways, criteria for, 308–9
Friction between tyre and carriageway
  and unacceptable deceleration rates, 239
  at horizontal curves, 247–50
  when braking, 239–40
Fringe parking, *see* Peripheral car parks
Fuel tax
  differential, to control congestion, 358
  levied, 34

Geometry of surface car parks, 178–80
Glare
  and vision, 296
  anti-glare fences, 314–15
  from highway lanterns, 372–4
Gradients, *see* Vertical alignment
Gravity model methods of trip distribution, 122–3
Gridiron street patterns, 69–70
Growth factor methods of trip distribution, 121–2
G-turn at an intersection, 320
Guide posts, *see* Post delineators

Headlight glare, *see* Glare
Hearing and accidents, 296
Hesitation time, 242
Highway, derivation of term, 4
Highways Act of 1959, 28
Highways and Locomotive (Amendment) Act of 1878, 19
Highways (Miscellaneous Provisions) Act of 1961, 29
Home-interview O–D surveys
  evaluation of accuracy, 107–9
  method, 102–7
  usage, 111, 118–19
Horizontal curvature
  and accidents, 245
  centrifugal force effects, 246–7
  properties of circular curves, 245–6
  sight distance design, 250–1
  superelevation, 247–50
  transition curves, 254–8
  widening, 251–4
Housing, Town Planning, Etc., Act of 1909, 24

Identification of highways, *see* Nomenclature
Induced traffic, 115
Interchanges, design criteria, 281–5
Internal rate-of-return method of economic analysis, 212–13
Intersections at-grade
  accidents at, 273, 275, 290–2
  basic forms, 275
  channelized, 277–9
  non-channelized, 275–7
  restrictions at, 319–21
  roundabouts, 279–80
  See also Capacity of intersections *and* Delay
Intersections, grade-separated
  interchanges, 281–5
  justification for, 281
  without slip-roads, 281
Interstate and Defense Highway System, American, 20

Journey speed
  studies, 86–8
  See also Running speeds *and* Delay

Kerb parking, *see* On-street parking
Kiss-and-ride parking, 159–60, 161

Land-use planning, importance of, 63–5
Land value effects on highway improvement costs, 199
Lane-lines, 234, 341–3
Lay-bys, 273
Legislation, highway, in Britain
  between the World Wars, 24–7
  post-World War II, 27–9
  pre-World War I, 23–4
Lettering in traffic signs, 332–5
Licences, differential, to control congestion, 359–60
Lighting installation terms, 368–9
Lighting of highways
  arrangement of lanterns, 368, 374–5
  benefits from, 366
  British practice, 377–81
  glare problem, 372–4
  installation terms, 368–9
  lantern spacings, 375
  light distribution on the carriageway, 370
  locations requiring usage, 366

Lighting of highways—(*contd*)
means of discernment, 369–70
objectives, 369
outreach and overhang, 376
photometric terms, 367
types of lamp, 376–7
systems, 373–4
See also Accidents
Linear street pattern, 70–1
Local Government Acts of 1888, 1889, 1894 and 1929, 19, 23, 25
Locomotive on Highways Act of 1896, 19, 23–4
London Traffic Survey, 118–19
Luminance and luminosity, *see* Lighting for highways

Macadam (McAdam), John Loudon, 16–18
Magnetic detectors, 92
Mandatory signs, 337
Marking materials, reflectorized, 339–40
See also Carriageway delineation
Mechanical parking garages, 180–1
Mechanical travel, development of, 53–4
Median speed, 81
Mercury lamps, 376–7
Metcalf, John, 13
Middle Ages, roads during the, 8–9
Ministry of Transport
founding of, 24–5
See also Department of the Environment
Modal speed, 81
Modal split models, 124–5
Motor car and its effect on
accidents, 55
environment, 55–6
public transport, 57–8
Motor fuel tax, 34
Motor vehicle growth in Britain, 61
Motorist taxation concepts
area-occupied, 39–40
differential-benefits, 39
differential-cost, 38–9
operating-cost, 39
weight-distance, 39
Motorways
anti-glare fences on, 314–15
crash barriers on, 315–19
design speed, 217–18
grade-separated intersections on, 280–5
in Britain, 31–2
surveillance systems on, 356
See also Capacity of highways *and* Cross-section elements of a highway
Moving observer method of measuring speed and flow, 87–9, 95
Multi-way interchanges, 283
Multiple regression analysis of trips generated, 120

National highway network in Britain, characteristics of, 68–78
Net present value method of economic analysis, 211–12
Night vision, 296
Nomenclature, highway, 32–3
Non-British roadways, ancient, 2–3
Normal traffic growth, in highway design volume estimations, 114–15

Off-street car parks, design of
audit control, 181–2
mechanical and electrical services, 182–5
parking garages, 180–1
surface parks, 178–80
Off-street car parks, factors influencing the location of
access streets, 151–5
business, 150
intended usage, 155
origins of travellers, 150
parking generators, 150
topography, 155
Off-vehicle recording systems, 360–1
On-street parking, management of
facilities for commercial vehicles, 148–9
prohibited parking, 140–1
resident parking, 147–8
time-limit parking, 142–7
One-way street systems
advantages, 321–5
disadvantages, 325–6
introduction of, 326
signing on, 326
Opportunity model methods of trip distribution, 123–4

Origin and destination surveys
functions of, 95
methods of carrying out, 95–109
selection of method, 109–11
zoning for, 96
Outreach, 368
Overhang, 368

Park-and-ride, 157–9, 160
Park-and-walk, 66–8, 156–7, 160
Parking, detrimental effects of, 129–33
Parking discs and meters, *see* On-street parking
Parking plan, preparing the town centre
developing the plan, 166–77
number of spaces, 161–6
Parking policy, 66–8, 155–61
Parking surveys
concentration, 135–6
duration, 136
information provided by, 138–40
interview, 136–8
supply, 133–4
usage, 134–5
Parking tax, to control congestion, 358–9
See also Parking policy
Passenger car units, 224
Pedestrian accidents, *see* Accidents
Pedestrian channelization, 304–5
Pedestrian crossings
criteria for, 308
in car parks, 180
on one-way streets, 323
risk at, 307
segregated, 305–6
zebra, 306–7
Perception-reaction time, 238–9
Peripheral car parks
kiss-and-ride, 159–60, 161
park-and-ride, 157–9, 160
park-and-walk, 66–8, 156–7, 160
Phillips, Robert, 13
Photo-electric detectors, 93
Planning for parking, *see* Parking plan
Poll tax, to control congestion, 360
Population, future
of persons, 59–60
of vehicles, 61–2
Postcard O–D surveys
method, 99–101
usage, 111
Post delineators, 314
Principal roads, classification, 32
Priority intersection, capacity of a, 226–9
Purchase tax, 35–6

Queue delays
at roundabouts, 207–8
at traffic signals, 355–6
Q-turn at an intersection, 320

Radar speed-meter, 80
Radial acceleration, rate of gain of, 256
Radial street pattern, 71–3
Rate-of-return methods of economic analysis
internal, 212–13
short term, 212
Rating revenues, 36–7
Registration number method
of carrying out O–D surveys, 101–2
of measuring speed, 86–7
Regression analysis, *see* Multiple-regression analysis
Restriction of Ribbon Development Act of 1935, 26
Restrictions of turning movements
closing side streets, 329–30
left turns, 321
right turns, 319–21
Revenue, main sources of
equitableness of, 37–41
from central government, 34–6
from local government, 36–7
Reverse-flow operation, *see* Tidal-flow operation
Ring roads, *see* Radial street pattern
Road Act of 1920, 25
Road and Rail Traffic Act of 1933, 26
Road Fund, *see* Road Improvement Fund
Road Improvement Act of 1925, 25
Road Improvement Fund, 24
Road pricing
direct methods of charging, 360–3
indirect methods of charging, 358–60
operational requirements, 357–8
Road Safety Act of 1967, 29
Road studs, 340–2
Road Traffic Acts of 1930, 1934, 1956, 1960 and 1962, 26, 28–9

Road Traffic and Road Improvement Act of 1960, 28
Road-user benefits, evaluation of
  accident costs, 197, 198, 209–10
  labour, vehicle-time and operating costs, 193, 195–8
  land value effects, 199
  speed and delay, 204–8
  traffic flow, 199–204
Roads, historical development of
  ancient, 1–3
  between the 16th and 18th centuries, 9–10
  during the Middle Ages, 8–9
  into the 20th century, 10–20
  prior to the Middle Ages, 7–8
  Roman, 4–7
Roadside interview O–D surveys
  method, 96–9
  usage, 109 11
Roman roads, *see* Roads
Roundabout interchange, 283–5
Roundabout intersections, at-grade
  accidents at, 292
  capacity of, 229–32
  general design criteria, 279–80
Running speed
  formulae for economic studies, 204–7
  studies, 86–8

Safety fences, *see* Crash barriers
Safety, highway
  accident studies, 300–2
  motor vehicle accidents, 287–300
  pedestrian accidents, 303–4
Safety time, in passing, 242
Sag vertical curves
  designing for sight distance, 266–8
  types of, 261–2
Sampling
  for home interview surveys, 106
  for roadside interview surveys, 97–8
  of spot speeds, 84–5
  traffic flow for economic studies, 199–204
Saturation flow at traffic signals, 235, 354
Screen-line comparisons in O–D surveys, 107
Segregated crossings for pedestrians, 305–6
Semi-cut-off lighting systems, 373–4
Shifted traffic, 115
Shoulders
  effect on capacity, 219
  function and design, 272–3
Side-slopes, 274
Sight distances, safe passing
  definition, 242
  eye-height criterion, 309
  for multi-lane roads, 243–4
  for three-lane roads, 243
  for two-lane roads, 242–3
  frequency of sections, 244–5
  in relation to commercial vehicles, 244
Sight distances, safe stopping
  braking distance component, 239–241
  definition, 238
  in relation to commercial vehicles, 241
  in relation to speed, 309
  perception-reaction time component, 238–9
  See also Vertical alignment
Signing on highways, *see* Traffic signs
Skidding
  and accidents, 293–4
  and speed, 310
Simultaneous system of linked traffic signals, 349
Sodium vapour lamps, 376–7
Space-mean speed, 85
Spacing of lanterns, 368, 375
Special roads, *see* Motorways
Special Roads Act of 1949, 27
Speed
  and delay data used in economic studies, 204–8
  effect of, on accidents, 309–11
  /flow relationships, 219–20
  limits, 311–14
  method of measuring, 79–80, 86–9
  types of, 78, 85, 86
  See also Design speed; Journey speed; Running speed *and* Spot speed studies
Speed-change lanes, 275–7
Speed restrictions
  advisory speed signing, 314
  consequences of high speeds, 309–11
  enforcement, 313–14
  mandatory speed limits, 311–14
Speed studies, 78–89
Spiral transition curves, *see* Transition curves
Spot speed studies, 79–86

Stage-coach to motor vehicle, 17–19
Standard deviation, determination of, 83–4
Standard error of the mean, 83–4
Statute of Winchester, 8
Summit vertical curves
  designing for sight distance, 262–6
  properties of, 260–2
  usage, 268–9
  See also Vertical alignment
Superelevation at curves
  application, 256–8
  Department of Environment recommendations, 249
  development of equations for, 247–9

Tag (or sticker) surveys, 102
Tee intersections, 226–9, 276, 282, 290–1
Telford, Thomas, 13–16
Three-lane roads
  capacities of, 225
  passing sight distance on, 243
  usage, 269–70
Tidal-flow operation
  controlling the reversible lanes, 328–9
  introduction requirements, 327
Time-lapse camera, 80, 85
Time-mean speed, 85
Town and Country Planning Act of 1947, 27
Trackways, British, 3
Traffic-actuated signals, 234, 347–8
  See also Traffic signals
Traffic assignment, 126–7
Traffic characteristics
  composition, 78
  daily patterns, 75–6
  directional distribution, 77–8
  effect of weather on, 77
  hourly patterns, 74–5
  monthly and yearly patterns, 76–7
Traffic data required for economic studies, 199–209
Traffic flow, *see* Volume; Speed *and* Traffic characteristics
Traffic lanes
  number used in a specific situation, 269–70
  width of, 219, 270
  See also Capacity of highways
Traffic restraint, 66–8
  See also Parking policy
Traffic signal settings
  effect on queue length, 355–6
  for a flexible progressive linked signal system, 350–1
  Transport and Road Research Laboratory method for an individual intersection, 353–5
Traffic signal-controlled intersections, capacity of, 233–7
Traffic signals
  advantages and disadvantages, 344–5
  area control systems using, 352–3
  basic features of, 346
  effect of closing side streets on, 330
  for pedestrians, 305
  linked systems, 348–53
  pre-timed, 346–7
  terminology, 345–6
  traffic actuated, 347–8
  usage in tidal-flow operation, 328–9
  See also Traffic signal settings
Traffic signs
  colour, 334
  definition, 331
  determination of letter height, 333
  effect on amenity, 336
  informatory signs, 332–6
  international developments, 330–1
  layout, 334
  location, 335–6
  mandatory signs, 337
  on one-way streets, 326
  on streets with tidal-flow operation, 328
  principles of signing, 331
  prohibitory signs, 337–8
  reflectorized, 335
  types of, 332
  warning signs and signs giving definite instructions, 337
Transition curves, 254–8
Transport Acts of 1947, 1953, 1962 and 1968, 27–30
Transport demand studies in urban areas, 116–27
Trésaguet, Pierre, 12–13
Trip distribution, 121–4
Trip generation, 119–21
Trumpet interchange, 282
Trunk Roads Acts of 1936 and 1947, 27
Trunk roads, classification, 30–1
Tungsten lamps, 376

Tunnel vision, 295–6
Turnpike Trusts, 11–12, 18
Turban interchange, 283

Uncontrolled crossings, *see* Zebra crossings
Uncontrolled intersections
  accidents at, 290–2
  capacity, 226–9
  See also Priority intersections *and* Tee intersections
United States, motor vehicle population, 20
Urban road patterns
  gridiron, 69–70
  linear, 70–1
  radial, 71–3
Utilization of parking spaces, 143

Variation of speed with flow on two-way British streets, 222–223
Vehicle-actuated signals, *see* Traffic-actuated signals
Vehicle Excise Act of 1962, 28–9
Vehicle metering systems, 361–3
Vehicle operating costs, *see* Road-user benefits
Vehicle registration duties, 34–5
Vehicle separations on the highways, 309–10
Vertical alignment
  creeper lanes, 259
  gradients, 258–9
  maximum grades, 259
  meeting sight distance requirements, 262–6
  minimum grades, 260
  sag curve design, 260–6
  summit curve design, 260–6
Vision and accidents, 295–6
  See also Lighting of highways
Volume, traffic, studies, 90–5
  See also Capacity of highways *and* Capacity of intersections
Volumetric counting, methods and devices, 92–5
  See also Speed studies

Wade, General, 10–11
Widening of horizontal circular curves, 251–4
Winchester, Statute of, 8

Zebra crossings, 306–7